C000097337

1 MONTH OF
FREE
READING

at

www.ForgottenBooks.com

By purchasing this book you are eligible for one month membership to ForgottenBooks.com, giving you unlimited access to our entire collection of over 700,000 titles via our web site and mobile apps.

To claim your free month visit: www.forgottenbooks.com/free8559

* Offer is valid for 45 days from date of purchase. Terms and conditions apply.

ISBN 978-0-428-71947-0
PIBN 10008559

This book is a reproduction of an important historical work. Forgotten Books uses
state-of-the-art technology to digitally reconstruct the work, preserving the original format
whilst repairing imperfections present in the aged copy. In rare cases, an imperfection in
the original, such as a blemish or missing page, may be replicated in our edition. We do,
however, repair the vast majority of imperfections successfully; any imperfections that
remain are intentionally left to preserve the state of such historical works.

Forgotten Books is a registered trademark of FB &c Ltd.
Copyright © 2017 FB &c Ltd.
FB &c Ltd, Dalton House, 60 Windsor Avenue, London, SW19 2RR.
Company number 08720141. Registered in England and Wales.

For support please visit www.forgottenbooks.com

APR 8   1929

ZOOLOGY OF THE INVERTEBRATA

ZOOLOGY

OF THE INVERTEBRATA

A TEXT-BOOK FOR STUDENTS

BY

ARTHUR E. SHIPLEY, M.A.

FELLOW AND ASSISTANT TUTOR OF CHRIST'S COLLEGE AND DEMONSTRATOR OF
COMPARATIVE ANATOMY IN THE UNIVERSITY OF CAMBRIDGE

LONDON

ADAM AND CHARLES BLACK

1893

c

PREFACE

In this book I have tried to give such an account of the Invertebrata as might be useful to students in the upper forms of Schools and at the Universities, who are already acquainted with the elementary facts of Animal Biology. The volume is in no sense a work for advanced students, and hence no references to original sources have been given, and the names of the various investigators who have promoted our knowledge have been mentioned as sparingly as possible.

In order to keep the book within reasonable limits, I have not described fully certain types which are dealt with in the admirable elementary text-books of Huxley and Martin, and Marshall and Hurst; but, with this reservation, I have endeavoured to describe some one example of each of the larger groups, and then to give a short account of the most interesting modifications presented by other members of the group.

The last few years have witnessed a great extension in our knowledge of the structure and relationship of the Invertebrata. The earth has been ransacked for new forms, and improvements in microscopes and in technique have facilitated a more minute and thorough examination of these forms in the laboratory. This increase in our knowledge has necessarily been accompanied by a rearrangement of material; many intermediate forms have been discovered, and unexpected relationships have been revealed, and these have entailed a revised classification.

These facts have led me to treat the subject largely from a morphological standpoint, touching but lightly on the Histology, Embryology, and Natural History of the forms described. More space has been, as a rule, devoted to those animals which are regarded as intermediate between the larger groups than to the more specialised members of the groups.

Any system of classification is to some extent a matter of personal judgment. I do not suppose that adopted here has any finality, but I hope the tables given will be of use to the student as expressing the results of the most recent research.

In preparing the volume I have been much helped by numerous friends, to whom my best thanks are due. Dr. D. Sharp, Dr. Hickson, Mr. Beddard, Mr. J. J. Lister, Mr. F. G. Sinclair, Mr. C. Warburton, and Mr. MacBride, have all given me the most generous assistance, and, above all, I am most deeply indebted to my friend Mr. S. F. Harmer, who has in the most kind way read through the proof-sheets, and whose careful revision has saved me from many errors.

To the Delegates of the Clarendon Press I owe thanks for permission to use Fig. 133, taken from Rolleston and Jackson's *Forms of Animal Life*. Herr Fischer of Cassel has kindly given me leave to use some reductions from the admirable diagrams of Professor Leuckart; these occur in the groups Echinodermata and Arthropoda, and are acknowledged under each cut; similarly the firm of Wieweg and Son have been good enough to allow me to use four figures taken from Vogt and Yung's *Lehrbuch der Praktischen Vergleichenden Anatomie*. I am also indebted to Messrs. Macmillan and Co. for their kindness in allowing me to use Figs. 37, 89, and 90, all of them taken from Professor Parker's *Elementary Biology*.

ARTHUR E. SHIPLEY.

CHRIST'S COLLEGE, CAMBRIDGE,
March 1893.

CONTENTS

CHAPTER I

PROTOPLASM is the name given to that colloidal, jelly-like substance which forms the basis of all life on this globe. Every living organism consists of protoplasm and the products of protoplasm. Whilst life lasts it is continually renewed from food which passes into the organism, and which, by the action of the protoplasm already there, is built up into new protoplasm. At the same time other portions of the protoplasmic body of the organism are being broken down, and the products thus formed are either thrown out from the body as excreta, or remain in the body, either stored away as useless, or in most cases performing some useful function, such as that of protecting the organism by forming a cyst or shell or internal skeleton.

The protoplasm of living beings is arranged in a series of units or elements, termed *cells*, and with very few exceptions each cell contains one or more specialised portions of protoplasm which take up staining material more readily than the body of the cell, and which are termed *nuclei*. An organism may consist of but one cell with its nucleus or nuclei, but more commonly it is composed of an enormous number of cells, connected together, and each dominated by a single nucleus. In either case, whether the organism is unicellular or multicellular, the cell is capable of an extraordinary degree of differentiation, and may assume the most diverse forms. In the multicellular beings similar cells are massed together into aggregates which form the various tissues composing the body of the higher organisms. In unicellular forms the cells composing the body sometimes remain in connection with one another, but they never form definite tissues, and the cells of such an aggregate are physiologically distinct and independent of one another, the whole forming a colony of unicellular beings.

The organic world has developed in two diverging directions, one corresponding to the animal the other to the vegetable kingdom, and though there is no difficulty in distinguishing the higher forms of these two kingdoms, it is often by no means an easy matter to determine whether some of the lower forms should be grouped with the plants or with the animals; hence any scheme of classification is largely dependent on individual opinion. There are a number of characters which if met with in an organism would justify us in claiming it as an animal, but in many cases one or more of these animal features are absent, and again other features may be present which, as a rule, are only found in plants, so that it becomes at once evident that the line between animals and plants, at any rate in their lowest forms, represents no scientific frontier, but is an arbitrary boundary which is apt to be shifted, now forward now backward, according to the opinion of the various investigators.

The most important morphological difference between plants and animals is perhaps the presence of a cellulose coat which encloses, at any rate during some part of its life, the vegetable cell. Cellulose is a substance which has a definite chemical composition, and which, though practically universal in plants, is very rarely met with in animals. Another constituent found in all green plants, but rare in animals, is chlorophyll; the presence of this enables the plant in sunlight to take in carbon dioxide, which serves as part of its food; chlorophyll is, however, not found in all plants, the Fungi, an important section of the vegetable kingdom, being devoid of it.

The physiological differences between plants and animals are more striking than the morphological. Plants can live upon much simpler compounds than animals; they can absorb their nitrogen in the form of nitrates or simple compounds of ammonia, and their carbon in the form of carbonic acid, or some other soluble compound; thus they can live on liquid inorganic food which may enter the organism at any point, and consequently plants require no mouth or digestive cavity, or organs for the prehension of food. Animals, on the other hand, require more complex compounds; their nitrogen, with scarcely an exception, must be supplied in the form of proteids, and their carbon in the form of starch, sugar, or fat. Some of these compounds are not soluble, and hence an animal must ingest its food in a more or less solid state; and to that end it is usually provided with a mouth and digestive tract, with organs for the prehension of food, and with locomotor organs so that it may find its food. Since the food of animals does not exist in nature except as the products of living beings, it is obvious that animals are ultimately dependent on the plant world for their means of subsistence.

The broken-down products of the protoplasm are usually excreted by special organs set apart for this purpose in animals, but in plants the waste products are either diffused from the surface of the organism, or are stored away in the plant. There are no special excretory organs.

In both plants and animals the most lowly organised beings consist of one cell, and the unicellular organisms are termed the Protophyta and Protozoa respectively. The Metaphyta and Metazoa, or the multicellular plants and animals, consist of a number of cells arranged in more or less definite tissues, but even these multicellular beings pass through a unicellular stage, that of the ovum, whose repeated divisions after fertilisation give rise to the cells composing the body of the animal or plant.

The Protozoa are therefore the simplest and most primitive animals, and it is natural to place them at the bottom of the animal kingdom.

SCHEME OF CLASSIFICATION OF THE
ANIMAL KINGDOM

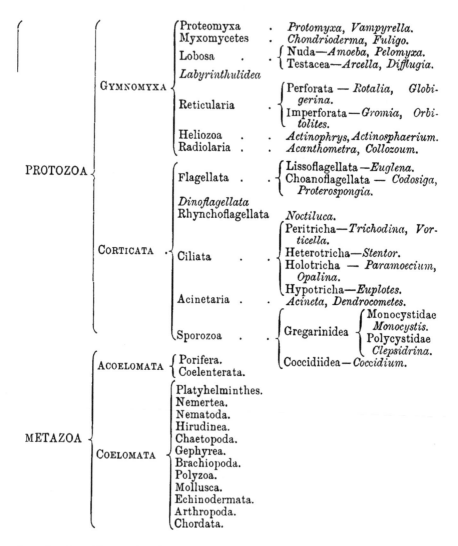

PROTOZOA	GYMNOMYXA	Proteomyxa	*Protomyxa, Vampyrella.*
		Myxomycetes	*Chondrioderma, Fuligo.*
		Lobosa	Nuda—*Amoeba, Pelomyxa.* / Testacea—*Arcella, Difflugia.*
		Labyrinthulidea	
		Reticularia	Perforata — *Rotalia, Globigerina.* / Imperforata—*Gromia, Orbitolites.*
		Heliozoa	*Actinophrys, Actinosphaerium.*
		Radiolaria	*Acanthometra, Collozoum.*
	CORTICATA	Flagellata	Lissoflagellata —*Euglena.* / Choanoflagellata — *Codosiga, Proterospongia.*
		Dinoflagellata / Rhynchoflagellata	*Noctiluca.*
		Ciliata	Peritricha—*Trichodina, Vorticella.* / Heterotricha—*Stentor.* / Holotricha — *Paramoecium, Opalina.* / Hypotricha—*Euplotes.*
		Acinetaria	*Acineta, Dendrocometes.*
		Sporozoa	Gregarinidea { Monocystidae *Monocystis.* / Polycystidae *Clepsidrina.* } / Coccidiidea—*Coccidium.*
METAZOA	ACOELOMATA	Porifera. / Coelenterata.	
	COELOMATA	Platyhelminthes. / Nemertea. / Nematoda. / Hirudinea. / Chaetopoda. / Gephyrea. / Brachiopoda. / Polyzoa. / Mollusca. / Echinodermata. / Arthropoda. / Chordata.	

Note.—In the tables of classification those groups which are not described in text are printed in italics. After the title of the minor sub-divisions of group, the name of one or more typical genera belonging to that sub-division ded in italics.

CHAPTER II

CHARACTERISTICS.—*Unicellular, or if composed of more than one cell, such elements not arranged in tissues. Food ingested by a special mouth or by any part of the cell substance. Reproduction never takes place by ova and spermatozoa. Some forms are colonial.*

Group A. GYMNOMYXA.

The Protozoa have been divided into two groups, the Gymnomyxa, corresponding with the old group Rhizopoda; and the Corticata, which comprise the Infusoria and Gregarinidea. The former group includes all those forms which, like Amoeba, have, during the dominant phase of their life-history, no limiting membrane. Their protoplasm is consequently exposed, at any rate at one portion of their surface, and tends to run into processes or pseudopodia, which vary in appearance in the different species. Food may generally be ingested at any point of the naked protoplasm.

Although the amoeboid condition is the one in which these organisms most frequently occur, they may pass through other phases, such as rounded spores enclosed in a membrane (*chlamydospore*), naked spores with a lash-shaped pseudopodium (*flagellula*), etc. Not infrequently two or more individuals fuse together, and this fusion may be the precursor of reproduction. When the bodies of numerous amoebiform individuals run together to form a large mass of protoplasm, the result is known as a *Plasmodium*.

The classification here adopted is taken from Lankester's article on Protozoa.

CLASS I. Proteomyxa.

The simplest forms of Gymnomyxa are grouped together in the class *Proteomyxa*. As an example of this class the life-history of *Protomyxa aurantiaca*, a minute organism found in 1867 by Professor Haeckel, living on the coiled shells of the Mollusc *Spirula*, in the Canary Isles, may be described. Many of these shells were found bearing on their white surface a minute globular mass of an orange - brown colour. Each globule or cyst consisted of a central mass of protoplasm, surrounded by a structureless membrane; in the older cysts the central protoplasm appeared to be segmented into a number of parts, each of which, on the bursting of the membrane, escaped in the form of a flagellula or pear-shaped swarm-spore. These moved actively about by the lashing of their whip-like pseudopodium, and soon underwent a change in form; instead of one pseudopodium which acted as a flagellum, they developed several, and then moved about like so many amoebae. After creeping about for some time, these amoeboid organisms fused together and formed a plasmodium, which in some cases attained such a size as to be visible to the naked eye. The plasmodium gave rise to many branching ragged pseudopodia, by whose aid it ingested great numbers of diatoms and other food particles. It was much vacuolated, although none of the vacuoles were contractile. After crawling over the Spirula shell for a time the plasmodium retracted its pseudopodia and became spherical; it then surrounded itself with a cell wall, and the contents of the cyst thus formed broke up into flagellulae in the way indicated above. No nucleus has yet been observed in any phase of the life-history of this organism.

Other genera have been described which live parasitically upon Spirogyra (*Vampyrella spirogyrae*) or Diatoms (*Archerina Boltoni*, described by Lankester). In the latter chlorophyll corpuscles are present, and seem to dominate the cell body in a manner suggestive of a nucleus, which is otherwise absent.

Class II. **Myxomycetes.**

The *Myxomycetes* differ from the Proteomyxa in their spores being always coated (chlamydospores), and in the fact that these are formed in definite cysts (Fig. 1, B), sometimes supported on columns, or in naked groups called *sori*. They

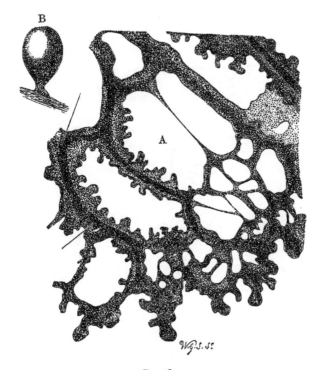

Fig. 1.

A. Plasmodium of *Didymium leucopus* (after Cienkowski).   × 350.
B. Spore cyst of *Arcyria incarnata* (after De Bary).

differ from all other Protozoa in being rarely aquatic; they usually live in the air, in damp places. Their plasmodia may attain a great size, several square inches in area, and form the largest masses of undifferentiated protoplasm to be met with. They live on organic particles; they are often of brilliant colour.

The life-history of most of the Myxomycetes is a repetition of that of Protomyxa: in some the flagellula phase is omitted, the chlamydospore giving rise directly to an amoeboid

organism, provided however with a nucleus. These *amoebulae*
may multiply by fusion, but ultimately they run together and
form the plasmodia (Fig. 1, A), which form the dominant and
characteristic phase in the life-history of the Myxomycetes.

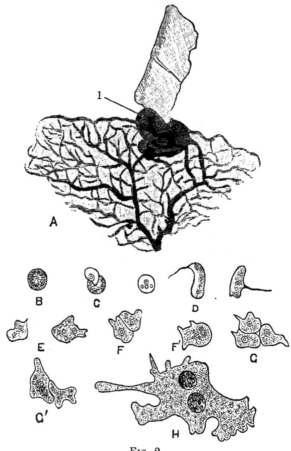

Fɪɢ. 2.

A. A portion of the plasmodium of *Bad-
hamia*, × 3½, showing a pseudopodium (1) commencing to enclose
a piece of mushroom stem. After
Lister.
B. Spore of *Chondrioderma*.
C. Spore of *Chondrioderma* dehiscing.
D. *Flagellulae* which have emerged from
the spores.

E. *Amoebulae* formed by meta-
morphosis of flagellulae.
F. Two amoebulae fusing to
form F'.
G and G'. Stages in the formation of a
three-celled plasmodium.
H. A small plasmodium.
(B-H., after Cienkowski.)

Chondrioderma difforme, the species illustrated in Fig. 2,
has a small plasmodium, easily visible to the naked eye. This
species occurs commonly on old bean-stalks. The plasmodia

may be easily obtained by soaking some dried bean-stalks in water for twenty-four hours, and then keeping them in a moist chamber for ten days or so; at the end of this time plasmodia may be observed crawling over the stems, etc.

The sulphur-coloured *Fuligo* (*Aethalium*) is a genus which is met with in considerable masses creeping over the tan in tan-yards; others occur in rotten wood, decaying bean-stalks, and dung. The spore cysts may or may not be stalked, and the protoplasm enclosed within them does not all become spores, but the remnant forms a meshwork of fibres differing in details in the various genera. This network, termed a *capillitium*, serves to support the spores, and possibly helps in their escape when the surrounding wall gives way. The walls of the cysts may be strengthened by the deposition of calcium carbonate. The coating of the spores is of a cellulose nature: a substance usually associated with the vegetable kingdom, but not unknown among animals, especially amongst the Protozoa. At times the plasmodia contract and surround themselves by a cyst, and pass through a quiescent period. This condition is known as the *sclerotium*.

Myxomycetes are capable of retaining their vitality for long periods of time in a dried-up condition; they resume their active life again when supplied with moisture. About 300 species of Myxomycetes have been described, chiefly by botanists, who regard these organisms as being allied to the Fungi.

Class III. Lobosa.

The individuals of this group are those Protozoa in whose life-history the amoeboid phase predominates. The pseudo-podia are lobose, thick, blunt processes of protoplasm, which are never filamentous and never anastomose. One or more contractile vacuoles are found, and it is stated that urates have recently been demonstrated in connection with these vacuoles in some *Amoebae*. The amoeboid individuals may conjugate from time to time, but do not form plasmodia. They some-times encyst, and the cyst is a resting one (*hypnocyst*) and not a reproductive one (*sporocyst*). The usual form of repro-duction is fission, which may pass into gemmation. The distinction between the endoplasm and ectoplasm found in these and other Gymnomyxa is more apparent than real, and depends only on the presence or absence of food and other granules, the actual protoplasm of the organism being of one consistency.

Some of the Lobosa have acquired the power of forming shells, and this affords a convenient character by which we can divide the class into two orders: (1) the Nuda, and (2) the Testacea.

Order 1. **Nuda.**—The most familiar example of the former order is the *Amoeba*, of which there are many species quite distinct from the amoeboid spores of the Myxomycetes, which are often taken for Amoebae. The various species differ one from another in the nature of their pseudopodia and in the character of their nuclei. In some species the former are little more than low eminences, standing out from the general surface, in others they are long finger-shaped processes which stream rapidly hither and thither. Some members of this order, as the *Amoeba princeps* and *Pelomyxa*, have numerous nuclei scattered through the body: in the first-mentioned form these arise by the gradual "fragmentation" of the original nucleus. Such a multinucleated condition is constant in some species. In some cases the soft protoplasmic body has been observed to contract away from, and to lie within, a very thin cuticular membrane, which maintains the outline that the Amoeba possessed the moment before contracting; this cuticle is not usually visible, except in *Lithamoeba*, when it exists it must be very attenuated and elastic.

Pelomyxa is one of the largest of the Lobosa, the species *P. palustris* having a diameter of more than 2 mm. The external protoplasm is clear and produced into pseudopodia (Fig. 3). The inner mass is crowded with vacuoles, and contains in addition to the numerous nuclei (5, Fig. 3) a number of refringent bodies of unknown function (6, Fig. 3), and many food particles. It has been observed to set free minute amoeboid spores, which probably grow into new Pelomyxas.

Order 2. **Testacea.**—The shell which encloses the proto-plasmic body of these Lobosa may be soft and cuticular, and may then be strengthened by grains of sand adhering to it, or it may be hard. In either case the protoplasm can be extruded from an aperture in the shell. *Arcella* (Fig. 4) is a common genus found in the soft debris at the bottom of clean ponds and ditches, and on the surface of aquatic plants. The shell is chitinoid, and arched on the upper surface, flat on the lower, so

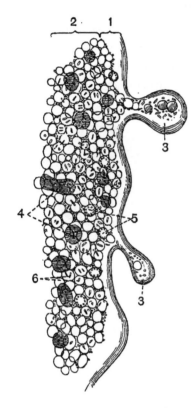

Fig. 3.— Portion of a *Pelomyxa*, highly magnified.

1. Clear external layer of protoplasm, ectoplasm.
2. Central protoplasm, crowded with granules, etc.—endoplasm.
3. Pseudopodia.
4. Refractive bodies.
5. Nuclei.
6. Cylindrical crystals scattered through the protoplasm.

that it is somewhat dome-shaped or hemispherical in form. In the centre of the flat surface is a circular opening through

Fig. 4.—*Arcella vulgaris*, Ehr.

1. Shell.
2. Protoplasm within the shell.
3. Protoplasm without the shell—pseudopodia.
4. Nucleus ; there is more than one.
5. Contractile vacuole.
6. Aperture of shell.
7. Space where the protoplasm has withdrawn from shell.
8. Gas vacuoles.

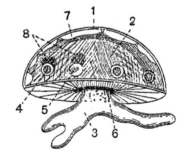

which the protoplasm protrudes in the form of blunt pseudopodia. The protoplasm within the shell encloses more than one nucleus, several contractile vacuoles, arranged round the circular

border, and numerous food vacuoles.   In addition we find one
or more contractile vacuoles which enclose no liquid, but a
gas, possibly CO_2.   This gas vacuole serves as a hydrostatic
balance ; when it disappears the Arcella sinks.   Two individuals
are sometimes found lying with their flat surfaces applied to
one another in the process of conjugating.   This has been in
some cases observed to precede reproduction, which takes place
by the constriction of small portions of protoplasm, either from

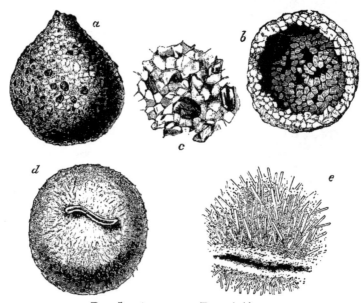

FIG. 5.—Arenaceous *Foraminifera*.
a. Exterior of *Saccammina.*   *b.* The same laid open.
c. A portion of the test more highly magnified.   *d. Pilulina.*
e. Portion of test more highly magnified.

the protruding pseudopodia or from the protoplasm enclosed in
the shell.   In the latter case the abstricted portions escape
through the shell mouth and grow up into new Arcellae.

Difflugia is a genus with a soft shell strengthened by the
presence of sand particles and diatoms ; the various species have
various shapes, some being flask or urn shaped, and one is
slightly coiled.

CLASS IV. **Reticularia** (Foraminifera).

In these Protozoa, the pseudopodia are filiform, and anastomose into a fine reticulum, along the strands of which gran-

ules may be seen streaming, evidence of the active movement of the protoplasm. They are never entirely naked, but are enclosed in a shell, which may be chitinous, calcareous, or composed of agglutinated sand grains (Fig. 5). There may be one or many nuclei, and a contractile vacuole has not been observed in most cases. Their method of reproduction is not very well known; it may take place by fission, or by the formation of buds. There are both marine and freshwater representatives, of this class. The enormous variety of forms under which the shells of the Reticularia present themselves, and their importance in building up large masses of chalk, limestone, etc., have always attracted the attention of naturalists. The class was formerly divided into two groups: the Perforata, those whose shell is pierced by numerous fine pores all over its surface, through which the filiform pseudopodia find exit; and the Imperforata,

FIG. 6.—Globigerina, as captured in the tow-net near the surface.

without the minute pores, but with one or more larger openings, for the exit of the protoplasm. This division is, however, tending to be obliterated. Many of the shells consist of one chamber only (monothalamia, Fig. 5), others, as they grow in size, accommodate their increased bulk by the addition of more chambers (polythalamia, Fig. 7), and it is chiefly the marvellous variety of ways by which the new chambers are added which produces the great divergency of forms. In some cases the protruding pseudopodia deposit a secondary shell, which obliterates the outline of the primary shell, and usually masks

FIG. 7.—*Globigerina bulloides*, as seen in three positions.

its form. The mud at the bottom of the Atlantic and other seas is composed to such an extent of the calcareous shells of *Globigerina bulloides* (Fig. 7), which, when the protozoan dies, sink to the bottom, that it is usually known as Globigerina ooze.

FIG. 8.—Globigerina ooze from 1900 fathoms.

The living Globigerina (Fig. 6) floats at the surface of the sea, the protoplasm extending round the shell and forming a much vacuolated envelope to it. Some slight idea of the enormous number of these organisms which must have lived to build up the foraminiferous rocks which extend from the Palaeozoic times onward may be formed from the fact that D'Orbigny estimated there were 160,000 shells in a gramme of sand from the West Indies, and Schultze gives 1,500,000 in 15 grammes of sand from the coast of Sicily. The nummulitic limestone of the Mediterranean basin is composed of the calcareous shells of

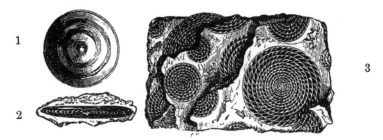

FIG. 9.

1. A piece of Nummulitic limestone from the Pyrenees, showing *Nummulites* laid open by fractures through the median plane.

2. Vertical section of Nummulites.
3. *Orbitoides.*

Nummulites, a Foraminiferan, which sometimes acquired the diameter of a shilling. Other species, such as *Fusulina* and

FIG. 10.—*Rotalia* with pseudopodia extended through the pores of the shell.

Rotalia (Fig. 10), also took a large share in building up the imestones of the Old World.

Gromia is a form found in both fresh and salt water; it has a membranous shell of the imperforate type, with an opening at

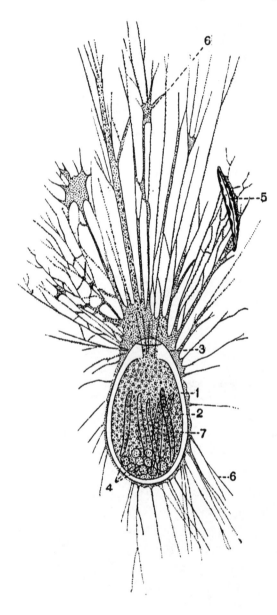

FIG. 11.—*Gromia oviformis*, Duj.

1. The shell.
2. Protoplasm inside the shell.
3. Protoplasm outside the shell.
4. Numerous nuclei.
5. A Diatom surrounded by pseudopodia.
6. Pseudopodia anastomosing.
7. Ingested diatoms.

one or both ends, from these the protoplasm passes out and forms a layer round the outside of the shell, from which the fine reticulating pseudopodia arise. The shell is thus completely imbedded in protoplasm, both inside and outside. In Lieberkühnia (Fig. 12), an allied form, the pseudopodia anastomose to a great extent and form a close reticulum.

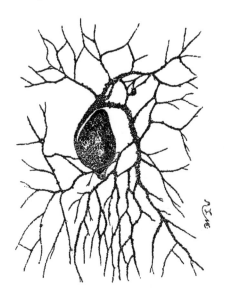

FIG. 12.—*Lieberkühnia*, with reticulate pseudopodia.

CLASS V. **Heliozoa.**

Mostly spherical in form, sometimes supported by a siliceous skeleton, and with radiating stiff pseudopodia. The protoplasm of the body is very vacuolated, and contains one or more nuclei. Near the surface of the body one or more contractile vacuoles may be observed. With few exceptions, they inhabit fresh water.

Actinophrys sol, the sun animalcule (Fig. 13), is one of the common microscopic objects found in still fresh water. It may be met with floating amongst the leaves of submerged plants, and presents a globular body which undergoes slight changes of outline, and is usually very vacuolated. The single nucleus occupies a central position, the contractile vacuole is on the surface, and food vacuoles containing portions of algae, infusoria, etc., may be seen throughout the body. The pseudopodia are stiff and hair-like, and are supported by an axial fibre; they can be withdrawn into the body. When they come in contact with a particle of food they bend slowly over it, and bring it near to the surface of the protoplasmic body, when it is swallowed with the surrounding drop of water, and thus a food

2

vacuole is formed.   Encystment rarely takes place.   *Actinophrys*
and another genus, *Rhaphidiophrys*, have been observed to form

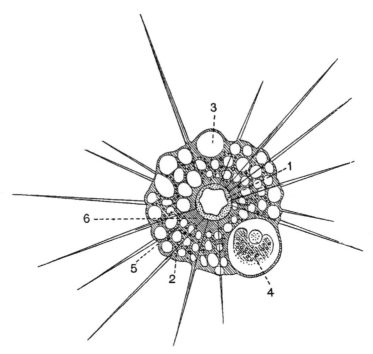

FIG. 13.—*Actinophrys sol*, Ehrb.   From Bronn.

1. Nucleus in centre of body.
2. Axis of pseudopodium extending into cell as far as nucleus.
3. Contractile vacuole.
4. A mass of food in a food vacuole.
5. Superficial vacuolated protoplasm.
6. Deep, finely granular protoplasm.

colonies by incomplete fission.   Reproduction commonly takes
place by fission, but in some cases spores have been observed;
those of *Actinosphaerium* being provided with a siliceous shell.
This last-named genus (Fig. 14) is much larger than *Actinophrys*; it contains numerous nuclei, situated in the deeper
protoplasm.   The pseudopodia are supported by an axial ray.
Rhaphidiophrys is usually found in colonies; it has a skeleton of
siliceous spicules, matted together round the body, each spicule
lying tangentially to the surface.   *Acanthocystis* has siliceous
rays arranged radially; they are of two kinds: short ones, which
are forked at their outer end, and long stout ones.   They are
attached to the body by a small disk.   Finally, leading on to
the condition found in the Radiolaria, *Clathrulina* (Fig. 15),

a stalked genus, has a spherical siliceous shell perforated by numerous openings.

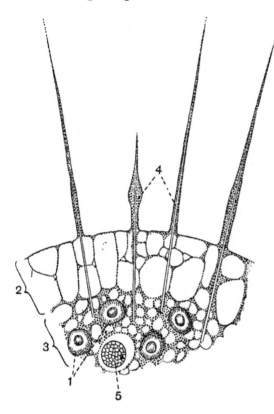

FIG. 14.—A portion of *Actino-sphaerium Eichhornii*, Ehrb., highly magnified, seen in optical section. From Bronn.

1. Nuclei.

2. Ectoplasm.

3. Endoplasm.

4. Pseudopodia with axis.

5. Food mass in food vacuole.

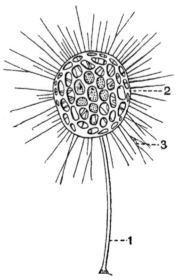

FIG. 15.—*Clathrulina elegans*, Cienk. × 150-200. From Bronn.

1. Stalk.

2. Shell.

3. Pseudopodia protruding through apertures in shell.

Class VI. **Radiolaria.**

Organisms which are either spherical or with one principal axis whose body is divided into a central mass containing one or more nuclei and a peripheral portion, by the presence of a membrane known as the "central capsule." This is perforated so that the intracapsular protoplasm is continuous with the extracapsular protoplasm. A well-developed skeleton, in most cases siliceous, is present. This consists either of loose siliceous spicules or of a continuous skeleton which may take the form of lattice-work spheres, arranged concentrically, and united to one another by radial spicules, which project beyond the surface of the body. The skeletons of Radiolarians occur in vast numbers on the floor of some seas, forming a layer of siliceous ooze (Fig. 16). The skeleton may be wholly outside the central capsule, or it may be partially within it. Numerous fine pseudopodia radiate around the body; these unite to some extent, nodes of protoplasm being found at the point of

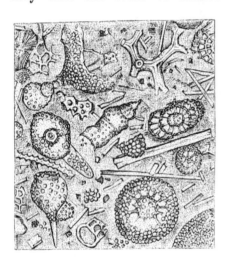

Fig. 16.—Radiolarian ooze from 4475 fathoms in Central Pacific.

union. A streaming of the protoplasm along the pseudopodia, as in Heliozoa and Reticularia, takes place, and granules have been seen to circulate between the intra- and extra-capsular protoplasm. No contractile vacuole has ever been observed. The protoplasm is much vacuolated: a condition commonly met with in those Protozoa which, like the Radiolaria, swim near the surface of the sea.

Some very remarkable bodies, known as *yellow cells*, are found widely distributed amongst the Radiolaria. These are small oval yellow bodies, only found in the extracapsular protoplasm. They were formerly regarded as part of the body of the Radiolarian ; more recent research has, however, shown that they continue to live after the death of the animal, that they multiply in more than one way, occasionally forming mobile swarm-spores, that similar cells occur in the tissues of many Coelenterates, and that they contain chlorophyll, although this colouring matter is masked by a yellow pigment. A nucleus and a cellulose cell-way are also present. These features have caused these yellow cells to be regarded as unicellular algae, living in a state of commensalism with the Radiolarian, and they have received the name of *Zooxanthella nutricola*. They are not found in all species, and are usually absent in *Acanthometra*, and in the other species with a horny skeleton.

The protoplasm contains, in addition to the yellow cells, numerous oil or fat globules, and crystals and concretions of unknown use.

No conjugation has ever been observed in this class ; reproduction is sometimes by simple fission, which commences first in the central capsules. Spore formation in the central capsules also takes place, and results in the formation of mobile spores ; but the details are complex, and the exact sequence of events not thoroughly understood.

Many form colonies by the fusion of their extracapsular protoplasm. That of *Collozoum*, the individuals forming which are devoid of skeleton, may be an inch or more long. The various members of the colony are held together by a gelatinous matrix.

Group B. CORTICATA.

The animals which are grouped together in this second division of the Protozoa have as a common feature a differentiation of the protoplasm into a more fluid central portion and a firmer cortical layer usually associated with a limiting membrane, which surrounds their body and gives it a definite shape. As a consequence of the presence of this cortical layer, these forms which take solid food have acquired one or more channels through which the nutriment is ingested, and usually a definite area whence the undigested remnants are extruded. The parasitic forms, which live in the nutritive fluids of their hosts, are usually devoid of any such cell mouth or anus. The presence of the cortical layer has also rendered locomotion by pseudo-podia rare, and those Corticata which move about actively do so as a rule by means of rows of cilia or by a single or paired flagellum.

CLASS I. **Flagellata.**

The bodies of the members of this class are usually very minute, and always contain a nucleus; they are moved by the lashing of one, sometimes by two or three flagella. A mouth

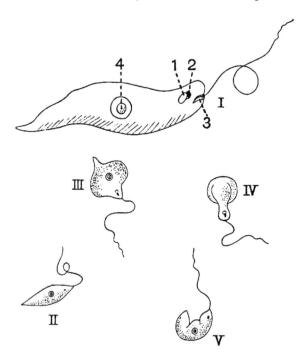

FIG. 17.—I. Typical form of *Euglena viridis*, Ehrb., after Sav. Kent.

1. Contractile vacuole.   3. Gullet and origin of flagellum.
2. Pigment spot.   4. Nucleus.
   II. III. IV. V. Four views of *Euglena viridis*, showing the change of shape consequent upon the euglenoid movement.

may be present, but in those forms which live in nutritive fluids the nourishment is usually imbibed by the whole surface of the body. One or more contractile vacuoles occur, and sometimes a pigment spot is situated at one end of the body. Conjugation sometimes occurs, and is followed by the breaking up of the body into spores, reproduction also takes place by simple fission.   Many Flagellata form colonies, the individuals of which are imbedded in a gelatinous matrix.

Euglena viridis (Fig. 17) is a minute oval Flagellate found in puddles by the wayside, or on roofs, etc.   It has a thin cuticle, and undergoes curious rhythmical changes of outline.   The elongated spindle-shaped body shortens, and becomes correspondingly thicker.   The thickening appears then to travel to the posterior end of the body and die out.   The animal has at this moment its elongated spindle-shaped form; it then shortens again, and the whole movement is repeated.   At the anterior end is a single long flagellum, which by its lashing drags the body swiftly through the water.   Lankester has dwelt on the difference between the action of such a flagellum (*tractellum*) and of one that propels an organism in front of it, as the tail of a spermatozoa or the flagellum of Bacteria (*pulsellum*).

Round the base of the flagellum is a depression, the mouth, which leads into the central protoplasm; and close to this, and apparently opening into it, is a reservoir communicating with a contractile vacuole.   A pigment spot also is found in the same region, but the reticulate nucleus occupies the centre of the body.   The whole body is coloured green, by chlorophyll granules.   Grains of paramylum, a body with the same composition as starch, are also found in the protoplasm.

Hypnocysts, or resting encysted forms, are frequently formed amongst the Flagellata, when they find themselves in unfavourable circumstances.   The encysted *Euglena* may emerge after a certain period of rest from the Hypnocyst, or it may whilst in the cyst divide into 2 or 4 spores each of which emerges as a young Euglena.   Reproduction by multiple fission has also been described in this species, a vast number of spores being formed, each of which grows into a new individual.   Conjugation also takes place in the Flagellata, and is usually followed by encystment and the division of the contents of the cyst (sporocyst).   At other times fission may occur in the free state.   Sometimes *macrogonidia* and *microgonidia* are produced, and the latter fuse with one another or with adult individuals (*Protococcus*).

The Flagellata are divided into two groups: the *Lisso-*

flagellata, with no collar round the base of the flagellum ; and the *Choanoflagellata,* in which the protoplasm is produced into a collar which surrounds the anterior end, from the middle of this the single large flagellum takes its origin. *Codosiga* (Fig. 18) is a colonial form of this kind, composed of long branching stalks, the end of each branch bearing an individual. These collared flagellates have a striking resemblance to the collar cells lining the flagellate chambers in a sponge ; and a genus, *Proterospongia* (Fig. 19), discovered by Saville Kent, in which the individuals of the colony are sunk in a jelly, lends some support to the view that Sponges may have originated from colonies of Choanoflagellata. In this genus the individuals near the surface are of the typical form ; but certain wandering amoeboid cells have sunk into the central jelly, and some of these have become spherical, and then divided up into microgonidia, in a manner recalling the formation of spermatozoa in a Sponge.

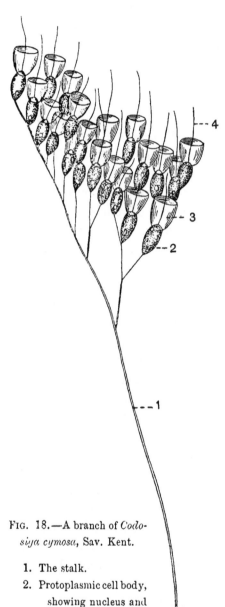

FIG. 18.—A branch of *Codosiga cymosa,* Sav. Kent.

1. The stalk.
2. Protoplasmic cell body, showing nucleus and granular protoplasm.
3. Collar.
4. Single flagellum.

Most Flagellata live in fresh water ; some are marine, and some parasitic, living in the alimentary canal or blood of

Vertebrates and Arthropods. *Euglena viridis* often exists in such numbers as to turn the water green, and a species of *Haematococcus* is responsible for the red snow of the Arctic regions. Many of them, such as *Polytoma*, thrive in putrid

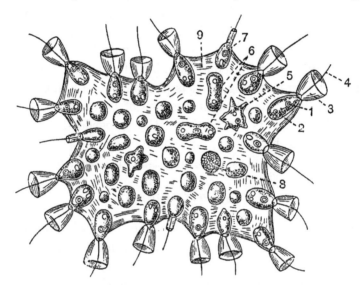

FIG. 19.—*Proterospongia Haeckeli*, Sav. Kent. × 800.

1. Nucleus.	6. Other individuals undergoing fission.
2. Contractile vacuole.	
3. Collar.	7. Individual with collar contracted.
4. Flagellum.	8. Individual divided up into number of spores (microgonidia).
5. Amoeboid individual sunk in supporting jelly.	
	9. Jelly-like supporting matrix.

liquids, such as the water of macerating tubs, and these forms are mostly saprophytic.

Some of the lower forms are apt to withdraw their flagellum and become amoeboid (*Ciliophrys*); in others (*Cercomonas*), the posterior end of the body is very apt to throw out pseudopodia, and the young of many species exhibit amoeboid movements. These facts support the view that the Flagellata are derived from the Gymnomyxa; and indeed there are certain forms which might equally well find a place in the class Proteomyxa.

CLASS II. **Rhynchoflagellata.**

This class contains but one or two genera. One of them, *Noctiluca*, attains a large size : $\frac{1}{25}$ of an inch in diameter. It is of a spherical form, and grooved on one side like a peach. From the bottom of this groove a very large transversely-striated flagellum takes its origin. Near the base of the flagellum is the mouth, opening into a sort of pharynx; a second smaller flagellum has its origin in the latter. The protoplasm of the globular body is very reticulate. No contractile vacuole has been observed. At times *Noctiluca* withdraws its flagella and passes into a resting condition, but it does not form a cyst. This animal is interesting, as it is phosphorescent, and gives rise to a large part of the phosphorescence of temperate seas. The seat of the light is said to be the superficial protoplasm. Reproduction is by fission, but motile swarm-spores are also formed in large numbers. Conjugation has been observed.

Class III. Ciliata.

This class is characterised by the possession of cilia as locomotor organs, arranged either in a perioral ring, or forming a more or less complete covering. A nucleus is always present; this may be single, and is then accompanied by a paranucleus, or it may be distributed in small fragments throughout the body. One or more contractile vacuoles are present. The shape of the body is very various. Some Ciliata are united into colonies, and some form gelatinous tubes in which they live; the majority, however, are free-swimming. Conjugation is common, but usually does not end in permanent fusion. Fission, usually simple, but sometimes multiple, is the usual method of reproduction.

This class has been divided into four Orders, characterised by the arrangement and nature of the cilia.

Order I. Peritricha.—*Cilia arranged in an anterior ring or spiral, to which a posterior ring may be added. The rest of the body unciliated.*

Torquatella has its cilia fused to one another, and so a vibratile membrane is formed which surrounds the anterior end. *Trichodina* is very common, crawling on the tentacles and body-wall of Hydra; it is pyramidal in shape, and has two circlets of cilia. On its sucker-like base a curious horny toothed ring is situated.

Vorticella is attached by a stalk to submerged water-weeds, etc. Up the centre of this stalk runs a muscle fibre, a differentiation of the protoplasm, attached at intervals to the cuticular sheath of the stalk. The differentiation of cortical and medullary protoplasm is well marked. The nucleus is a coiled loop. The animal sometimes encysts, but this is prob-

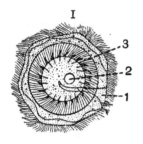

I

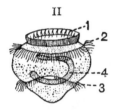

II

FIG. 20.—*Trichodina pediculus*, Ehrb. × 300.

I. View of the base.

  1. Mouth.

  2. Contractile vacuole.

  3. Corneous collar.

II. View from the side.

  1. Corneous collar.

  2 and 3. Ciliated rings.

  4. Nucleus.

ably only the formation of a hypnocyst, and has no reproductive significance. Binary longitudinal fission is the commonest form of reproduction, one half remaining on the stalk, the other (macrozooid) acquiring a ring of cilia and swimming away to settle elsewhere. At other times the *Vorticella* divides into eight microzooids, which conjugate permanently with the sessile individuals. These also occasionally produce microzooids by budding, and the colonies are also increased by the formation of buds.

Order II. Heterotricha.—*The body is covered uniformly with short cilia, and a circlet or spiral of long cilia is developed in relation to the mouth.*

Stentor has a moniliform or beaded nucleus, and a considerable number of paranuclei. In some cases these latter correspond in number with the beads of the nucleus. *S. polymorphus* is one of the largest Ciliata, reaching a length of $\frac{1}{20}$ in. In this genus conjugation takes place by the oral face,

as in *Paramoecium*, and fission is oblique.   Recent observation
on this and other forms shows that as long as the products of
artificial division contain part of the nucleus with its chromatin,
they are capable of regenerating the lost parts ; those portions
of the body which are without any portions of the nucleus die.

Balantidium is a genus which lives parasitically in the
human colon, and with *Nyctotherus*, is found in the rectum
of Anura, etc.   The latter is interesting, since it is provided
with a permanent anus with a cuticular lining.   In most Ciliata
the situation of the cell anus is constant, but there is nothing
to indicate its position, except when waste matter is being
expelled.

Order III. Holotricha.—*In this order the body is uniformly clothed with short cilia, arranged in regular rows.   Sometimes those on the adoral surface are slightly longer than the
others.*

Paramoecium is one of the commonest genera of this
order.   Close underneath its cuticle in the ectosarc, is a layer
of oval bodies, the *Trichocysts*; these, when the animal is
irritated, discharge threads, which have probably the same
functions as the stinging threads in the nematocysts of
Coelenterata.

Maupas has recently described the conjugation of *Paramoecium aurelia*, which takes place as follows.   As soon as two
individuals come together, the paranuclei, of which there are
two (Fig. 21, 1), separate themselves from the nucleus, increase
in size (Fig. 22, A), become spindle-shaped, and ultimately divide
into two (2 and B).   The two halves of each divide again (C), so
that a stage is found with eight similar portions of the original
paranucleus in each individual (3 and 4).   Of these eight
corpuscles, seven are absorbed and disappear (5) ; the eighth
alone, and this is always that one which lies nearest to the
mouth, undergoes further change.   This corpuscle increases
in size and divides into two (6 and D), thus giving rise to the
male and the female pronucleus.   The former of these passes out
of each conjugating individual into the other, and there fuses
with the female pronucleus (7 and E).   The conjugating animals
now separate.   The " fertilised " paranucleus now divides into

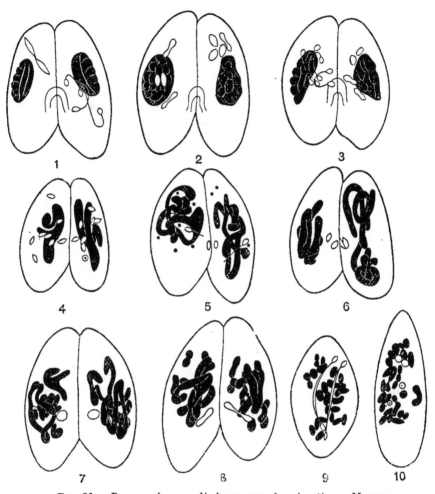

FIG. 21.—*Paramoecium aurelia* in process of conjugating. Maupas.

1. The two paranuclei in each individual becoming spindle-shaped and com-mencing to divide.

2. Stage with four paranuclei in each individual.

3. The four paranuclei again divide. The division is almost complete in the left Paramoecium.

4. Stage with eight paranuclei.

5. Seven paranuclei being absorbed, and the one nearest the mouth remaining and dividing in two, the male and female pronuclei.

6. Exchange of male pronuclei.

7. The male has fused with the female pronucleus to form the fertilised paranucleus.

8. The fertilised paranucleus divides.

9. The resulting halves divide again.

10. The two halves at the posterior end form the new paranucleus ; the two at the anterior end, the new nucleus.

The disintegration and dissolution of the nucleus are shown in the same series of figures.

two (8 and 9, and F), and each half divides again (G), so that
each individual contains four fragments of paranucleus : two at
one end of the body and two at the other (9).   The two at

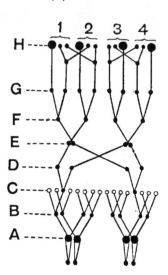

FIG.   22.—Diagram   of   changes   undergone   by
      paranuclei of *Paramoecium aurelia* during
      conjugation.   Maupas.

A. Increase in size of paranuclei.

B. 1st division of paranuclei.

C. 2nd division of paranuclei, and disappearance
of seven-eighths.

D. Division of remaining portion into male and
female pronuclei.

E. Fusion of male and female pronuclei.

F. Division of fertilised paranucleus into two
halves.

G. Division of these halves.

H. Of the four quarters shown in G, one forms
the new nucleus of young Paramoecium which is
formed by fission, the other divides into two and
forms two paranuclei, one in each of the new individuals.

the posterior end undergo no change, and form the paranuclei
of the new individuals which result from the subsequent
fission ; the two at the anterior end increase, and are destined
to form the nuclei of the new individuals (10).   Before this
is completed, however, the two paranuclei have again divided

FIG. 23.—*Opalina ranarum*, Ehrb.   From Bronn.

1. Nuclei.

2. Ectoplasm.

(H), so that after fission each new individual contains one
nucleus and two paranuclei.

Whilst the paranucleus of the original Paramoecium has
been undergoing these changes, the nucleus has first become

mammillated and then band-like, and ultimately undergoes fragmentation. The fragments persist some time, and in other species, *P. caudatum*, possibly take part in forming the new nucleus. In *P. aurelia* they ultimately disappear, the majority of them being in all probability extruded.

Opalina ranarum (Fig. 23) lives in the rectum of a Frog. As it grows, its nucleus divides, until a great number of nuclei are found. The animal then slowly segments, until each portion contains only one or two nuclei; these form a cyst, and in this condition leave the body of the frog. When eaten by a Tadpole, they emerge, grow, and the nucleus again begins to divide. In *Opalinopsis* this division of the nucleus has been carried further, and it exists in the form of a fine powder scattered through the endoplasm, the particles of which at times coalesce and form a single nucleus again.

Order IV. Hypotricha.—*These Ciliata have a flat ventral surface, completely ciliated, or provided with enlarged muscular cilia. The dorsal surface is convex, unciliated, but sometimes bears retractile setae. Both mouth and anus are well developed.*

In this group the cuticle is sometimes strongly developed, and forms a protective plate of some thickness; in the Euplo-

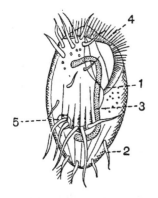

Fig. 24.—*Euplotes patella.* After Bütschli.

Mouth.
Hypotrichous processes.
Nucleus.
. Cilia of oral groove.
Contractile vacuole.

tidae this exists on the dorsal surface only. The distinction between endoplasm and ectoplasm is almost lost. Encystment is not uncommon, and one species (*Gastrostyla vorax*) has been kept alive in a hypnocyst for the space of two years.

Class IV. **Acinetaria.**

No cilia or flagella present, but a number of tentacular processes, which may be adhesive or may be tubular and suctorial. The Acinetaria are usually fixed, and most commonly stalked (Fig. 26). The nucleus is single, and often branched. One or more contractile vacuoles occur. Reproduction takes place by binary fission and gemmation; the latter is often internal.

The Acinetaria are either marine or freshwater, and they are all carnivorous, living chiefly on the soft parts of Infusoria.

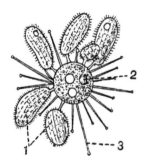

Fig. 25.—*Sphaerophrya magna*, Maupas.   × 300.

1. Individuals of *Colpoda parvifrons*, a ciliated Infusorian whose soft parts are being sucked up by the *Sphaerophrya*.

2. The nucleus.

3. One of the hollow tentacles ending in a knob.

The cuticle may be thin, but in some cases a definite membranous capsule is formed. Two kinds of tentacles may be found: one long and adhesive, whose function is to catch and hold the prey; the other is shorter, tubular, and ends in a sucker,—these latter are sometimes provided with a spiral thickening. The soft protoplasm of Infusoria, on which they prey, is sucked up through these hollow tentacles (Fig. 25). *Dendrocometes* has a round body, from which four to six many-branched stout arms project. Each branch ends in a point, which is said to be hollow, and by means of which food is sucked into the body. In this genus, and in one or two others, the contractile vacuole has an excretory duct.

Acineta (Fig. 26) is a stalked form with a membranous cup; the tentacles are arranged in two clusters at each side of the body.

Reproduction may be by fission, or by external gemmation, or by internal gemmation, in which case a brood-pouch is formed, which may be open, but is more commonly closed.

The bud, or buds, are formed in the floor of this cavity. When they leave the parent they are ciliated; a fact which lends support to the view that the Acinetaria are descended from the Ciliata. They at first lead a wandering life, but after a time settle down, lose their cilia, and acquire tentacles.

Class V. **Sporozoa.**

The members of this class are all parasitic; and correlated with this condition of life is the absence of locomotor organs, the absence of any mouth,—the nutriment being absorbed all over the body,—and a reproductive process which results in the formation of a very large number of young. The nucleus is always single even when the cell is divided into two chambers. No contractile vacuole is present. The cortical layer of protoplasm may show traces of fibrillation.

Of the four sub-classes which compose this class, that of the Gregarinidea is the most important. The true Gregarines, with very few exceptions, pass their early life as cell parasites, afterwards

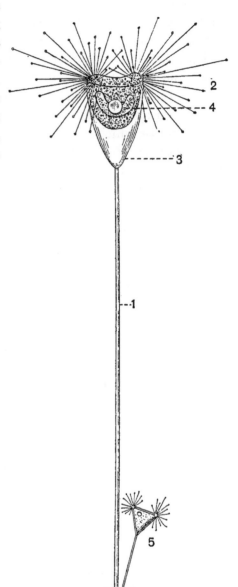

Fig. 26.—*Acineta grandis*, Sav. Kent.

1. Stalk.
2. Knobbed tentacles.
3. Membranous shell.
4. Nucleus.
5. An example of *Acineta lemnarum*, showing relative size of the two species.

emerging and living parasitically in the various cavities of their host. The Coccidiidea pass their whole life as cell parasites. The true Gregarines are divided into the Monocystidae, whose body is not divided into two chambers, and which inhabit Annelids, Gephyrea, Platyhelminthes, and Tunicata; and the Polycystidae, in which a transverse partition divides the cell into an anterior and posterior chamber, the latter invariably containing the nucleus. An anterior outgrowth, the epimerite, which serves to attach the Gregarine to the tissues of its host, is often present, but this is shed sooner or later. The Polycystidae have hitherto only been found in the alimentary canal of Arthropods.

Monocystis magna (Fig. 27) is frequently to be found with its anterior end embedded in one of the epithelial cells of the en-

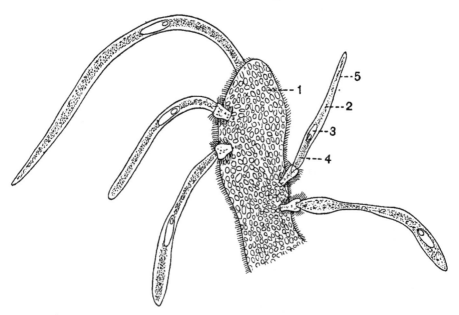

FIG. 27.—Five individuals of *Monocystis magna* (Schmidt), with their anterior ends embedded in the cells of the rosette-shaped inner end of the vas deferens of *Lumbricus terrestris.*

1. Part of the vas deferens of *L. terrestris.*
2. The endoplasm of the Gregarine.
3. The nucleus.
4. The ectoplasm.
5. The cuticle.

larged inner rosette-shaped openings of the vas deferens of the common earthworm. It is visible to the naked eye, and sometimes attains a length of 5 mm. The body is limited by a thin cuticle; within this lies the cortical protoplasm, which, though full of granules, is transparent. The medullary protoplasm is dark brown and opaque. Bütschli has shown that some of the granules are composed of a starch-like material. The body exhibits movements of a euglenoid character; waves of contraction passing down the elongated cell. The flowing about of the protoplasm is rendered visible by the granules. The nucleus is clear and vesicular, with few granules, and it lies in the centre of the cell.

At times two individuals come together and surround themselves by a spherical capsule; apparently no true fusion takes place, but the bodies of the Gregarines commence to form spores. This spore formation proceeds from without inwards in each cell, but the whole protoplasm is not always used up for this purpose. The spores are shuttle-shaped, they acquire a capsule (chlamydospores), and are often spoken of as *pseudonavicellae.* The pseudonavicellae escape from the cyst by its bursting, or in *Clepsidrina*, a Polycystid, by special sporoducts. Their contents divides into eight elongated bodies, known from their shape as *falciform* bodies; these leave the pseudonavicella, and probably grow up directly into the adult form. A recent observer has, however, stated that the contents of the pseudonavicella does not break up into falciform bodies, but the protoplasm becomes grooved, and thus the appearance of segmentation is produced. According to him, the whole contents of the pseudonavicella escapes and grows in a new Gregarine.

Many of the Polycystidae are more highly differentiated than the species described above; their cuticle may be ridged or tuberculated, and is frequently produced into hooks in the epimerite, and the cortical layer of protoplasm may show traces of fibrillation. When these septate forms conjugate, they usually lie side by side. *Gregarina gigantea*, which inhabits the alimentary canal of lobsters, attains the astonishing length of $\frac{3}{4}$ of an inch.

Coccidiidea are minute spherical cells which infest the epithelium of the intestine, the liver cells, etc., of Vertebrates, Mollusca, and Insects. Whilst still in their cell host, they give rise to chlamydospores and falciform young.

CHAPTER III

METAZOA

CHARACTERISTICS.—*Multicellular animals, which pass through a unicellular stage, the ovum or egg. This multiplies by division, and the cells thus formed, instead of remaining equivalent to one another, become differentiated and are arranged in tissues. Reproduction takes place by means of ova and spermatozoa.*

ACOELOMATA.

Metazoa in which a two-layered condition is the predominant one. The ectoderm and the endoderm may constitute the whole animal, but in many cases an intermediate layer (the mesoderm or mesogloea) lies between them. This middle layer may be homogeneous, but is more usually invaded by cells from one of the two layers, or from both. The cavities of the Acoelomata, except certain ectodermal pits, are in all cases continuations of the primary central cavity lined by endoderm, and no cavities exist lined by mesoderm comparable to a coelom. Radial symmetry about an axis passing through the mouth is a primitive and common feature of this subdivision. The animals which constitute it are exclusively aquatic, and almost entirely marine.

PORIFERA
- CALCAREA
  - Homocoela—*Ascetta.*
  - Heterocoela—*Grantia, Leucandra.*
- NON-CALCAREA
  - Hyalospongiae—*Hyalonema, Euplectella.*
  - Spiculispongiae—*Halisarca, Oscarella, Geodia.*
  - Cornacuspongiae—*Euspongia, Velinea, Spongilla.*

PORIFERA (the Sponges).

CHARACTERISTICS.—*Animals of very varied size and shape. Numerous minute pores allow the passage of water into the interior of the sponge, and the water is discharged through larger openings known as Oscula. The current of water is maintained by certain flagellate cells, which are usually aggregated in what are known as flagellate chambers. The mesoderm is well developed, and usually gives rise to a skeleton of calcareous, siliceous, or horny material; it also gives origin to the reproductive cells. Sponges may be unisexual or hermaphrodite. They are aquatic, and, with the exception of the Spongillidae, they are marine. They are devoid of tentacles and of nematocysts.*

The simplest type of Sponge is that of *Ascetta primordialis*, described by Haeckel. It is a hollow vase-like structure borne on a stalk with its free end open. This opening is the *osculum*. The walls are perforated by a series of small circular apertures, the "pores," and its cavity is lined by a layer of flagellate collared cells, whose activity keeps a current of water entering the pores and finding an exit through the osculum. The flagellate cells are endodermal. The outside of the vase-like body is covered with ectoderm, and between these two layers is a mesodermic tissue which produces triradiate calcareous spicules. The flagellate endoderm cells are said to possess contractile vacuoles. In the more highly organised Sponges the endodermic lining of the central cavity has lost its flagellate character and become a flat epithelium.

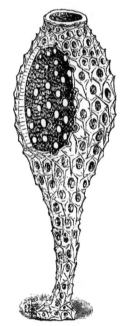

FIG. 28.—*Ascetta primordialis*, Haeckel. After Haeckel.

Grantia compressa is a sponge with a calcareous skeleton, which is frequently met with attached to rocks and stones round our coast. It is of a whitish colour, seldom more than an inch long, and rather variable in outline; a

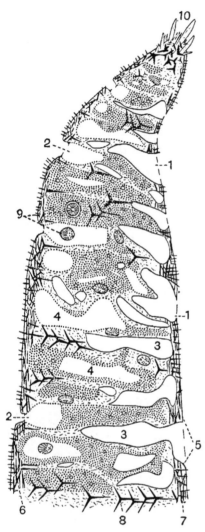

FIG. 29.—Part of a section through *Grantia labyrinthica*, vertical to the margin and to the two surfaces of the wall of the cup. After Dendy.

1. Inhalent pore.
2. Exhalent canal.
3. Inhalent canal.
4. Cavity of flagellate chamber.
5. Pore area.
6. Gastral skeleton.
7. Dermal skeleton.
8. Tubar skeleton.
9. Embryos.

common form is somewhat flask shaped. At its free end are situated one or more slit-like openings, the oscula. The body is compressed from side to side, and its wall is pierced by numerous minute inhalent pores, which lead by a system of branching tubes into the central cavity.

The substance of the sponge is composed of three layers the ectoderm, the endoderm, and between them the mesoderm. The ectoderm consists of flattened cells covering the outside of the sponge, and lining certain pits or depressions which are pushed into the substance of the sponge, and are termed intercanal spaces. The openings from the exterior into the intercanal spaces are termed "pores" (1, Fig. 29). Several pores are usually grouped together, and form the pore area. The intercanal spaces open on their inside by numerous apertures, called by Sollas "prosopyles," into the flagellate chambers (4, Fig. 29). These flagellate chambers are the most characteristic feature of the Sponges. They are lined by collared flagellate cells similar to those of the Choanoflagellata. Their flagella keep up a constant current of water, which passes in at the pores through the intercanal spaces and flagellate chambers, and into the central cavity, thence it leaves through the oscula.

Embryological research shows that we must regard the collar cells of the flagellate chambers, the cells lining the tubes which lead from them to the central cavity, and the cells lining the latter cavity, as endoderm. The mesoderm is a gelatinous tissue in which certain cells are found embedded; some of these form ova or break up into spermatozoa, whilst others give rise to the skeleton of calcareous spicules.

In sections of *Grantia* the intercanal spaces (3, Fig. 29) may be seen lying between the flagellate chambers, but quite distinct from these. They are lined by flat epithelial cells, and they ultimately open by a more or less wide mouth on to the exterior on the one hand, and by a series of circular pores, the *prosopyles*, into the flagellate chambers. These intercanal spaces are formed by the pushing in of the outer coating of the sponge, and are lined by ectodermal cells similar to those covering the outside of the sponge. In some sponges they reach a great degree of complexity.

The ectoderm of sponges is, as a rule, composed of flat cells

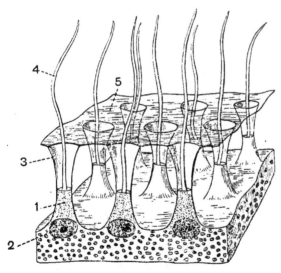

Fig. 30. — *Stelospongus flabelliformis*. Diagram of a portion of a flagellate chamber, showing the various parts of the collared cells and their relation to Sollas's membrane, which connects together the margins of the collars. After Dendy.

1. Body of flagellate cell.
2. Its nucleus.
3. Its collar fusing at its edges to form 5.
4. The flagellum.
5. Sollas's membrane.

in a single layer; rarely these cells become columnar, and bear flagella (*Oscarella lobularis*). The endoderm consists, in all but the Homocoela, of two kinds of cells: (i.) flat pavement cells, which line the central cavity and the ducts opening into it ; and (ii.) collared cells, which line the flagellate chambers, and in the Homocoela the general cavity. In *Grantia* the flagellate chambers do not open directly into the central cavity, but into a short exhalent canal (2, Fig. 29), the entrance to which is guarded by a sphincter diaphragm. The endoderm lines the whole canal system from the prosopyles to the osculum. In several genera distributed among several orders the collars which surround the base of the flagella are at their outer ends fused to form a membrane, which was first described and figured by Sollas (Fig. 30, 5). Bidder has shown that if a *Leucandra* be placed in water with carmine suspended in it, the water which comes from the oscula is always free from carmine granules, thus showing the presence of a very efficient filter, presumably the fused collars of the flagellate cells. It is still an open question whether the space within this membrane, between the body of the flagellate cells, is empty, or occupied by a transparent gelatinous substance.

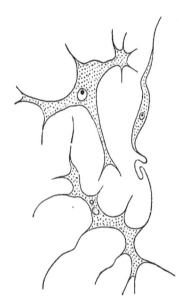

The mesoderm is a gelatinous layer, with branched or stellate cells scattered through it. Amoeboid cells (Fig. 31) wander through its substance, and convey nutriment from one part to another, and when occasion arises assist in removing irritant foreign matter from the body of the sponge (*phagocytes*). The branching mesoderm cells have been traced into direct protoplasmic continuity with both ectoderm and endoderm. The reproductive cells are also mesodermal, and the fertilised ovum develops in a space in the gelatinous mesoderm (9, Fig. 29), which is lined by a layer of flat endothelial cells, also mesodermal in origin.

Fig. 31.—Branching connective tissue cells from the mesoderm of *Thenea muricata*. After Sollas.

Certain mesodermal cells in the neighbourhood of the

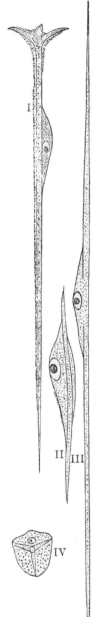

diaphragms in the exhalent canals of the flagellate chambers are believed to be muscular, and a nervous nature has been attributed to others situated near the inhalent pores.

The skeleton of *Grantia* consists of calcareous spicules, most of which are of a triradiate shape (6, 7, and 8, Fig. 29). Each spicule commences to appear in one of the mesodermal cells, but as it increases in size it may protrude from this. The spicules are said to gradually work towards the exterior of the sponge, and to be discharged as waste matter.

The character of the skeleton is made use of in classification. Only a very few sponges—*Halisarca*, *Oscarella*, and *Chondrosia*—are devoid of any kind of supporting structure. Those which possess calcareous spicules have been grouped together as the Calcarea, and opposed to all the other sponges, or Non-calcarea. The skeletons of this latter group may consist of siliceous spicules, or of a fibrous substance—spongin. The siliceous, like the calcareous spicules, originate in a single mesodermic cell (Fig. 32); both assume a great variety of size and shape, and the former may be articulated or fused to one another. The organic skeleton, found in the Order Cornacuspongiae, consists of spongin, a substance chemically allied to silk. It is secreted by a number of mesodermic cells termed spongoblasts, which form a layer all round the fibre, and a multicellular cap covering the ends (Fig. 34). The fibrous skeleton of *Euspongia*, devoid of spicules, and characterised by the regular arrangement of the network and the smallness of the meshes, forms the bath sponge of commerce.

In *Grantia* some of the amoeboid mesodermic

FIG. 32.—Spicules originating in single cells. After Sollas.

I. From *Stelletta*. II. From embryo of *Craniella cranium*. III. The same from adult *Craniella cranium*. IV. A four-rayed spicule from *Theonella Swinhoei*.

cells may be noticed withdrawing their pseudopodia, and passing into a spherical resting condition. These form the ova. Others

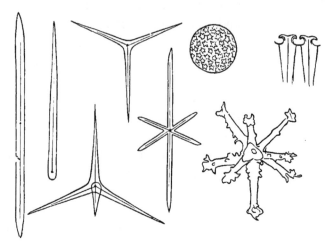

FIG. 33.—Various forms of Sponge spicules.

divide up into an immense number of spermatozoa (Fig. 35), each with a head and long vibratile tail. In *Grantia labyrinthica*

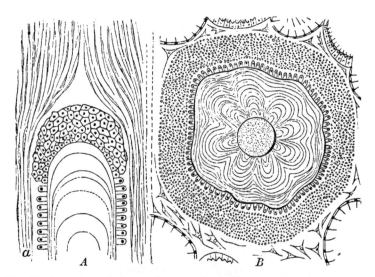

FIG. 34.—Sections through horny fibre of *Dendrilla*. *A*, Longitudinal section; layers of spongin, surrounded by a layer and surmounted by a cap of spongo-blasts; *a*, fibrous sheath. *B*, The same in transverse section. After Von Lendenfeld.

Dendy has described the ripe ova making their way through the mesoderm to the epithelium of the inhalent canals, and passing through the epithelial lining of these ducts. They then hang freely, supported by a short peduncle, into the lumen of the canal (Fig. 36), and are doubtless fertilised in this position by

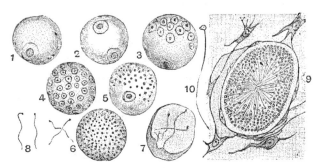

FIG. 35.—1-7, Development of spermatozoa in *Sycandra raphanus*; 8, mature spermatozoa, after Polejaeff (× 792); 9, a sperm morula in *Oscarella lobularis* (× 500); 10, an isolated mature spermatozoon, after Schulze (× 500).

spermatozoa carried in by the inhalent current from another sponge. After fertilisation they must return back into the mesoderm, where the larva develops. A similar migration of ova has been described by Weismann in many Hydroids.

FIG. 36.

1. Ovum (*Grantia labyrinthica*) hanging in lumen of inhalent canal.
2. Epithelium lining inhalent canal.
   After Dendy.

The early stages of development take place whilst the ovum is in the mesoderm; and in the Non-calcarea a special investment of epithelial cells lines the cavity in which the embryo lies. Many sponges are hermaphrodite, and then the spermatozoa usually mature before the ova; others are unisexual. Reproduction by gemmation occurs in some species, buds being formed which, as a rule, include portions of all the three layers; these separate from the parent form, and grow into new sponges. In Spongillidae, the only freshwater family, resting buds or gemmules are produced, and their production involves the death of the sponge. The gemmules consist of certain yolk-bearing cells enclosed in a complicated capsule, which serves to protect the cells until the external circumstances permit them to emerge and grow up into a new sponge.

Artificial fission has been successfully adopted in the cultivation of sponges for commercial purposes.

All sponges are marine, with the exception of the family Spongillidae, the commonest species of which is *Spongilla fluviatilis*, an incrusting mass with crater-like oscula, commonly found on the woodwork of locks, weirs, etc., in our rivers. It may be coloured green by the presence of chlorophyll. Many other sponges are brightly coloured, and they may assume a very great variety of shapes. Some emit a powerful and unpleasant odour, which may, like the presence of the spicules, tend to prevent their being devoured.

The classification of the Porifera cannot be regarded as settled. That at the head of this chapter is suggested by Vosmaer.

A. The CALCAREA *includes those forms which possess a skeleton of calcareous spicules, generally triradiate in shape. The collar cells are large.*

This class comprises two orders :

(i.) Homocoela, *in which the endoderm consists wholly of collared cells, which line the central cavity.* Ex. Leuco-. solenia.

This group includes all those forms which were formerly known as Ascones.

(ii.) Heterocoela, *in which the endoderm is differentiated into (a) flat epithelial cells, lining the central cavity and excurrent canals, and (b) collar cells, confined to the flagellate chambers.* Ex. Grantia, Sycon, Leucandra.

B. The NON-CALCAREA *possesses a skeleton of siliceous spicules or spongin ; very rarely none at all. The spicules may be isolated, articulated, or fused. The collar cells are markedly smaller than those of the Calcarea.*

The group includes three classes: (i.) Hyalospongiae (Hexactinellidae), (ii.) Spiculispongiae, and (iii.) Cornacuspongiae. The first of these three classes is much more clearly defined than the other two; indeed the latter tend to run into one another.

(i.) Hyalospongiae.—*Skeleton of siliceous spicules, which are usually sexradiate, isolated, or ²fused into a trellis-work. Usually deep-sea ²forms. Many ²fossils.* Ex. Hyalonema, whose spicules may reach a length of two feet; Euplectella, or Venus's flower-basket.

(ii.) Spiculispongiae.—*Skeleton absent in a few ²forms; in the great majority consisting of siliceous spicules, usually independent, but sometimes articulated together, or united by organic material.* One of the sub-orders, the Myxospongiae, is devoid of skeleton. Ex. Halisarca, Oscarella. In the Tetractinellidae, one of the largest subdivisions, the spicules are to a great extent four-rayed. Ex. Geodia, Tetilla.

(iii.) Cornacuspongiae.—*Skeleton of uniaxile spicules united by spongin, or of spongin. Inhabit the sea, brackish or ²fresh water.* Ex. Euspongia, the sponge of commerce; Velinea, Spongilla (the freshwater sponge), etc.

CLASSIFICATION.

COELENTERATA

HYDROZOA
- HYDROMEDUSAE.
  - Gymnoblastea - Anthomedusae — Tubularia, Clava.
  - Calyptoblastea-Leptomedusae.
    - Campanularidae— Campanularia, Obelia.
    - Plumularidae— Plumularia, Antennularia.
    - Sertularidae— Sertularia.
- SCYPHOMEDUSAE—
  - Hydrocorallinae { Milleporidae—Millepora. Stylasteridae—Stylaster.
  - Siphonophora—Velella, Physalia, Diphyes.
  - Trachomedusae, Carmarina (Geryonia).
  - Narcomedusae, Cunina.
  - Aurelia, Rhizostoma.

ACTINOZOA
- HEXACTINIA or ZOANTHARIA.
  - Actiniaria—Actinia, Cerianthus.
  - Antipatharia—Antipathes.
  - Madreporaria { Perforata—Madrepora. Aporosa—Oculina, Astraea.
- OCTACTINIA or ALCYONARIA—Tubipora, Pennatula, Alcyonium.

CTENOPHORA . Cydippe, Beroe, Cestus.

CHAPTER IV

COELENTERATA

CHARACTERISTICS.—*Acoelomata with a definite shape. The body is usually radially symmetrical about an axis which passes through the mouth. The ectoderm is separated from the endoderm by a middle layer—the mesogloea—which may be structureless and devoid of cells, or may contain numerous cells. The tissues are not pierced by a series of pores, and there are no collar cells. Protective organs known as nematocysts are characteristic of the group, with the exception of almost all the* CTENOPHORA. *Alternation of generations is common, and also the formation of colonies by budding. All the members of the group are aquatic, and most of them marine.*

This group is divided into four classes :

    A. HYDROMEDUSAE $\left.\right\}$ = Hydrozoa.
    B. SCYPHOMEDUSAE
    C. ACTINOZOA.
    D. CTENOPHORA.

CLASS A. HYDROMEDUSAE.

CHARACTERISTICS.—*The Hydroid form may be free or sessile, single or colonial. It is rarely without tentacles, which are nearly always solid. A horny perisarc or a calcareous skeleton may be developed. Asexual reproduction by gemmation usually takes place, the hydroid form budding off a medusiform sexual individual. The Medusa has a velum, and a double nerve ring. The sensory organs are ocelli and otocysts or modified tentacles. The Hydromedusae are bisexual, and the sexual cells are typically ectodermic, sometimes endodermic. No gastral filaments are present. The medusa may arise directly from the egg, but this is rare.*

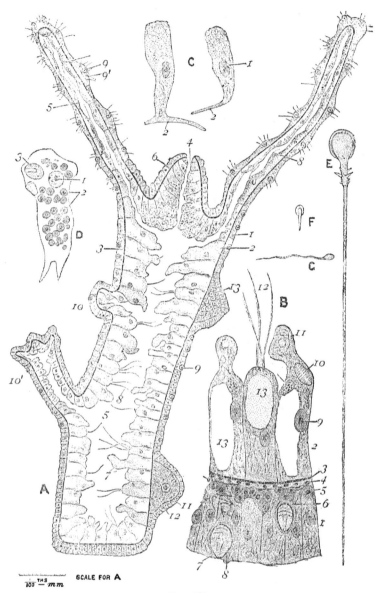

Fig. 37.

A. Vertical section of *Hydra*.
1. Ectoderm.
2. Endoderm.
3. Mesogloea.
4. Mouth.
5. Coelenteron.
6. Hypostome or oral cone.
7. Pseudopodia of endoderm cells.
8. Flagella of endoderm cells.

9 and 9′. Large and small nematocysts.
10 and 10′. Two buds in different stages of development.
11. Ovary.
12. Single ovum.
13. Testis.
B. Portion of transverse section through body wall of *Hydra*, more highly magnified.

The simplest form of Hydromedusa is represented by Hydra, which exists only in the hydroid form. Its reproductive organs show no trace of a medusoid nature; and although the medusae can be traced in other and more specialised species through stages of degeneration till they become little more than protuberances on the body-wall full of sexual cells, still there is nothing in the ovary and testis of Hydra to warrant the view that they are not simply sexual organs.

In some Hydromedusae a distinct alternation of generations is present; that is, the hydroid person produces asexually, by budding, a medusoid person which produces sexually, by means of ova and spermatozoa, the hydroid person again. Thus asexual and sexual modes of reproduction alternate in the life-history of these animals, and each mode is associated with a distinct kind of animal: the *asexual* with the *hydroid*, usually a fixed form; the *sexual* with the *medusoid*, a free-swimming form. This kind of alternation of generation—budding alternating with the sexual method—has been termed *metagenesis*; it occurs in many of the lower animals.

Although the fixed hydroid differs a good deal in appearance from the free-swimming medusa, they can both be reduced to a common type. In both forms of person very considerable complexity of form is often combined with great simplicity of ultimate structure. Many forms are colonial, and the individuals composing the colonies are commonly modified to subserve various functions, and thus may become degraded to the level of organs.

1. Ectoderm.
2. Endoderm.
3. Mesogloea.
4. Layer of muscular processes of ectoderm cells cut across just outside mesogloea.
5. Interstitial cells.
6. Cnidoblast containing nematocysts.
7. Nematocyst.
8. Cnidocil.
9. Nucleus of endoderm cell.
10. Ingested diatom.
11. Pseudopodium.
12. Flagella.
13. Vacuole.
C. Two ectoderm cells.
1. Nucleus.
2. Muscular tails.
D. An endoderm cell of *H. viridis*.
1. Nucleus.
2. Chromatophores.
3. Ingested nematocyst.
E. A large nematocyst with extended barb.
F. A small nematocyst.
G. A spermatozoon.

A, B, and C, after Jeffery Parker; D, after Lankester; F and G, after Howes.

Tubularia indivisa is a marine colonial Hydromedusan borne on a branching anastomosing stolon attached to sub-

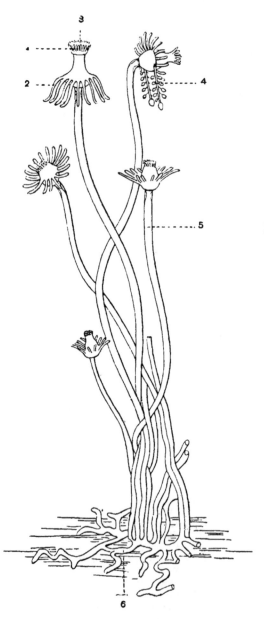

Fig. 38.—A colony of *Tubularia indivisa*. After Allman (natural size).

1. Circumoral tentacles.
2. Aboral row of tentacles.
3. Mouth.
4. Blastostyle.
5. Hydrocaulus.
6. Hydrorhiza.

merged objects. At intervals the stolon sends up straight unbranched stems, each of which terminates in a hydroid person, generally termed a *hydranth*. The branching stolon may be termed the *hydrorhiza*, and the single stem the *hydrocaulus*. Both the hydrorhiza and the hydrocaulus are protected by a thin chitinoid investment, the *perisarc*. This tubular covering, however, stops short at the base of each hydranth. The living tissue within the perisarc is known as the *coenosarc*; it consists externally of a single layer of ectoderm, which secretes the perisarc, and internally of an axis of endoderm traversed by ciliated canals; these serve to place the cavities of the various hydranths in communication with one another.

Each hydranth is a somewhat flask-shaped structure bearing two rings of tentacles; one composed of short tentacles situated round the mouth, the other, of much larger tentacles, arises from the middle of the body, where the diameter is greatest. The bases of the circumoral tentacles are visible as slight projections on the body-wall for some little distance below their point of emergence. Their ectoderm is crowded with nematocysts, and contains muscle cells, glandular cells, ganglion cells, usually bipolar, and four kinds of nematocysts. Their endoderm consists of several rows of cartilaginoid cells, which serve as a stiffening skeleton.

The digestive cavity of the hydranth is spacious, it is lined by endoderm cells, which in many species digest intracellularly. There is a thickened layer of endoderm forming a ring for the support of the larger tentacles. The middle layer or mesogloea is structureless and thin.

The relationship of the hydroid to the medusoid person is best explained by means of diagrams (Fig. 39). The oral axis of the hydroid is shortened, and the circumference of the middle part of the body is correspondingly increased. The ring of large tentacles is thus carried out to the edge of what is known as the umbrella; the oral cone, with its ring of tentacles, remaining in the position of the umbrella handle. This change of external shape is accompanied by the obliteration of certain parts of the *coelenteron* or central cavity. Along certain areas in the umbrella the endoderm cells come in contact and fuse, forming the endodermal lamella (Fig. 39, II. 12). This fusion takes place in such a way as to leave certain tracts open, the most important of which are the circular vessel running round the edge of the umbrella, and the radial canals, which communicate with the cavity of the oral cone on the one hand and the circular canal on the other. On the aboral surface of the endodermal lamella the middle layer usually becomes much

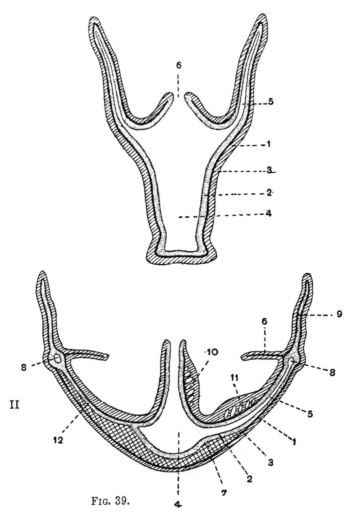

FIG. 39.

thickened, its gelatinous substance may remain structureless, or cells may invade it.

Tubularia is one of those *Hydromedusae* which do not give rise to free medusae. The medusa is produced, but does not break away, it remains connected with the hydranth by the aboral pole. In a mature individual a racemose extension of the body-wall is formed between the two rows of tentacles.

This is the *blastostyle* (Fig. 38, 4), and it bears a series of buds called *gonophores*. These buds increase in maturity as they approach the end of the blastostyle, so that a single blastostyle may illustrate the various stages in the development of the buds. The gonophores or rudimentary medusoid persons arise as elevations of the ectoderm, into which the endoderm projects. At the free end of the bud the ectoderm then thickens,

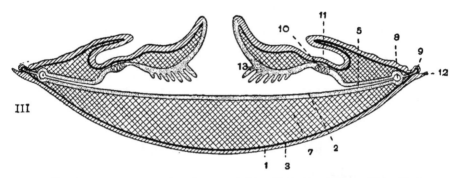

FIG. 39.—Diagrams to show the relations of (I.) a Hydroid polype to (II.) a Hydromedusa and to (III.) a Scyphomedusa.

1. Ectoderm, shaded.
2. Endoderm, dotted.
3. Structureless lamella, black line.
4. Coelenteron.
5. Extension of coelenteron into tentacles in I., and forming radial canals in II. and III.
6. In I., Mouth.
   In II., Velum.
7. Gelatinous tissue, cross-hatched.
8. Circular canal.
9. Tentacle into which the endoderm extends.
10. Generative organs. In II. they occupy Anthomedusan position.
11. In II., Generative organs in Leptomedusan position.
    In III., Sub-genital pit.
12. In II., Endodermal lamella.
    In III., Fold of umbrella over tentacle.
13. The gastric filaments, remains of the taeniolae.

and in this thickened portion a split appears in such a way as to separate a central portion, the manubrium, from a circular portion which surrounds the former and forms the umbrella. The space thus constituted, which is lined everywhere by ectoderm, is the sub-umbrella cavity (Fig. 40). The endoderm projects both into the manubrium and the umbrella; in those forms which break off as free medusae, the manubrium acquires a mouth, which leads into a cavity—the coelenteron—lined by endoderm. The endoderm of the umbrella in *Tubularia* is a flat plate of tissue enclosing no cavity except round the rim, where slight remnants of a circular canal can be found, and from this rudiments of radial vessels lead to the cavity of the manubrium. Rudimentary tentacles may occur on the gonophore. As stated above, the medusoid persons in *Tubularia* do not break away or develop further. The generative organs are found on the walls of the manubrium (Fig. 39, II.), the medusoid persons bearing either male or female cells, but not both.

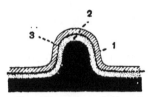

FIG. 40.—Diagrams of the formation of Medusae by budding. The ectoderm and endoderm are represented by the same shading as in Fig. 39. The Coelenteron is black. After Korschelt and Heider.

I.—Commencement of bud.

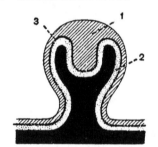

II.—Thickening of ectoderm and cupping of endoderm.

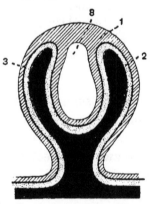

III.—Formation of sub-umbrella cavity in the ectoderm.

In those species in which the medusoid person is completed, other structures develop; tentacles and sense organs arise round the edge of the umbrella, a muscular fold, the *velum*, is formed inside the rim of the umbrella (Fig. 40, IV.) and functions as a swimming organ, and the medusa breaks away and swims freely through the water. The existence of these two phases in the life-history of one organism has rendered necessary a double classification. The order to which *Tubularia* belongs is the **Gymnoblastea-Anthomedusae :** the first word referring to the nature of the hydroid form, the second to that of the medusoid.

Order 1.

Gymnoblastea-Anthomedusae.

CHARACTERISTICS.—*These Hydrozoa have probably always a hydroid stage, which is very various in appearance, sometimes non-colonial, but more usually the hydroids are united into colonies. The ectoderm may form a perisarc, but this is never continued into a cup or hydrotheca surrounding*

the hydranth, nor are there capsules surrounding the gonophores.

The medusae may be well developed and free, or they may be
permanently attached to the hydranth (gonophore), or they may
not develope beyond rudimentary
buds (sporosacs), and finally,
in Hydra they are not formed
at all. When free, they may
be provided with ocelli at the
base of their marginal tentacles,
usually on their outer surface.
The sexual cells are arranged
round the manubrium, either
uniformly or in bands. Most
Anthomedusae are small.

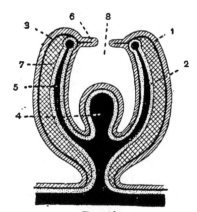

FIG. 40.

IV.—Sub-umbrella cavity opened,
manubrium arising, and the development of gelatinous tissue.

1. Ectoderm.
2. Endoderm.
3. Structureless lamella.
4. Manubrium.
5. Radial canal.
6. Velum.
7. Gelatinous tissue.
8. Sub-umbrella cavity.

The tentacles of the hy-
droid are nearly always solid.
Hydra, however, is an exception
to this rule; they are usually
arranged in one circle as in Hydra,
or in two as in Tubularia; in
the family Clavidae, however,
they are irregularly scattered over
the surface of the hydranth.

There is no special aggregation of nerve cells in the
hydroid stage, but in the medusa a nerve ring is present round
the edge of the umbrella; it is split into an upper and lower
nerve ring by the insertion of the velum. The ocelli consist
of a collection of pigment spots, and a cuticular lens is present
in Lizzia. The sexes as a rule are separate, the genus Hydra
again forming an exception. Reproduction by fission is rare,
but sexual reproduction by budding is common in both the
hydroid and the medusoid stage. When it takes place in the
latter, a medusa is invariably the result.

Order 2. **Calyptoblastea-Leptomedusae.**

CHARACTERISTICS.—*The hydroid forms of this genus never have
more than one row of tentacles round the mouth. The*

tentacles are solid. The ectoderm secretes a perisarc which is continued into a cup, the Hydrotheca, surrounding the Hydranth. The groups of medusiform buds are similarly enclosed in a capsule, the Gonotheca.

The medusae of this group are all Leptomedusae, and are characterised by having four, eight, or more radial canals, on which the generative glands are always situated. The tentacles on the rim may be few or very numerous. Ocelli are found in two families; more commonly the sense organs take the form of hollow vesicles, the otocysts, in which the otolith cells are formed from the ectoderm.

The hydroids belonging to this order are colonial, the colonies being arranged in branching filaments, which have a superficial resemblance to some of the branching colonies of Polyzoa. Certain families are distinguished by the way in which the hydranths are placed on the branch. In the CAMPANULARIDAE *each hydranth is stalked;* in the PLUMULARIDAE *the hydrothecae are sessile on one side of the branch only;* in the SERTULARIDAE *they are sessile, and on both sides of the branch.*

Certain tentaculoid structures, termed *nematophores,* occur in relation to the hydrothecae of the PLUMULARIDAE. They are solid, with an endodermic axis, and knobbed at the end, and the knob contains nematocysts and sense cells. Their ectoderm has been seen to ingest carmine granules, and they have been observed to bend into the hydrotheca and eat up the remains of dead hydranths of the same colony.

The medusae are often rudimentary, and remain attached to the blastostyle, as the hydroid individual which gives rise to them is called. The free medusae arise from the Campanularian hydroids; the gonophores of the Plumularians and Sertularians do not become detached.

Order 3. **Hydrocorallinae.**

This is a very well-marked order, in which the hydroid stage only has hitherto been found.[1] The hydrorhiza in these animals deposits a copious secretion of carbonate of lime instead of a chitinoid perisarc. By this means considerable masses

---

[1] A very primitive form of medusa bearing male organs only (spermospores) has been recently described in *Millepora Murrayi.*

of lime are built up, simulating some of the true corals. Certain tubular spaces are left for the accommodation of the hydranths, which can be withdrawn into them. The hydranths are colonial, and of two kinds : *gastrozoids*, provided with mouths, which are the nutritive persons of the colony ; and *dactylozoids*, elongated mouthless persons well armed with nematocysts.

The order is divided into two families : (i.) the MILLE-PORIDAE and (ii.) the STYLASTERIDAE.

(i.) THE MILLEPORIDAE.—*This family may have an arborescent or branching skeleton, consisting of an outer living part which surrounds an inner and older dead part. The younger parts of the pores, in which the living hydranths are found, are separated from the older ones, wherein their predecessors lived, by horizontal partitions known as tabulae. The colonies are hermaphrodite.*

(ii.) THE STYLASTERIDAE.—*These are always arborescent, never encrusting, and their colonies are unisexual.*

The hydranths may be scattered irregularly, but more usually the gastrozoids are surrounded by a regular circle of

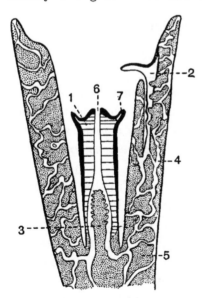

FIG. 41.—Longitudinal median section through a Stylasterid. After Hickson.

1. Gastrozoid.
2. Dactylozoid.
   Style.
3. Coenosarcal canals.
4.
5. Coenosarc, calcareous.
6. Mouth.
7. Tentacle.

dactylozoids. The gastrozoid (Fig. 41) is usually provided with tentacles, but in some of the STYLASTERIDAE these are wanting. The dactylozoids are long tentacular structures with no mouth, they bear short knobbed branches, or tentacles, in the Milleporidae.

The generative cells of *Millepora*, which is hermaphrodite, arise in the ectoderm of the coenosarcal tubes, which connect the gastrozoids and the dactylozoids. The spermospores or cells which are destined to break up into spermatozoa, migrate through the mesogloea into the endoderm, and then travel along the tubular coenosarc to a hydranth, usually a dactylozoid. They pass into the cavity of the hydranth, and then re-enter the endoderm, and push out the mesogloea into a diverticulum between the tentacles. The number of the diverticula or sporosacs on each dactylozoid varies from one to five. The spermatozoa mature in the sporosacs. The ova make their way in a similar manner through the mesogloea into the endoderm; here they attach themselves by a stalk to the inside of the mesogloea and grow considerably. The ovum may wander about to seek a more favourable position for fertilisation, which probably takes place in the lumen of the canals. Similar wanderings of the sexual cells take place among the Hydromedusae.

Order 4. **Siphonophora.**

CHARACTERISTICS.—*These are* HYDROMEDUSAE *which live together in colonies floating at or near the surface of the sea* (PELAGIC). *The members of the colonies are hydriform individuals alone* (VELELLA),*or hydriform and medusiform; they have undergone great modifications, so that in each colony a great number of persons are present performing various functions, and exhibiting a great diversity of form.*

The chief modifications are as follows : (i.) the *Gastrozoid*, whose function is to absorb nutriment. It has a mouth, and communicates at its aboral end with the tubular coenosarc connecting the various persons. (ii.) *Hydrophyllia* : these are protective covering-pieces, which have a central endodermic canal, and are usually borne on the coenosarc. (iii.) *Tentacles* : these may be short and tubular, or may attain the length of many feet (*Physalia*). They are usually branched, and each branch is provided with a battery of nematocysts. (iv.) *Dactylozoids* : hydroid structures without a mouth, and usually well armed with nematocysts. (v.) *Nectocalyces*: swimming bells. They consist of the umbrella of a medusoid, with four radial canals, but without a manubrium or mouth. (vi.) *Pneumatophores*: these form floats, and the air secreted within them serves to keep the whole colony in a vertical position. (vii.) The sexual persons: these may become free medusae of the *craspedote* type—that is, provided with a velum; or they may remain undeveloped as *sporosacs*. The medusae have their generative organs on the radial canals (*Velella*), or on the manubrium (*Physalia*). The colonies are usually hermaphrodite.

The seven modifications described above are not all to be met with in every Siphonophoran, but usually four or five of them coexist in each colony, thus giving rise to a form of extreme complexity; various combinations of these persons also permit great variety amongst the different species. Many of them are brilliantly coloured in parts, and are amongst the most beautiful marine objects which float along the surface of the sea. They are common in the Mediterranean and open seas, *Velella* and *Diphyes* being occasionally found on our shores.

The two remaining orders of the HYDROMEDUSAE—the TRACHOMEDUSAE and the NARCOMEDUSAE—have no hydroid stage, but the ova in most cases give rise immediately to the medusae. They are characterised by the possession of sense organs formed by modifications of the tentacles on the rim of the umbrella. They are termed *tentaculocysts*, and into them alone of the sense organs of the Hydromedusae does the endoderm enter. One or more otoliths formed from the endoderm cells, which correspond with the axis of the ordinary tentacle, contain crystalline concretions, and form the auditory organ.

Order 5. **Trachomedusae.**

CHARACTERISTICS.—The **Trachomedusae** *have their tentaculocysts free or enclosed in capsules. Coecal radial canals may be present, which open into the circular canal, but never reach the central stomach.*

The generative glands are on the sub-umbrella surface of the radial canals. *Limnocodium*, the only freshwater medusa with whose anatomy we are acquainted, is placed by Lankester in this order. It forms an exception to the rule of the absence of a hydroid form. A small hydranth has been described as budding off the medusae, which are found at intervals in tanks of the Victoria regia in the Botanical Society's Garden, Regent's Park.

Order 6. **Narcomedusae**.

CHARACTERISTICS.—*The* **Narcomedusae** *always have their tentaculocysts free. The tentacles arise some way up the aboral face of the umbrella, not from its edge. Their bases are connected with the edge of the umbrella by stiffening rods of tissue termed Peronia. These may also be found in the Trachomedusae.*

The generative organs arise from the sub-umbrella surface of the stomach, not on the radial canals.

CLASS B. SCYPHOMEDUSAE.

CHARACTERISTICS.—*The hydroid stage of this sub-class is a polype, which may give rise to the medusa, not by budding but by transverse fission. It is termed the Scyphistoma. The mouth of the Scyphistoma is squarish, and the coelenteron is sub-divided by four ridges, termed taeniolae, which project from the sides into its lumen.*

The medusa has also a square mouth, whose angles may be produced into four oral processes. The edge of the umbrella is lobed. The sense organs are modified tentacles, into the base of which the coelenteron is produced. No continuous nerve ring exists, but scattered nerve centres are found round the margin, and no true velum is present. CHARYBDAEA *forms, however, an exception to this statement. The generative cells are endodermic in origin, and the medusae are unisexual.*

In the Scyphomedusae a primary series of four tentacles corresponding to the four angles of the mouth are the first to be formed. This affords a convenient method for mapping out the regions of the body. The radii on which they are formed are termed *perradii,* the next four are situated half-way between the perradii, and are known as *interradial tentacles*; and eight more may be added, one between each perradius and interradius, and these are called *adradial.* The same convention is made use of in describing the position of the various organs in medusae of other classes.

The most striking features in the Scyphistoma are the taeniolae, which project into the enteric cavity and divide its outer portion into four chambers (Fig. 39, 13). The taeniolae are four in number, and are inter-radial in position. They are comparable with the mesenteries in the **Actinozoa**.

Towards the end of spring the Scyphistoma passes into the Strobila stage, it then becomes constricted on the aboral side of its tentacles, and thus a disk is partly cut off, behind this a series of furrows arise, cutting the Scyphistoma into a series of segments (Fig. 42). These separate off and form the immature medusae or the *Ephyrae.* Eight bifid lobes have grown out, which give to the Ephyrae their characteristic eight-rayed appearance. Each lobe encloses in its notch either a perradial or interradial tentacle destined to become a tentaculocyst. The gastric cavity extends into each lobe, and the taeniolae become detached from the aboral side of the disk and give rise to gastric filaments. The same processes of growth go on in each of the segments of the Scyphistoma.

FIG. 42.—Strobila of *Aurelia aurita.* After Haeckel.

1. Base of Strobila.
2. Sense organ.
3. Marginal tentacle.
4. Lappet on side of sense organ.

The Ephyra becomes a medusae chiefly by the filling out of the adradial spaces, and it thus acquires a circular outline.

The position of the tentaculocysts, however, always marks the position of the eight original lobes.

Aurelia is a Scyphomedusan very commonly met with round our coasts swimming at the surface of the sea. If the adult medusa be examined, it will be seen that the original four-lobed chamber of its Scyphistoma has given off a number of branching enteric canals which ultimately open into a circular vessel. Between these canals the aboral and oral endoderm has given

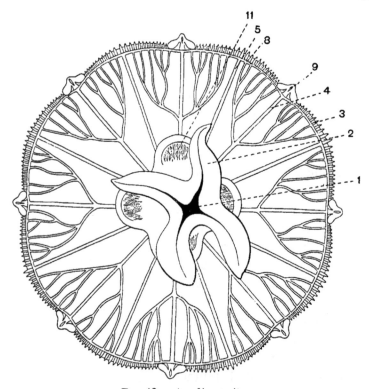

FIG. 43.—*Aurelia aurita.*

1. Mouth.
2. Circumoral perradial processes.
3. Tentacles on the edge of the umbrella.
4. One of the branching perradial canals. There are four of these, and four similar interradial canals.
5. An adradial canal.
8. The circular canal.
9. Marginal lappets hiding tentaculocysts.
11. Gastral filaments.

rise by concrescence to the endodermal lamella (Fig. 39). Eight adradial canals are given off from the central chamber between these branching canals, and these pass directly to the circular canal without giving off any branches.

The mouth of *Aurelia* is surrounded by four perradial

processes, and the manubrium is short. The coelenteron is ciliated. In *Rhizostoma* the four oral processes are divided into eight, and have fused together in such a way as to occlude the mouth. This is compensated for by the presence of fine canals which open at one end into the gastric cavity, and at the other on the frilled edges of these processes. The food is absorbed through these canals.

The gelatinous layer of the mesogloea in *Aurelia* is traversed in all directions by fibrous cells and by wandering amoeboid cells. The muscles are in some cases formed of transversely-striated processes of epithelial cells; in the oral processes of some medusae, however, distinct nucleated muscle cells occur.

The sense organs of *Aurelia* are modified tentacles, which bear endodermal otocysts and ectodermal pigment spots or eyes. An aboral and an oral pit, both lined by specialised epithelium on the surface of the disk, are regarded as olfactory. The coelenteron is prolonged into the modified tentacles.

No regular nerve ring exists in Scyphomedusae, with the exception of one genus, *Charybdaea*; but nerve fibres occur in the region of the sense organs, and are connected with scattered ganglion cells.

The ova and spermatozoa arise from endodermal cells, and escape through the mouth. Four large pits may be excavated in the sub-umbrella ectoderm, and these form sub-genital pouches, the skin of which is thin, and forms only a slight membrane between the sea-water and the genital cells (Fig. 39, III.). This arrangement may promote the respiration of these parts.

In some Scyphomedusae, *e.g. Pelagia*, the medusa rises directly from the ovum without the intervention of a Scyphistoma.

Class C. ACTINOZOA.

CHARACTERISTICS.—*Single or colonial Coelenterata. The mouth leads into an ectodermic invagination, the stomodaeum or oesophagus; this is attached to the walls of the body by a series of radial mesenteries, so that the coelenteron is divided into a central portion and a series of radiating intermesenteric chambers. The generative cells are endodermic, and lie in the mesentery. There are no special sense organs. Nemato-cysts occur, and are more complicated than those usually found in the Hydrozoa.*

The most fundamental distinction between the **Actinozoa** and the **Hydrozoa** lies in the presence of the mesenteries and the consequent division of the coelenteron into a central chamber and several radiating lateral chambers, which all open below into the central chamber. In the Hexactinia these chambers are put into direct communication with each other by one or two pores, which pierce the mesenteries near their upper end. The existence of these mesenteries is to some extent foreshadowed by the taeniolae of the Scyphomedusae. In two forms recently described by Danielssen, *Fenja* and *Aegir*, the oesophagus is continued to the aboral disk, where it opens to the exterior by an anus; thus a body cavity is shut off from the digestive cavity. In *Aegir*, however, these two cavities communicate by a series of pores which lead from the inter-mesenteric chambers into the alimentary canal near the anus. In *Fenja* each intermesenteric chamber opens to the exterior by a pore close to the anus. Thus a ring of genital pores is formed. In both genera the various intermesenteric cavities communicate with one another by pores at the oral end of the mesentery.

The Actinozoa have an apparent radial symmetry which does not hold true for all parts; but a genuine bilateral symmetry exists.

The **Actinozoa** are sub-divided into two orders: (i.) the HEXACTINIA (ZOANTHARIA) and (ii.) the OCTACTINIA (ALCY-ONARIA).

Order 1. HEXACTINIA.

CHARACTERISTICS.—*Tentacles simple; they and the mesenteries are very generally some multiple of six. Single or colonial; when colonial a continuous organic or calcareous skeleton is usually deposited by the ectoderm.*

Actinia mesembryanthemum is a beautiful red sea-anemone common in rock pools round our coasts. When in the expanded condition it may be seen that the animal is in the form of a short cylinder, with one end firmly attached to a stone or the rocks, the other end free. The mouth is an elongated slit in the middle of the peristome or free end, it is surrounded by several rows of tentacles (Fig. 44). At the ends of the elongated mouth are special grooves which are continued down the oesophagus and are lined with especially long cilia (Fig. 46).

FIG. 44.—Colony of sea-anemones (*Sagartia parasitica*) on shell of hermit crab.

When the mouth is closed the central parts are in apposition, but the grooves, called *Siphonoglyphs*, remain always open, and through them a current of water may be kept circulating in and out of the animal even when it is in its most contracted condition. *Cerianthus* has only one siphonoglyph. The oesophagus ends with a free edge, and never reaches the base of the sea-anemone.

The mesenteries are vertical radial partitions which extend from the peristome to the base. The outer edge is continuous with the inner side of the body-wall. The inner edge in the primary mesenteries is divided into two parts (Fig. 45). The part nearest the peristome is continuous with the outside of the oesophagus, but below the lower edge of the oesophagus the inner edge of the mesentery is free. The secondary mesenteries have not reached the oesophagus in Fig. 46, so that the whole extent of their inner edge is free. The mesenteries are grouped in pairs, and the members of each pair are separated by an

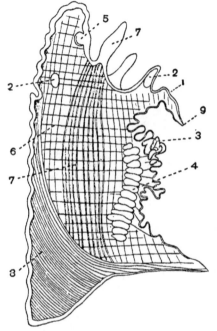

Fig. 45.—A mesentery of *Tealia crassicornis*. After the Hertwigs.

1. Edge of mouth.
2. External and internal pores in the mesentery.
3. Mesenterial filament.
4. Testis follicles.
5. Circular muscle.
6. Radial muscle fibres.
7. Longitudinal muscle fibres.
8. Parietal longitudinal muscle fibres.
9. Lower edge of the stomodaeum.

intramesenteric space, while the various pairs have an *inter-septal* chamber between them. The secondary mesenteries always occur between two primary ones. Each mesentery is pierced by a round hole, and in some species by two, so that the various intermesenteric chambers open into one another above, as well as all communicating with the central chamber below.

Along one side of each mesentery runs a longitudinal bundle of muscle fibres from the peristome to the base. These assist the sea-anemone to contract. They are as a rule facing one another in each pair. But on each side of each siphono-glyph is a mesentery known as the *directive* mesentery (Fig. 46), and on these the longitudinal muscles turn away from each other. Transverse muscle fibres are found on the mesenteries on the side where the longitudinal muscles do not exist. There are also sphincter muscles running round the peristome, and parietal muscles running obliquely from the walls to the base of the body. The nervous system consists of scattered ganglion cells, chiefly in the peristome and tentacles.

The free edge of the mesentery below the level of the oesophagus is divided into three lobes; the middle one is crowded with nematocysts and glandular cells, the outer lobes

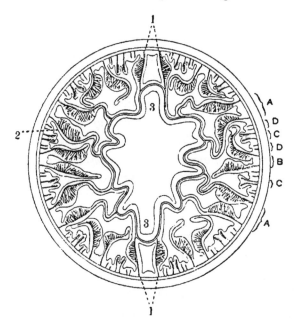

FIG. 46.—Transverse section through the body of *Adamsia diaphana*, in the region of the stomodaeum. After the Hertwigs.

1. The directive mesenteries.
2. Longitudinal muscle fibres in the mesenteries, cut across.
3. The siphonoglyphs.
A. Primary mesenteries, 12.
B. Secondary mesenteries, 12.
C. Tertiary mesenteries, 24.
D. Quaternary mesenteries, 48.

are ciliated. When a piece of solid food is swallowed, the edges of several of the mesenteries come together and surround it, and the secretion of the glandular cells helps to digest it.

The sexes are usually separate, the generative cells arising from the endoderm of the mesenteries (Fig. 45); the young escape through the mouth, *A. mesembryanthemum* being viviparous.

In some species the free edges of the mesenteries are produced into long whip-like processes, called *acontia*, armed with nematocysts. These are protruded through the mouth or through special pores in the body-wall called *cinclides*. They are found in *Sagartia*. Occasionally there is a central pore in the base of the body, and as a rule the tentacles are perforated at their ends; and in some of the deep-sea forms the tentacular pores are large, in others the tentacles are reduced, the pores only remaining as a circlet of holes surrounding the mouth.

The Hexactinia are divided into three groups :

(i.) *The* ACTINIARIA, *which are devoid of any kind of skeleton, are usually single, and are mostly adherent to some foreign body ; occasionally they live half embedded in mud or sand.* Cerianthus, Actinia.

(ii.) *The* ANTIPATHARIA : *they possess a horny axial skeleton secreted from their ectoderm ; they are colonial, and form large branching structures.* Antipathes.

(iii.) *The* MADREPORARIA *are solitary or colonial ; their most remarkable characteristic is their power of secreting a calcareous skeleton. The skeleton is often very massive, and recent research has shown that it is entirely formed from the ectoderm.* Oculina, Astraea, Madrepora, Fungia, Caryophyllia.

The MADREPORARIA are of the greatest importance in the history of the earth. They are the true corals, and their skeletons form by far the greater part of the coral rock which has built up a considerable portion of the earth's crust. Reef-forming corals do not as a rule grow below the forty-fathom line, and are not usually found north or south of a belt extending 30° each side of the equator.

As the coral grows, large masses of the *coenenchyma* or common skeleton become covered over by the younger formations. This skeleton may be quite solid and dead, or it may be pierced by canals which shelter coenosarcal tubes of living matter, connecting one individual with another. This enables us to divide the Madreporaria into two divisions: (i.) *Perforata, with the skeleton perforated by the cavities which lodge the coenosarcal tubes*, and (ii.) the *Aporosa, in which no such perforations exist.*

The form and shape of the skeleton is extremely varied, and often complicated by the colonial habits of the actinozoan. But whatever its shape, and however deeply it may have penetrated into the body of the soft gelatinous-looking animal, it is always formed by *ectoderm*, and is consequently always outside the animal, whose tissues are, as it were, moulded over it.

The skeleton commences to appear by the ectoderm of the base secreting a flat plate between it and the substance on which the young actinozoan is fixed. From this plate a number of radially-arranged vertical ridges grow up (Fig. 47, 2). These are,

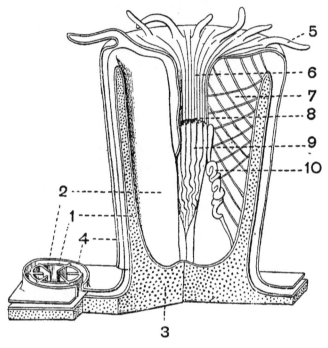

FIG. 47.—Diagram to illustrate anatomy of *Galaxea esperi.* After Hickson.

1. The theca.
2. The septa.
3. The basal plate.
4. Flesh covering the theca.
5. Tentacles.
6. Mouth.
7. Mesentery.
8. Lower edge of stomodaeum.
9. Free edges of mesenteries beneath the stomodaeum.
10. The mesenteric filaments.

like the rest of the skeleton, secreted by ectoderm ; and as they grow they push the ectoderm and the skin of the base up into the coelenteron. These vertical plates are termed *septa.* Partly by the fusion of the external edges of the septa, and partly by the upgrowth of a circular rim from the basal plate, a circular ridge is formed. This ridge forms the wall of the cup or *theca,* and like the septa it projects into the coelenteron, pushing the body-wall before it; occasionally a second circular rim, external to but concentric with and close to the first, is formed, the *epitheca,* and this forms a ridge round and outside the base of the animal. Some of the septa may coalesce in the centre of the body, and thus the *columella* is

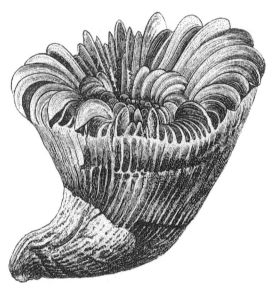

FIG. 48.—*Caryophyllia borealis*, Fleming, a simple coral, twice the natural size. After Sir Wyville Thomson.

produced; this is usually continued as a little pillar pointing towards the mouth. The outer wall of the theca frequently

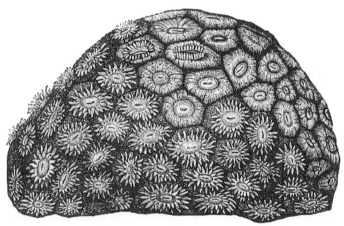

FIG. 49.—*Astraea pallida*, Dana, a compound coral in its living condition. After Dana.

bears ridges, the *costae*, which do not necessarily correspond with the septa; the latter may be connected with their neighbours by horizontal shelves, termed *dissepiments*.

Neither the septa nor the costae correspond with the mesenteries, but are situated in intermesenteric spaces. Like the mesenteries, they may increase in number, secondary septa arising between the primary. Most of the corals which form massive skeletons increase the number of individuals in the colony by budding; in some the theca of each member of the colony remains distinct (*Madrepora, Oculina, Astraea*) (Fig. 49), in others complications arise by the fusion and obliteration of the walls of the theca, etc. (*Meandrina*).

Order 2. OCTACTINIA.

CHARACTERISTICS.—*Colonial Actinozoa with eight pinnate tentacles and eight mesenteries, which bear the longitudinal muscles on their ventral surface, that is, on the face which looks towards the single siphonoglyph.*

The well-known skeleton of the organ-pipe coral, *Tubipora*, consists of a stolon or encrusting lamina which attaches the colonies to some foreign body, and of a series of tubes in which the polyps live, termed *corallites*, which arise from the stolon. The corallites are externally connected together by horizontal plates, forming the *platforms* or *exothecal tabulae* (Fig. 50); and within each corallite is a series of *tabulae*, the top one of which

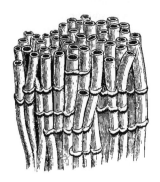

FIG. 50.—A portion of the corallum of *Tubipora musica* of the natural size, showing the tubular corallites and the exothecal tabulae or platforms.

cuts off the polyp from the dead skeleton below. New corallites constantly arise from the platforms, so that the whole coral increases in circumference as well as in height as it gets older.

The platforms are formed by outgrowths of the lips of the living coral, and are at first very thin; they are traversed by many branching canals.

Within, the cavity of the corallites is divided into a series of chambers by the presence of partitions termed tabulae; these may be simple flat plates, but more often they are cup-shaped, and in some cases are drawn down in the centre into funnel-shaped structures. The tabulae often give off tubes, which run out into the platforms.

The whole skeleton is so compact that it appears to be formed of a continuous homogeneous deposit of calcium carbonate. In reality, however, it consists of a number of spicules, each with minute serrations which fit into other serrations on the neighbouring spicules.

The polyp wall is built up of the ectoderm, mesogloea, and endoderm. The mesogloea contains a few scattered cells and fibres, as well as the skeletal spicules, which, however, in those cases where the embryology is known, originate in certain ectodermal cells which wander afterwards into the mesogloea.

The tentacles, eight in number, have about fifteen pinnae on each side in a single series. They are hollow, their cavities

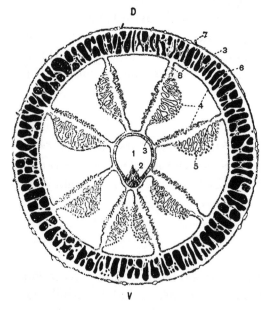

Fig. 51.—Transverse section of a polyp of *Tubipora purpurea.* After Hickson.

1. Stomodaeum.
2. Cilia of siphonoglyph.
3. Ectoderm of stomodaeum and of outer surface of body-wall.
4. Endoderm lining mesenteric chambers.
5. Longitudinal muscle fibres in mesentery, cut across.
6. Calcareous spicules.
7. Parasitic diatoms on the ectoderm.
8. Mesogloea.
D. Dorsal surface.
V. Ventral surface.

opening into the intermesenteric chambers, over which they are placed. Their ectoderm is ciliated. The whole body of the polyp, including the tentacles, can be withdrawn into the corallite. The stomodaeum is ciliated throughout, and on one side, the ventral, there is a special groove (Fig. 51) lined with long cilia, the siphonoglyph.

The stomodaeum is supported by eight mesenteries; on the ventral side of each of these are placed the powerful retractor muscles, which draw the polyp swiftly into the corallite at the approach of danger. The two dorsal mesenteries alone are continued far below the lower edge of the stomodaeum in the form of mesenteric filaments, their edge being much thickened, bilobed, and covered with cilia. The endodermal cells are probably some of them amoeboid, and digestion may be intracellular.

The ova are found attached to the dorsal and dorsolateral mesenteries, immediately below the stomodaeum.

The tabulae are formed by the mesogloea splitting into two layers: the outer remains attached to the ectoderm, the inner layer with the endoderm; the latter shrinks away from the outer layer, and then in this contracted condition begins to secrete a fresh layer of spicules, which ultimately stretches across the corallite. Hence the space below each tabula is morphologically a space in the mesogloea.

The polyps of the *Tubipora* remain free and distinct one from another; in other groups of the OCTACTINIA, however, they may be sunk in a well-developed coenosarc, as in *Alcyonium*, commonly known as "dead men's fingers," or they may be arranged side by side, their lateral surfaces fusing in the form of a leaf-like plate, as in the Pennatulidae (Fig. 52). In *Alcyonium* the skeleton is not continuous, but consists of spicules scattered loosely through the coenosarc. The leaf-like plates of *Pennatula* are borne on each side of a rachis, this is continued into a stalk free of polyps. Both stalk and rachis are traversed and supported by a long calcified horny rod, secreted by an epithelium whose origin is uncertain.

Among the Alcyonidae and Pennatulidae the individual zooids are often of two kinds. In *Pennatula*, for instance, the leaf-like expansions are composed of a single layer of polyps (*autozoids*) fused side by side, whilst the zooids (*siphonozoids*) cover that surface of the rachis on to which the bases of the leaves do not extend, and pass up between the leaves. The zooids differ from the polyps, having no tentacles or retractor muscles, only two of their mesenteries are continued below the level of the oesophagus and bear mesenteric fila-

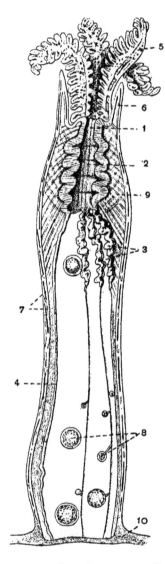

Fig. 52.—A longitudinal section through a polyp of *Pennatula phosphorea*. After Marshall.

1. Mouth.
2. Walls of stomodaeum.
3. Free edges of mesenteries below the level of the stomodaeum, showing short mesenterial filaments.
4. Long mesenterial filaments.
5. Feathered tentacles.
6. Calyx.
7. Spicules in body-wall.
8. Spermatospheres.
9. Mesentery showing protractor and retractor muscle fibres.
10. Coenenchyme.

ments, they have a well-developed siphonoglyph, but no reproductive system.

The cavities of all the individuals are put into communication by means of a number of ramifying coenosarcal tubes lined by endoderm.

The Pennatulidae are phosphorescent, and emit a bright glowing light when disturbed, this is said to originate from eight cords of fatty cells attached to the oesophagus in both kinds of polyps.

Class C. CTENOPHORA.

CHARACTERISTICS.—*Pelagic Coelenterata, usually spheroidal, more rarely band-shaped in form. Eight meridional rows of vibrating plates composed of fused cilia form the locomotor apparatus. A large pair of retractile tentacles are usually present, which can be withdrawn into pouches. At the aboral pole is a special sense organ. They are all hermaphrodite, and pelagic.*

This group of animals possesses considerable importance from a phylogenetic point of view. Haeckel has described a

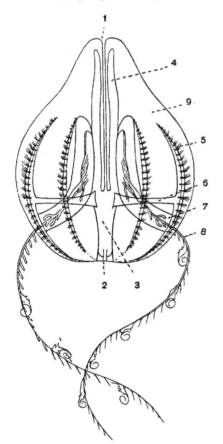

FIG. 53.—*Hormiphora plumosa.* After Chun. Side view.

1. Mouth leading into stomach.

2. Aboral pole with sense organ.

3. Funnel.

4. Recurrent canals running back towards oral pole.

5. One of the eight bands of fused cilia.

6. One of the eight canals running towards 5.

7. A tentacular pouch.

   A tentacle.

8. Gelatinous tissue.

remarkable Anthomedusan named *Ctenaria ctenophora* which has many of the characteristic features of a Ctenophor; whilst on the other hand two curious organisms, the *Coeloplana* of Kowalevsky and the *Ctenoplana* of Korotneff, have lately been discovered which unite some characters of the Ctenophora with others of the Turbellaria.

In a typical Ctenophor, one of the Cydippidae, such as *Hormiphora,* the mouth leads into a flattened stomach lined with ectoderm, which in its turn opens into the funnel, lined with endoderm (Fig. 53). The funnel gives off two gastric canals, which pass out towards the base of the long tentacles. These two primary canals give off a secondary canal on each side, which forks, and forms tertiary canals (Fig. 54), of which therefore there are eight. Each of these eight canals opens into one of eight meridional canals which lie under the rows of vibratile plates, and which end blindly both at the oral and aboral pole.

The two long tentacles can be completely retracted into their pouches; they bear peculiar adhesive cells.

The characteristic vibratile plates are formed by a number of very large cilia fused together side by side; an arrangement also met with in some of the Hypotricha.

The central nervous system consists of an area of ciliated

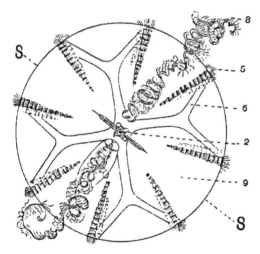

Fig. 54.—*Hormiphora plumosa.* After Chun. Aboral view.

2. Aboral pole with sense organ.

5. One of the eight bands of fused cilia.

6. One of the eight canals running towards 5.

8. A tentacle.

9. Gelatinous tissue.

SS. sagittal plane.

cells at the aboral pole. This area is sunk into a depression which contains certain otoliths, and from which nerve cells pass to the meridional rows of paddles. This specialised sensory apparatus possibly serves as a balancing organ, and helps to keep the floating animal the right way up.

The reproductive organs are arranged along the meridional canals, the male cells on one side and the female cells on the other of each canal. The ova and spermatozoa escape through the canals, and eventually leave the body through the mouth.

The main features of the anatomy of the Ctenophora have been indicated in order to render intelligible their possible relationship on the one hand to the Anthomedusae and on the other to the Turbellarians. There is a mass of interesting detail with reference to these animals which cannot be referred to here.

The Anthomedusan *Ctenaria* has the mouth of its umbrella very much contracted, and the edges have grown round and over the manubrium, which is small. The opening into this sub-umbrella cavity corresponds with the opening into the stomach of the Ctenophor; and the lumen of the latter, lined as it is with ectoderm, corresponds with the sub-umbrella cavity of *Ctenaria*. The shape of the medusa is very like that of Cydippe, and its surface is provided with eight rows of modified ectodermal cells, which correspond in position with the eight rows of vibratile plates in Ctenophors. The arrangements of the enteric canals also approaches that of Cydippe, and the resemblance between the two animals is further increased by the presence in both of two long fringed tentacles which project from pouches as in the Ctenophora.

The *Cestus veneris*, or Venus's girdle, is a Ctenophor in which the spherical form has been replaced by a flattened band-like shape. It is found swimming at the surface of warm seas, and moves through the water by a series of graceful undulations. *Beroe*, in which the stomach attains a very great size, has no tentacles.

The group is a carnivorous one, the chief food being pelagic organisms. Many of them are phosphorescent.

CHAPTER V

COELOMATA

In the Acoelomata there is a common cavity, the Coelenteron, which is lined by endoderm cells, and which pervades various parts of the body. Whatever cavities exist in these animals, with the exception of certain ectodermic pits, are all diverticula of this one primitive cavity, and remain in connection with it.

The Coelomata, on the other hand, start with an Archenteron, probably the equivalent of the Coelenteron; but this is replaced by two distinct and separate cavities—that of the alimentary canal, and that of the body. The latter is termed the Coelom, and is entirely shut off from the cavity of the digestive system. The cavity of the alimentary canal is lined by endoderm ; between the endoderm and ectoderm a new layer of cells has appeared, the mesoderm, and it is in this layer, and lined by it, that the coelom appears.

The Coelomata have typically a bilaterally symmetrical form, which is however in many cases lost. In front of the mouth there is a *prostomium*, or preoral lobe. The mouth and anus, when present, are lined by invaginations of the ectoderm, termed the *stomodaeum* and *proctodaeum* respectively.

The very important cavity, the coelom, which distinguishes the Coelomates from the lower animals, is characterised by the following peculiarities. (i.) It develops as one or more diverticula from the primitive archenteron (*enterocoel*), or it arises as a space or spaces in the mesoblast (*schizocoel*), or it is a remnant of the segmentation cavity, into which a lining of mesoblast grows (*archicoel*) ; it is consequently always lined by mesoblast. (ii.) Its walls give rise to the reproductive cells, which are set free into a portion of the coelom, and leave the body either through the nephridia or by special ducts or apertures. (iii.) It communicates with the exterior through the nephridia, or the organs which excrete waste nitrogenous matter.

It is probable that the body-cavities or coeloms which originate in different manners may not be homologous throughout the Coelomata; on the other hand they all possess the characteristics enumerated above.

There are other cavities which arise in the mesoblast to which the above characteristics do not apply. These are the vascular and lymphatic systems; they contain blood and lymph. In some Coelomata these systems are composed of vessels with certain muscular differentiations to propel the contained fluid; in others, they form large spaces which simulate the appearance of the body-cavity of other animals. Such spaces are termed *pseudocoels* or *haemocoels*; they occur in Arthropods and Molluscs, and when they are present the true coelom is very much reduced. It is not impossible that the vascular system is but a part of the general coelom; in many cases the two systems are distinct, but in others—as the Leeches, where they communicate through the botryoidal tissue, and in Vertebrates through the thoracic duct—there is an indirect connection, whilst in Nemertines the coelom and vascular system appear to be one.

The mesoderm, which is such an important feature in the Coelomata, occupies the same position between the ectoderm and endoderm as the mesogloea of the Coelenterata; it is not, however, always regarded as homologous with the mesogloea, partly because the latter appears originally as a clear gelatinous layer which is devoid of cells, and may remain so throughout life; but also because cells, when they do wander into the mesogloea, do not arrange themselves in definite tissues.

CHAPTER VI

PLATYHELMINTHES

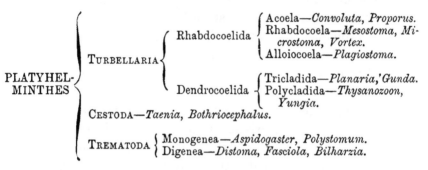

The three classes included under this heading—the TURBEL-
LARIA, the TREMATODA, and the CESTODA—include those animals
which stand in the nearest relationship to the Coelenterata.

CHARACTERISTICS.—*They are not very typical members of the
sub-grade Coelomata, because, although they possess a distinct
mesoderm, the coelom itself is reduced to mere splits and
irregular cavities between the mesodermic cells. The absence
of any well-developed coelom may, however, be due to degeneration. The PLATYHELMINTHES are further characterised
by the absence of a distinct vascular system, and the alimentary canal has no anus. The nervous system consists
of a plexus of nerves, mostly on the ventral side of the
animal. As we ascend in the group the nerves become
more concentrated. A central nerve ganglion, the brain,
is found which in the higher forms gives off two longitudinal
cords which may be connected by transverse commissures.
An excretory system, consisting of fine branches, often intracellular, which at their inner end terminate in flame cells,
at their outer end open to the exterior, is very characteristic
of the group.*

The PLATYHELMINTHES *are hermaphrodite ; in the female organs a vitellarium is often found. Reproduction may be asexual at times, and then an alternation of generations takes place.*

CLASS I. TURBELLARIA.

CHARACTERISTICS. — *Free-living Platyhelminthes, whose oval, usually flattened body is covered with cilia. A mouth, a muscular pharynx, and an alimentary canal exist, but no anus. No special respiratory system or vascular system, or hooks, are present. Nervous system, a paired cerebral ganglion and two lateral nerve cords. With few exceptions hermaphrodite.*

The Turbellaria are mostly aquatic, inhabiting the sea and fresh water, but a few live on the earth amongst damp

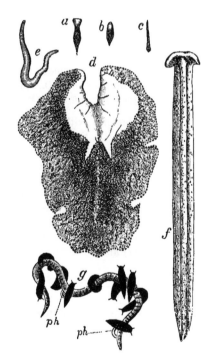

FIG. 55.—Various forms of Turbellaria, all natural size, and viewed from the dorsal surface. After Von Graff.

a. *Convoluta paradoxa*, Oe.

b. *Vortex viridis*, M. Sch.

c. *Monotus fuscus*, Gff.

d. *Thysanozoon brocchii*, Gr., with elevated anterior extremity (after Joh. Schmidt).

e. *Rhyncodemus terrestris*, O. F. Müller (after Kennel).

f. *Bipalium ceres*, Mos. (after Moseley).

g. *Polycelis cornuta*, O. Sch., attached by the pharynx (*ph*) to a dead worm (after Johnson).

surroundings (Fig. 55); a very few are parasitic, and then occur in Mollusca or Holothurians. Several are common in ditches and standing water in England; *Planaria lactea*, a Dendrocoel, especially so. One of the Rhabdocoela, which are simpler in their organisation, *Mesostoma*, is, however, not infrequently found in ponds and streams, and the transparency of its body permits a more thorough examination of its internal organs than is the case with most of the members of the group.

Mesostoma Ehrenbergii is a flat leaf-shaped Rhabdocoel, its dorsal surface being but very slightly vaulted. Its length does not exceed 15 mm., and its greatest breadth 5 mm. Another member of this genus has a square cross-section, with the angles somewhat produced. The body is very transparent, and sometimes nothing of it can be seen but the brown stomach and occasionally the red eggs.

The epidermis is composed of flat cells, with irregular outlines and conspicuous nuclei. The whole surface of the body is covered with fine cilia, somewhat longer on the ventral side than on the dorsal. Within the layer of ectoderm cells, but separated from it by a basement membrane, is a thin sheet of

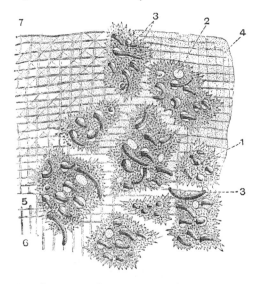

Fig. 56. — Integument of *Mesostoma lingua*, O. Sch. After Von Graff.

1. Epidermis with perforations (2) through which the rhabdites (3) project. Beneath is the basement membrane (4), and beneath this again the muscular layers consisting of circular (5), longitudinal (6), and diagonal (7) fibres.

circular muscle fibres, and within this a layer of longitudinal fibres, crossing the outer layer at right angles, and a few oblique fibres (Fig. 56). These tissues form the integument, and surround a mass of tissue, the parenchyma, in which the various organs of the body are embedded. The coelom is broken up into irregular spaces or splits by the presence of numerous dorso-ventral muscle fibres and connective tissue strands. The lacunae left between these branching fibres contain a fluid in which certain cells float. Some of these cells contain the yellow pigment which gives the animal its yellowish colour. Unicellular slime - glands, with longer or shorter ducts, which open into the surface of the integument, are comparatively numerous, and connective tissue cells are also to be met with in the parenchyma.

The *rhabdites* are homogeneous refractive little rods which occur in most Turbellaria, and are usually found in great numbers either embedded in, or protruding from, the epidermis. They escape from the body, and often occur in great numbers in a slimy deposit the animal leaves on its track ; when they are expelled they leave a round hole in the ectoderm cells. In some species they are secreted in the ectoderm cells, in others, as in Mesostoma, from special cells beneath the integument (Fig. 57); these cells, however, arise from the ectoderm and remain connected with it by fine processes along which the rhabdites travel.

The connective tissue cells and the rhabdite-forming cells,

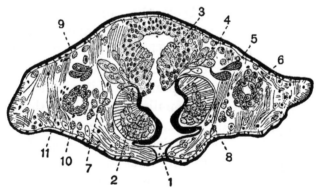

FIG. 57.—Transverse section through the pharynx of *Mesostoma Ehrenbergii*, O. Sch. After Von Graff.

1. Mouth.
2. Pharynx walls crowded with glandular cells.
3. Walls of alimentary canal.
4. Salivary glands.
5. Rhabdite-forming cells.
6. Yolk glands.
7. Testes.
8. Nerve cord.
9. Integument beneath cuticle.
10. Pigment cell.
11. Dorso-ventral muscle fibres.

with some strands of muscle fibres, constitute the parenchyma. In Mesostoma this tissue is not so well developed as in most Turbellaria, and the spaces containing fluid are correspondingly large, in other forms the parenchyma occupies, with the exception of a few small lacunae full of fluid, the whole space within the integument; the digestive, nervous, excretory, and reproductive systems are embedded in it.

The mouth opens on the ventral surface a little in front of the middle of the body. It leads into a pharynx whose wall, just within the mouth, is drawn out into two pharyngeal pouches. The next portion of the pharynx has a very thick muscular wall, and internal to this are two more pharyngeal pouches. The pharynx leads directly into the stomach, the demarcation between them being shown by the change in the lining epithelium, and by the presence of the salivary glands (Fig. 57). These consist of numerous pear-shaped glandular cells, with ducts which open at the junction of the pharynx and stomach. The stomach of Mesostoma is a simple sac, of uniform diameter. Owing to the position of the mouth near the middle of the body, the stomach may be divided into a short preoral and a longer postoral region. The stomach is lined by a single layer of cells, which vary much in shape, and appear to exhibit amoeboid movements similar to those of the endoderm in Hydra, and they ingest nutritious food-particles in the same intracellular way. No anus is present, and undigested matter is ejected through the mouth.

The excretory system consists of (i.) main trunks which open to the exterior, (ii.) secondary branches given off from these, and (iii.) the excretory cells.

In Mesostoma the main ducts open one on each side close to the external pharyngeal pouches in the mouth (Fig. 58). From each opening a transverse duct runs dorsalward on each side of the pharynx and intestine; when it reaches the dorsal side of the alimentary canal it divides into two, one branch running

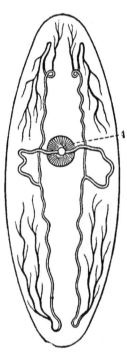

Fig. 58.—Main trunks of the excretory system of *Mesostoma Ehrenbergii,* O. Sch. After Von Graff. They open on to the exterior close to the mouth.

1. Pharynx.

forward and one running backward on each side of the body. These branches give off the very numerous secondary ducts which ramify all over the body. The secondary branches give off numerous minute side branches, each of which ends in a flame cell. The flame cells are pear-shaped cells, the stalk being formed by a flagellum which projects into the lumen of the minute duct. The body of the cell is often branched; it contains a large nucleus, and a vacuole which is continuous with the lumen of the duct. The fluid contained in the excretory system is clear and free from corpuscles, it is kept in motion by the flagella of the flame cells.

The nervous system consists of a central organ, the brain, and certain nerves running from it. The brain is somewhat

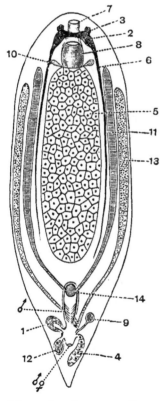

FIG. 59.—Plan of a Rhabdocoelous Turbellarian. Von Graff.

1. Bursa copulatrix.
2. Brain.
3. Eye.
4. Germarium.
5. Intestine.
6. Longitudinal nerve trunk.
7. Mouth.
8. Pharynx.
9. Receptaculum seminis.
10. Salivary glands.
11. Testis.
12. Uterus, containing an egg.
13. Yolk gland.
14. Vesicula seminis.
♂. Chitinous copulatory organs.
♂♀. Common sexual aperture.

oblong in shape, and separated into two halves by a slight furrow. It is situated between the pharynx and the anterior end of the animal, and is embedded in the parenchyma. From each corner a strong nerve is given off. The anterior nerves run to the anterior end of the body, and break up into a number of fine branches. The nerves arising from the posterior angles run backwards on each side of the pharynx and stomach, rather ventral to the latter. Von Graff figures a transverse commissure connecting these two posterior nerves just behind the pharynx.

A pair of eyes are situated directly upon the dorsal side of the brain, they are composed of two triangular aggregations of pigmented cells.

Turbellaria are, with few exceptions, hermaphrodite, and

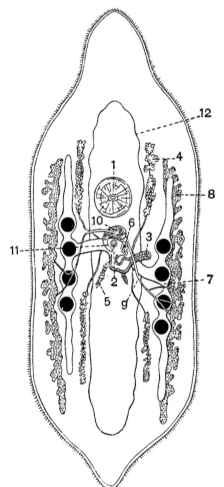

FIG. 60.—The reproductive organs of *Mesostoma Ehrenbergii*. After R. Leuckart, slightly altered.

1. Pharynx.

2. Common male and female genital opening.

3. Ovary.

4. Yolk gland.

5. Shell gland.

6. Receptaculum seminis.

7. Uterus with eggs.

8. Testis.

9. Vas deferens.

10. Vesicula seminis.

11. Penis.

12. Outline of alimentary canal.

their female reproductive organs are complicated by the egg being formed from two sets of glands. The ovum, with its nucleus, arises in the *germarium*, but the yolk or food-material which serves to nourish the developing embryo has its origin in another gland, the *vitellarium.*

In *Mesostoma* the male organs comprise a testis running along each side of the body near the surface; it is of a clear white colour, and follicular nature (Fig. 59). The testis communicates with a vas deferens, arising by one or two branches from the middle of the gland. The vasa deferentia of each side unite just behind the pharynx, and form a short ductus seminalis which traverses a coiled penis. At the base of the latter is a vesicula seminalis. The penis lies in a genital recess, into which the female organs also open; the opening of this recess lies behind the pharynx.

The ovary consists of a number of oval follicles, grouped together on each side of the stomach. The ovary of each side opens separately into the uterus near the opening of the latter into the atrium genitale. The yolk glands (Fig. 60) lie one on each side of the body, and their ducts unite into a common channel, which opens into the genital recess. The two uteri lie between the testes and the ovary, one on each side of the body; they are a pair of tubes which open right and left in the genital recess, and when they are full of eggs they are very conspicuous objects. Other unpaired glands open into the genital recess, one of these is the bursa copulatrix, another forms the receptaculum seminis, and the third is the shell gland.

Mesostoma produces two kinds of eggs : summer eggs, with a thin, transparent shell, and winter eggs, with thick, brown, horny shells. The developement of the summer eggs takes place whilst in the uterus. The winter eggs are laid in the autumn, and can withstand a considerable amount of desiccation. Copulation is said to be reciprocal, and fertilisation always takes place in the atrium genitale.

As a rule, *Mesostoma Ehrenbergii* is found living in still or slowly-flowing clear water, especially where rushes and sedge grow, and where the bottom is of clay. It swims quietly through the water by means of the undulations of its body, or glides with its cilia along the stems of water-weeds. It feeds on the minute freshwater crustacea, small worms, and insect

larvae.   When it has caught a Daphnia, for instance, it will
wrap its body round it and attack it with its proboscis; very
frequently it makes prisoners of its prey by ensnaring them
with threads of mucus.   In the autumn these animals become
of an opaque white colour, lose their power of movement, and
sink to the ground.   The survival of the species through the
cold of winter is ensured by the existence of the well-
protected winter eggs, a device frequently met with amongst
fresh-water creatures.

Mesostoma is not a large Turbellarian; the marine forms
often measure more than an inch in length, and are usually
flattened and leaf-like (Fig. 55); the land Planarians, on the

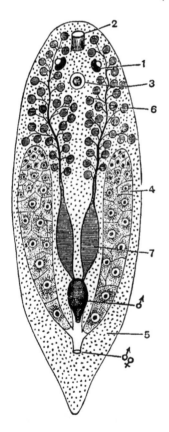

FIG. 61.—Plan of an Acoelous Turbellarian.
After Von Graff.

1. Eye.
2. Mouth.
3. Otolith.
4. Ovary.
5. Digestive parenchyma.
6. Testicular follicles.
7. Vesicula seminalis.
♂. Male copulatory organ.
♂ ♀. Common sexual aperture.

other hand, are elongated and linear: one of this has been
described which attains a length of nine inches.   The marine
forms are often brilliantly coloured, and their dorsal surface
beset with papillae.

The position of the mouth in the different species of TUR-
BELLARIA varies greatly; in some it is anterior, in others it is
ventral and median, and it may be nearly at the posterior end
of the body. True nematocysts,
such as are found in the COELEN-
TERATA, exist in the ectoderm of
a few genera, amongst others in
Microstoma. Some of the ectodermic cells are also modified, and
form adhesive or glutinous areas
which make up for the want of
suckers.

The pigment may be in the
ectodermal cells, or, as in
Mesostoma, in the parenchyma;
Vortex viridis and *Convoluta
Schultzii* contain cells coloured
with chlorophyll and with starch
in them. These are probably
symbiotic algae.

The arrangement of the
muscle fibres in *Mesostoma* is
very simple; the diagonal layer
exists in many genera, and in
some of the Dendrocoels there
may be as many as six separate
layers.

The disposition of the pharynx and stomach in Turbellarians
is very various, and has served as
a basis for the classification of
the class. The muscular part of
the pharynx may be very much
enlarged, and capable of being
protruded through the mouth,
and acting as a powerful sucker.

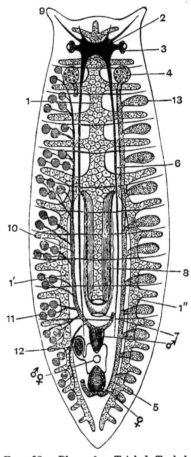

FIG. 62.—Plan of a Triclad Turbellarian. After Von Graff.

1. Anterior.
1', 1''. Posterior branches of alimentary
canal.
5. Oviduct.
9. Tentacle.
10. Vas deferens.
2, 3, 4, 6, 7, 8, 11, 12, 13, as in Fig. 59.
♀. Female copulatory organs.
♂♀. Common sexual aperture.

In one division of the Rhabdocoelida, the Acoela (Fig. 61),
the pharynx ends in the solid parenchyma, into which the food
passes, and is there digested; in the Rhabdocoela the alimentary

canal is a straight and unbranched cavity, such as exists in *Mesostoma*. The other great division of the class, the Dendrocoela, have a branched stomach, the main division of which in the Polycladida is ciliated, the branches only being lined with an amoeboid epithelium, which digests intracellularly. The branches may anastomose, and in rare instances, *Yungia* and *Cycloporus*, they open on to the exterior.

The excretory system is absent in Acoela; in Rhabdocoela the main trunk may be single and open posteriorly, or double, and then the two ducts may, as in *Mesostoma*, open separately and near the mouth, or together at the posterior end. In the Dendrocoels the main trunks open by paired apertures situated on the dorsal surface. The main trunks may be ciliated, the finer branches are probably always intracellular, piercing the cells and not lying between them.

The two main nerve cords are rarely connected by transverse commissures in the Rhabdocoels, but in the Triclads this is the usual arrangement. In the Polyclads there are many, usually eight, nerve cords which diverge from the central cerebral ganglion (Fig. 63). Sensory cells provided with tactile hairs occur in the ectoderm. The eyes are usually two or four in number, but they may be more numerous, and in the Polyclads they increase by division. Auditory vesicles also occur, though they are rare in the Dendrocoels; they are often single, and consist of a vesicle full of fluid in which a calcareous otolith floats. The anterior end of the body is remarkably sensitive, and in some genera forms a tactile proboscis which can be retracted into a sheath; this, together with a pair of lateral ciliated grooves which lie one on each side of the brain in many Rhabdocoels, affords matter for a comparison with the members of the class Nemertea.

Turbellarians, with the exception of two genera, *Microstoma* and *Stenostoma*, are hermaphrodite, but many of them are protandrous—that is to say, their male reproductive cells mature before the female. Most Rhabdocoels and all Triclads have a common genital opening for both male and female ducts; others have separate apertures, and then the male is usually anterior to the female. Self-fertilisation is said to occur in the summer eggs of some Rhabdocoels.

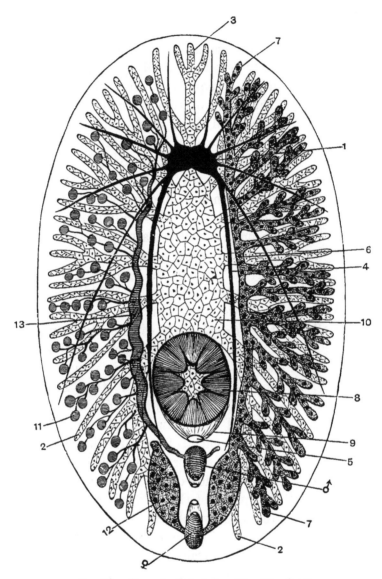

Fig. 63.—Plan of a Polyclad.   After Von Graff.

1. Brain.
2. Intestinal branches.
3. Anterior unpaired branch.
4. Longitudinal nerve cord.
5. Mouth.
6. Oviduct.
7. Ovarian follicle.
8. Pharynx.
9. Pharyngeal pouch.
10. Stomach.
11. Testicular follicle.
12. Uterus.
13. Vas deferens.
♂. Male copulatory organ, with the male aperture behind.
♀. Female copulatory organ, with the female aperture before it.
The eyes are omitted.

The family *Microstomida* is unique in the group in possessing the property of asexual reproduction, which takes place by transverse fission. The posterior third of the body becomes cut off by a septum, and grows until it equals in size the anterior half. Then each portion again cuts off its posterior third, and these again grow (Fig. 64). By this process of segmentation and equalisation chains of 4, then 8, then 16, and sometimes even 32 individuals are formed. Each forms a mouth, and for some time the chain persists, but the individuals ultimately become sexually mature and then separate. They lay eggs during the autumn, which persist through the winter, and these give rise in the spring to the budding individuals.

Two curious animals, the *Coeloplana* and the *Ctenoplana*, have recently been described by Kowalevsky and Korotneff respectively, which are intermediate in structure between the CTENOPHORA and the TURBELLARIA. *Coeloplana* is minute, its flattened body covered with cilia, creeps along the surface on its ventral face. In the middle of its back is a sensory vesicle enclosing an otolith.

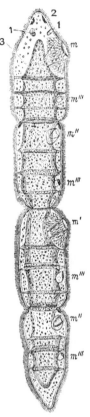

FIG. 64.—*Microstoma lineare*, Oe., undergoing division. There are 16 individuals, 8 with mouth apertures, showing the buds of the first (*m*), second (*m'*), third (*m''*), and fourth (*m'''*) generation. The fifth generation has not yet acquired a mouth aperture. After Von Graff.

1. Ciliated grooves.
2. Eye spots.
3. Intestine.

On each side of its body, at right angles to the plane of the stomach, is a long feathered tentacle which can be withdrawn into a pouch. The mouth in the middle of the ventral surface opens into a stomach which gives off numerous anastomosing canals, two of which pass dorsally, one in front and one behind the otocyst, and there end blindly.

Ctenoplana has much the same shape as *Coeloplana*. In addition to the uniform covering of cilia, there are on the dorsal surface eight bands of ctenophoral plates corresponding to those of the Ctenophors. The alimentary canal resembles that of *Coeloplana*. The sense organ, situated in the middle of the dorsal surface, is a structure resembling the aboral sense organ of Ctenophors ; on each side of the body is a tentacle with short side branches.   In the neighbourhood of the tentacles is the opening of a branching system of canals, which have been compared with the excretory system of Turbellarians. Beneath the ectoderm is a basal membrane which encloses an outer layer of longitudinal muscles and an inner of transverse muscle fibres ; there are also dorso-ventral muscles.

From this short description it will be seen that these two animals resemble Ctenophors in (i.) their aboral sense organ ; (ii.) the presence of ctenophoral plates (Ctenoplana) ; (iii.) the pair of solid, retractile, feathered tentacles ; (iv.) the general arrangement of their body.   On the other hand they resemble the Polyclads in (i.) their flat shape, uniform ciliation, and creeping motion ; (ii.) their basal membrane (Ctenoplana) and disposition of muscle fibres ; (iii.) the branched stomach with ventral mouth ; (iv.) the possession of an excretory system (Ctenoplana ?).

These two animals serve to show that there may have been a connection between the Ctenophora and the Polyclads ; and the resemblance is emphasised by the fact that both groups are hermaphrodite.   On the other hand, it is not easy to homologise the nervous system of the two groups.   The Ctenophors have nothing corresponding with the excretory system of Turbellarians, and the mesogloea of the former does not necessarily represent the mesoderm of the latter.   Further, if the view that there is a connection between the CTENOPHORA and the TURBELLARIA be accepted as sound, it must be through the Polyclads, in many respects the most complicated tribe, and in this case the other members of the group must be regarded as degenerate.

The Turbellaria are classified as follows :

A. RHABDOCOELIDA.—*Of small size, body cylindrical or depressed, with or without an intestine ; when it is present it is simple and unbranched.   Female genital glands compact, not follicular.   Genital orifice single or double.*

Tribe 1. **Acoela.**—*No intestine ; digestion takes place in general parenchyma.   No nervous system or excretory*

system.   *Hermaphrodite ; follicular testis and paired
ovaries.   Marine.   Frequently with brown or green
parasitic algae.*   Convoluta, Proporus.

Tribe 2. **Rhabdocoela.**—*Stomach distinct from parenchyma,
and simple.   Pharynx complicated.   Compact testes and
ovary, the latter divided into a germarium and vitellarium.   Otocysts rare.   Marine, freshwater, and land
forms.*

Prorhynchus inhabits damp earth.   The MESO-
STOMIDA and VORTICIDA are the chief freshwater families ;
the latter includes two parasitic genera—*Graffilla* found
in Gastropods, and *Anoplodium* in Holothurians.

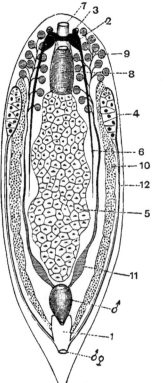

FIG. 65.—Plan of an Alloiocoelous Turbellarian.
After Von Graff.

1. Bursa copulatrix.
2. Brain.
3. Eye.
4. Germarium.
5. Intestine.
6. Longitudinal nerve trunk.
7. Mouth.
8. Pharynx.
9. Testis.
10. Yolk gland.
11. Vesicula seminis.
12. Oviduct.
♂. Chitinous copulatory organs.
♂♀. Common sexual aperture.

Tribe 3. **Alloiocoela.**—*Stomach lobed or irregular, pharynx
present, testes follicular.   Otocysts occur in the family
Monotida.*

All marine, except *Plagiostoma Lemani*, which inhabits the deep water of certain Alpine lakes.

B. DENDROCOELIDA. — *Forms with flattened body and branched intestine. Testes and ovaries follicular. Otocysts very rare.*

Tribe 1. **Tricladida.**—*Body elongated. Intestine in three main branches, which diverge from a cylindrical retractile proboscis. Testes and vitellarium follicular. Germarium compact. Genital aperture single.*

Planaria and *Dendrocoelum* live in fresh water. *Gunda*, which shows traces of metameric structure, and *Bdelloura*, which is an ectoparasite of *Limulus*, are marine. The leech-like land planarians are chiefly tropical. *Rhynchodemus terrestris* and *Geodesmus bilineatus* are European. The former is found in England amongst damp bark, etc., in gardens and woods.

Tribe 2. **Polycladida.**—*Body thin and broad. Numerous branched intestinal caeca open into the stomach. Testes and ovaries follicular. Genital apertures separate, the male anterior to the female. A ventral sucker sometimes present. Marine.* Thysanozoon, Yungia.

CLASS II. CESTODA.

CHARACTERISTICS. — *Endoparasitic Platyhelminthes, devoid of mouth and alimentary canal. Organs of adhesion, in the shape of chitinoid hooks or suckers, enable the parasite to attach itself to its host. The nervous system is composed of a cerebral ganglion and two lateral nerve cords. The coelom is represented by irregular splits in the mesodermic parenchyma. The excretory system is of the Platyhelminthine type, opening to the exterior by one or more pores, which are sometimes furnished with a pulsating vesicle. The Cestoda are hermaphrodite, the generative organs being usually repeated many times.*

One of the commonest Tapeworms found in man is *Taenia saginata (mediocanellata)* (Fig. 66), and it will be a convenient form to describe.

The worm in question may attain a length of five or six yards, and may be composed of many hundred segments, often

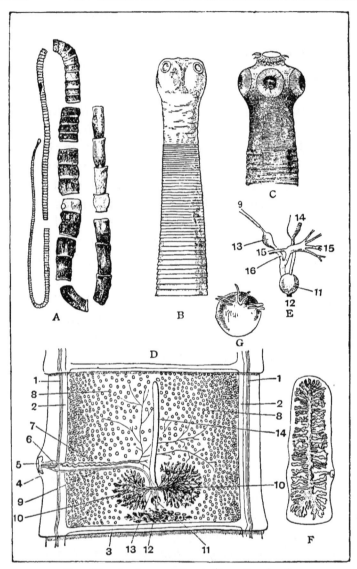

FIG. 66.—Anatomy of Taenia (from Leuckart).

A. Portion of *Taenia saginata*. × ½.
B. Head of the same. × 8.
C. Head of *Taenia solium*, showing the crown of hooks. × 22.
D. A segment of *Taenia saginata*, showing the generative organs. × 7.
   1. Nervous system.
   2. Longitudinal excretory tubes.
   3. Transverse vessels.
   4. Genital papillae.
   5. Genital cloaca.
   6. Cirrus pouch.
   7. Vas deferens.   8. Testes.

   9. Vagina.
  10. Ovaries.
  11. Shell glands.
  12. Yolk glands.
  13. Receptaculum seminis.
  14. Uterus.
E. The connections of the generative organs ; figures as above. × 30.
  15. Oviducts.
  16. Fertilising canal.
F. Detached segment of *Taenia saginata*, showing ripe uterus. × 2.
G. Six-hooked embryo, highly magnified.

termed *proglottides*, which increase in size as they recede from
the head.   The body of a tapeworm may be divided into three
regions: (i.) the head, (ii.) the neck, and (iii.) the segmented
trunk.

The head of *T. saginata* (Fig. 66) is spherical in shape, and
bears on its sides four well-developed suckers.   Other species,
as *T. solium*, in addition to the suckers, are provided with a
double circlet of chitinoid hooks, which assist the suckers to
attach the worm to the inner surface of the alimentary canal
(Fig. 66, C) of its host.   The head is mobile, and can shift its
point of attachment with ease.

The neck is the region immediately succeeding the head ;
its most anterior half is not segmented.   The first trace of
division into segments is the appearance of shallow grooves
which separate the various proglottides one from another.
As they grow backward the proglottides increase in size, and
those situated a foot or more behind the head are sexually
mature.

The surface of the body is covered by a thin clear structureless cuticle, the layer of cells beneath this, corresponding
with the ectoderm of other animals, is composed of long-tailed
cells, the tails running down into the parenchyma.   The body
of the tapeworm is practically solid, the coelom being represented by poorly-developed splits in the parenchyma.

The muscle fibres are arranged in longitudinal transverse and dorso-ventral bundles.   The outermost layer, the
longitudinal, is not a very definite layer, but consists of a
number of unstriated fibres scattered through the parenchyma.
The transverse muscles lie immediately within the longitudinal ;
they serve to divide the parenchyma into a central and a
cortical portion.   The dorso-ventral fibres run from one surface to the other, and are very irregularly arranged.   The
muscle fibres are non-striated, and often branched at their
ends.   The animal has very considerable powers of extension
and contraction.

The parenchyma is composed of ill-defined connective
tissue cells, amongst which are scattered, especially in the
cortex, a number of ovoid calcareous corpuscles (Fig. 67),
about whose function little is known.   They have been vari-

ously regarded as excretory in nature, and as material for counteracting the acidity of the digestive fluids of their host.

The excretory system consists of an annular ring in the head, from which four ducts corresponding in position with the suckers pass backward. Two of these soon disappear, and the other two pass down one on each side of the proglottides, just inside the longitudinal muscle layer (Fig. 67). These

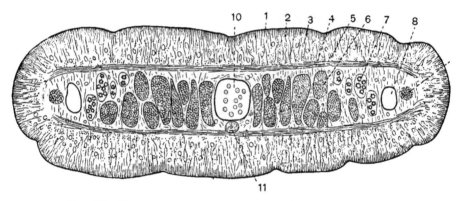

Fig. 67.—Transverse section through a mature proglottis of Taenia.

1. Cuticle.
2. Tailed cells of ectoderm.
3. Longitudinal muscle fibres cut across.
4. Layer of circular muscles.
5. Split in the parenchyma which lodges
   a calcareous corpuscle.
6. Ovary.
7. Follicle of testes with sperm morulae.
8. Longitudinal excretory canal.
9. Longitudinal nerve cord.
10. Uterus.
11. Oviduct.

two lateral ducts coalesce at the posterior end of the last proglottis, and open there by a common vesicle. A transverse vessel at the posterior end of each proglottis serves to place the two longitudinal ducts in communication. The main ducts are provided with valves, which only permit the flow of fluid towards the external opening. A series of secondary ducts arise from these main ones and ramify all over the body. The secondary ducts give off still finer tubules, each of which ends in a flame cell. The cilium of this cell hangs down into the lumen of the tubule, which is here slightly enlarged and funnel-shaped. Some observers maintain that this funnel-shaped end of the tubule opens by a pore into the splits in the parenchyma which represent the coelom, but this is a disputed point.

There is a central nervous system in the head. This gives

rise to some nerves which pass forward and supply the suckers, and to two stout nerve cords which pass back along each side of the proglottides, lying just outside the longitudinal ducts of the excretory system (Fig. 67).

Both male and female reproductive organs are repeated in each proglottis (Fig. 66). The testis is composed of very numerous vesicles in which spermatozoa arise, and which are each attached to one of the branches of the ramifying vas deferens. Thus the testis is dispersed all over the proglottis. The branches of the vas deferens unite and form a common tube, which is slightly coiled, and which runs from the centre of the proglottis to the common genital pore situated in the middle of one side (Fig. 66). The portion of this common vas deferens lying next the orifice has very muscular walls, and can be extruded. It functions as a penis.

Self-fertilisation takes place, and is probably the usual method, there is no evidence to show that one proglottis fertilises another.

The vagina opens into the genital pore a little behind the penis, it then passes backward to a small swelling, the receptaculum seminis. In this the spermatozoa are stored until the ova are ripe for fertilisation. The ovaries are two in number, and are composed of numerous tubules on a branching duct. Each ovary gives off an oviduct, and the two unite, and then receive a small duct which comes from the receptaculum seminis and conveys the spermatozoa; it is therefore called the fertilising canal (Fig. 66). The oviduct then receives a duct from the yolk gland, which lies between and behind the ovaries, and passing through a small spherical shell gland, which deposits the shell, enters the uterus.

The uterus is at first an inconspicuous simple sac, whose only opening is that from the oviduct. As it becomes full of eggs it increases greatly in size, and becomes much branched; eventually it occupies almost the whole of the interior of the proglottis, the reproductive organs of both sexes having atrophied (Fig. 66, F). A proglottis which is ripe for separation consists of little more than a sac—the uterus—crowded with minute spherical eggs, which eventually escape by the rupture of its walls.

The terminal proglottis breaks off from the rest of the Tapeworm owing to the contraction of its muscles, its rupture affects the transverse excretory duct of the segment in front, which is put into communication with the outer world, and thus forms the new terminal vesicle, which opens to the exterior.

The ripe proglottis passes out of the alimentary canal of its host, and the eggs with which it is crowded escape by the rupture of its walls. In some species the free proglottides,

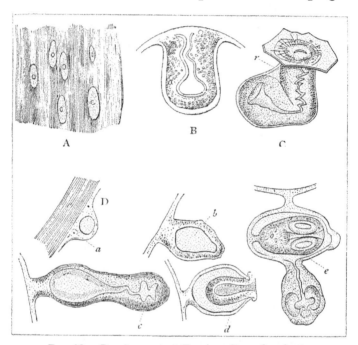

FIG. 68.—Development of *Taenia*.   From Leuckart.

A. *Cysticercus bovis*, from beef. Natural size.
B. Invaginated head of *Cysticercus* before the formation of the suckers. × 20.
C. Invaginated head of *Cysticercus cellulosae*, showing the bent head. × 30.

D. Stages in the development of the brood capsules in *Echinococcus*.
a. The thickening in the parenchyma of the bladder.
b. Subsequent formation of a cavity in it.
c. Development of the suckers.
d. A capsule with one head invaginated.
e. A capsule with two heads.   × 90.

after they have left their host, retain a considerable power of movement, and crawl over blades of grass, etc. In this way their ova are disseminated, and stand a better chance of being eaten by their second host, in whose body alone they can develop. Some of them succeed in making their way, either by means of the food or water, into the stomach or intestine of an ox. Here the egg shell is dissolved by the action of the digestive juices, and a small embryo, the *proscolex*, emerges (Fig. 66, G). The proscolex is a minute spherical embryo provided with three pairs of hooks, by the aid of which it burrows its way through the wall of the alimentary canal, and eventually passes into some blood-vessel, or into the body-cavity. If the former is the case, it is carried either to the liver, brain, lungs, muscles, connective tissue, or eye (Fig. 68). Its presence causes a cyst to be formed round it, pathological changes being induced in the surrounding tissues of the host.

The embryo soon loses its hooks and begins to grow. Some of the cells in its interior liquefy, and a vesicle full of fluid results. These vesicles attain a length of 4 to 8 mm., and when found between the muscle fibres of beef or pork, the meat is technically termed measly. At one side of the vesicle the head commences to be formed ; it arises inside out as a projection into the lumen of the vesicle (Fig. 68). This projection is hollow, and on its walls, facing its own lumen, the suckers and rudimentary ring of hooks arise, whilst in the substance of its walls the excretory system is being formed. The head then turns inside out, and the result is a *Cysticercus* or bladder worm. The Cysticercus consists then of the head and short neck, termed the *scolex*, and of the bladder or vesicle. That formed by the embryos of *Taenia saginata* is known as *Cysticercus bovis*, from the host it inhabits (Fig. 68). As long as it remains in the body of the ox, the Cysticercus is incapable of further developement ; if, however, it is swallowed by a man, the vesicle is digested, whilst the head fixes itself to the walls of the intestine, and commences to divide into proglottides.

LIFE-HISTORY OF TAENIA SAGINATA.

Scolex............in man gives rise by strobilisation to the
|
Proglottides.........these leave their host carrying with them the
|
  Ova...............which, when found on grass or in water, contain the
  |
Six-hooked.........embryo or proscolex. This passes into alimentary canal of an
                  ox, and working its way into the tissues of this intermediate
  | host, becomes the
Cysticercus... ... ...or bladder-worm. When this is eaten by man it develops
  | into the
  Scolex.

The question whether the proglottides which are cast off one by one by the tapeworm should be regarded as individuals, is one which has given rise to much difference of opinion. The comparison which has been instituted between a single proglottis and the body of a Trematode gives some support to the view that the scolex produces by strobilisation a number of individuals (the proglottides), which in their turn reproduce sexually. If this view be retained, the Taeniidae exhibit a well-marked alternation of generations.

On the other hand, the nervous, excretory, and to some extent the muscular systems are common to all, and the detached proglottides are not capable of individual life or

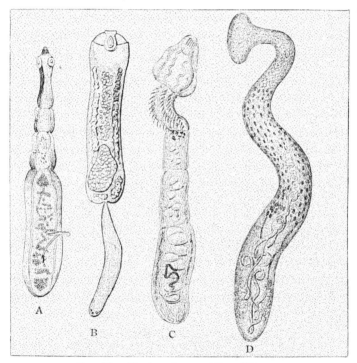

FIG. 69.—Various forms of Tape-Worms.
A. *Taenia echinococcus.*   × 12.   From Leuckart.
B. *Archigetes Sieboldii.*   × 60.   From Leuckart.
C. *Echinobothrium typus.*   × 10.   From Van Beneden.
D. *Caryophyllaeus mutabilis.*   × about 5.   From Carus.

maintenance. It is true that they are stated to increase in size when separated, but this may be due to imbibition of fluids, and it is doubtful how far it corresponds with growth.

Further, the separation of parts of the body containing reproductive cells occurs in the hectocotylisation of the Cephalopods, and this is no longer regarded as an instance of
alternation of generations.

It is an interesting fact that, with the exception of *Archigetes*,
which inhabits the Oligochaet *Tubifex rivulorum*, the sexual form
of the Cestoda are entirely confined to Vertebrate hosts. The
Cysticercus form, although usually found in Vertebrata, has also
been observed in the bodies of Molluscs, Arthropods, and Worms.

Some interesting divergences from the arrangements of the

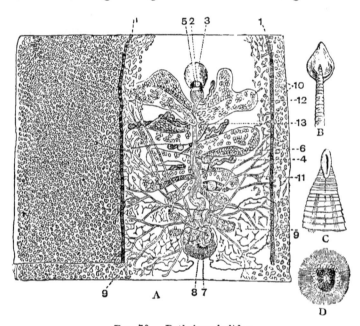

FIG. 70.—Bothriocephalidae.

A. A segment of *Bothriocephalus latus*,
  showing the generative organs from
  the ventral surface.
1. Excretory vessels.
2. Cirrus.
3. Cirrus pouch.
4. Vas deferens.
5. Vaginal opening.
6. Vagina.
7. Shell gland.
8. Oviduct.
9. Ovary.
10. Yolk glands.
11. Their duct.
12. Uterus.
13. Uterine opening. The testes are not
  visible from this side. × 23.
  From Sommer and Landois.
B and C. Marginal and lateral view of *B.
  cordatus*, showing the cephalic
  grooves. × 5. From Leuckart.
D. Ciliated larva of *B. latus*. × 60.
  From Leuckart.

organs in *Taenia* are met with in *Bothriocephalus latus*, the
largest Cestode which inhabits the human intestine; it may

attain a length of nearly ten yards. The head is flattened, and bears two lateral slits; the genital pores are not on the side, but on one surface of each segment. The male orifice is distinct, and in front of the female, and the uterus has an aperture for the exit of the ova, behind the female orifice through which the spermatozoa are introduced. The asexual stage has been found in certain freshwater fish. This parasite is chiefly met with in Russia and Central Europe.

Certain genera, as *Amphilina*, found in the body-cavity of the Sturgeon, and *Amphiptyches* in the stomach of the Chimaera, have flat bodies with one anterior sucker. Their embryos are ciliated, and their female reproductive organs closely resemble

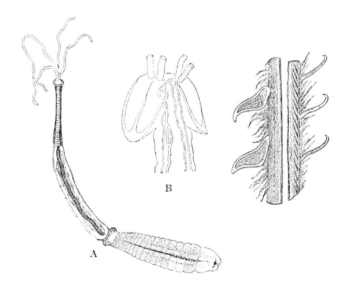

FIG. 71.—Tetrarhynchus.

A. General view of the worm.   × 4.
B. Head, showing the suckers, proboscides, and excretory canals.   × 25.
C. Portion of a proboscis showing the two forms of hooks, highly magnified.
   All after Pintner.

those of Trematodes. They are therefore looked upon as connecting the latter class with the Cestoda. *Ligula* is another genus in which the segmentation into proglottides does not occur.

Although, if the proglottides be not regarded as distinct individuals, there is no alternation of generations in *Taenia saginata*, in some other species this phenomenon is brought about by the formation of many scolices from one proscolex. Thus, in *Taenia coenurus*, the bladder-worm which gives rise to the "staggers" in sheep produces several heads, and thus several scolices are produced asexually, which in their turn, if they attain their right host, reproduce sexually.

The *Taenia solium*, which is common in man, has its bladder-worm, *Cysticercus cellulosae*, in the pig; it is, however, also found in man, and is the cause of considerable disturbance, often ending in the death of the host. *Taenia echinococcus*, which is found in the sexual form in dogs, exists in the Cysticercus condition in man.

The genus *Tetrarhynchus* is curiously modified, it appears to be flattened in a plane at right angles to that of other Cestoda. The head bears four long hooked proboscides, which can be protruded from two disks; each of the latter seems to be homologous with two of the suckers of Taenia.

The Cestoda are divided into seven families:

Family   I. Amphilinidae—*Amphilina*, found in Sturgeon.
,,   II. Caryophyllaeidae—*Caryophyllaeus*, found in intestine of Cyprinoid fishes.
,,   III. Pseudophyllidae—*Bothriocephalus*, *Ligula*, *Archigetes*, found in fishes.
,,   IV. Diphyllidae—*Echinobothrium*, found in Selachians.
,,   V. Tetrarhynchidae—*Tetrarhynchus*.
,,   VI. Tetraphyllidae—*Anthobothrium*, *Calliobothrium*, found in Selachians.
,, VII. Taeniidae—*Taenia*.

Class III. TREMATODA.

CHARACTERISTICS.—*Platyhelminthes with a cylindrical or flat leaf-like body, devoid of segmentation. Their adhesive organs are in the form of suckers. Hooks are rarely present. The excretory system ends internally in flame cells, and opens to the exterior by a contractile vesicle, or by two pores. They are parasitic and hermaphrodite; both reciprocal and self-fertilisation occur. The embryo may develop directly, or it may pass through a series of stages, in some of which asexual reproduction takes place.*

Fasciola (*Distoma*) *hepatica*, commonly known as the liver fluke, is found in the liver of diseased sheep. It is about $\frac{3}{4}$ of an inch long, and has a flattened leaf-like shape.

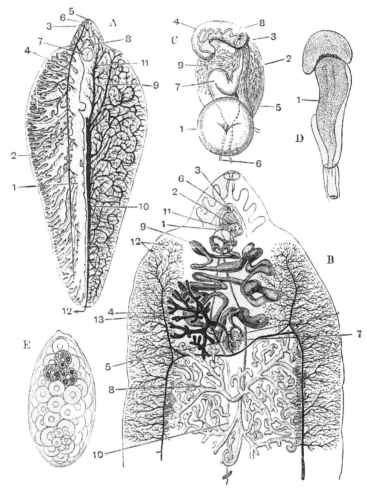

FIG. 72.

A. *Fasciola hepatica*, from the ventral surface (× 2), the alimentary and nervous systems only shown on the left side of the figure, the excretory only on the right.

1. Right main branch of the intestine.
2. A diverticulum.
3. Lateral ganglion.
4. Lateral nerve.
5. Mouth.
6. Pharynx.
7. Ventral sucker.
8. Cirrus sac.
9. Left anterior dorsal excretory vessel.
10. Main vessel.
11. Left anterior ventral trunk.
12. Excretory pore.

B. Anterior portion, more highly magnified.

1. Cirrus sac.
2. Ductus ejaculatorius.
3. Female aperture.
4. Ovary.
5. Oviduct.
6. Penis.
7. Shell gland.
8. Anterior testes.
9. Uterus.
10. Vasa deferentia.
11. Vesicula seminalis.
12. Yolk gland.
13. Its duct (from Marshall and Hurst, after Sommer).

C. Genital sinus and neighbouring parts.

1. Ventral sucker.
2. Cirrus sac.

The cuticle which covers its body is inflected at the mouth, and at the genital and excretory pores. The nature of the cuticle, like that of Cestodes, is a matter of dispute, some authorities looking upon it as a basement membrane, whilst others hold the view that it is a metamorphosed layer of cells. It is perforated by numerous fine pores, and is produced into many backwardly-directed spines.

The intestine, which consists of two main branches giving off numerous secondary ones, is lined by a layer of amoeboid cells, and the digestion of the solid food, such as blood corpuscles, is intracellular.

The reproductive organs are similar to those met with in the Cestodes, with the addition of a channel known as the Laurer-Stieda canal, whose function, although the subject of much debate, still remains uncertain. The canal is given off from the common duct of the yolk gland, just before it enters the substance of the shell gland to unite with the oviduct. The canal opens dorsally upon the surface of the animal. Two views have been held as to its use : it is regarded by some as the duct through which the spermatozoa enter the female organs, and in *Polystoma* (Fig. 74, B) this has actually been observed ; on the other hand it has been urged that in some species, notably in *Fasciola*, it is too minute for this purpose, and the second view has been propounded, that it acts as a valve, and permits the escape of an excess of the secretion of the yolk glands, or of spermatozoa.

The method of fertilisation in the liver fluke is not very definitely settled, probably self-impregnation takes place.

The life-history of the *Fasciola hepatica* may be briefly summarised as follows. The sexual form lives in the liver and bile ducts of sheep and other domesticated animals. Its fertilised eggs develop to some extent in the uterus, and

3. Genital pore.
4. Evaginated cirrus sac (? penis).
5. End of vagina.
6. Vasa deferentia.
7. Vesicula seminalis.
8. Ductus ejaculatorius.
9. Accessory gland (from Sommer).

D. A ciliated internal end from the excretory apparatus.

1. Orifice of the flame cell (highly magnified).

E. Egg of *Fasciola hepatica*, × 330 (from Thomas).

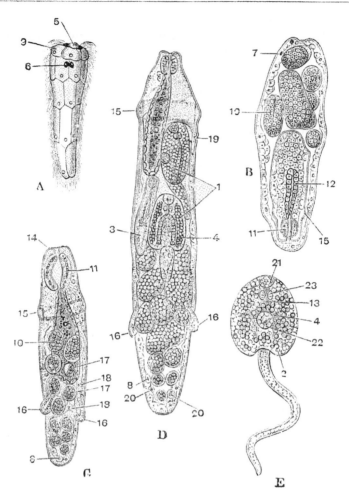

FIG. 73.—Five stages in the life-history of *Fasciola hepatica*, all highly magnified.

A. The free-swimming embryo.
B. The sporocyst containing young rediae.
C. The young redia, the digestive tract shaded.
D. An adult redia, a daughter redia, two cercariae, and germs.
E. The free cercaria. The figures have the same significance throughout.
 1. Nearly ripe cercariae.
 2. Cystogenous cells.
 3. Daughter rediae.
 4. Limbs of the digestive tract.
 5. Head papilla.
 6. Eye-spots.
 7. Same degenerating.
 8. Germinal cells.
 9. Cells of the anterior row.

10. Embryo in optical section, gastrula stage.
11. Pharynx of redia.
12. Digestive sac.
13. Oesophagus.
14. Lips of redia.
15. Collar.
16. Processes serving as rudimentary feet.
17. Embryos.
18. Trabeculae crossing body-cavity of redia.
19. Birth opening.
20. Morulae.
21. Oral sucker.
22. Ventral sucker.
23. Pharynx.   All from Marshall and Hurst, after Thomas.

are cast out from the body into the bile duct in such quantities that it has been computed that each fluke produces half a million eggs.

The further development of the embryo only takes place outside the body, and at a low temperature. If these conditions be present, at the end of two to three weeks the egg gives birth to a free-swimming *ciliated embryo*. This is a conical larva, provided with a double eye-spot, and rudiments of an excretory system in the form of ciliated funnels (Fig. 73, A). If this embryo is fortunate enough to be born in a pond or ditch, it swims about looking for a certain species of water snail, *Limnaea truncatula*. If it fails in its quest, it dies in eight or ten hours; on the other hand, if it succeeds, it immediately sets to work to bore into the soft tissues of the snail. This it effects by elongating its head papilla into a pointed structure, and revolving on its axis by means of its cilia. When once it has forced an entrance into the tissues of the snail, it loses its cilia and becomes a *Sporocyst*. This is an oval sac of cells, whose wall is covered with a cuticle and contains circular and longitudinal muscle fibres, and is lined by an epithelium (Fig. 73, B). The Sporocyst may multiply by transverse division. Within its body certain germinal cells arise, and these ultimately form a *Redia*, which bores through the walls of the Sporocyst, and makes its way to the liver of the snail (Fig. 73, C and D). The walls of the Sporocyst close up, and the process is repeated; but if too many Rediae are produced they may cause the death of the Snail. The Redia is a cylindrical larva with a terminal mouth, which leads through a pharynx into a blind stomach lined with a single layer of cells. A little way behind the mouth the surface of the body is raised into a circular ring, and posteriorly there are two projections which assist the larva in its movements. The excretory system is well developed, and the cells lining the body-wall give rise to the germinal cells. These latter may produce fresh Rediae, but as a rule they give rise to *Cercariae*: organisms which differ from their parent by the possession of a forked alimentary canal, two suckers, a tail, and certain cystogenous cells. The Cercariae escape from the Rediae through an opening just behind the collar; they are at first active, and make their way out of the body of the snail into the water (Fig. 73, E). They then swim about for a time, and ultimately settle on some water-plant, or during a flood on the grass, and by means of the cystogenous cells envelop themselves in a cyst. If they are then swallowed by the grazing sheep, they make their way to the bile duct, and there develop into the sexual *Fasciola hepatica*. It is thus evident that in the life-history of this Trematode there is an alternation of generations, during which there are several occurrences of asexual reproduction.

LIFE-HISTORY OF FASCIOLA HEPATICA.

Sexual Adult (Sheep).

Ciliated Embryo (Water and Snail).

Sporocyst (Snail).

Sporocyst (by division).   Redia (Snail).

Redia (by gemmation).   Cercaria (encysted on grass).

Sexual Adult (Sheep).

The disastrous effects which this internal parasite produces on its host are evidenced by the fact that it is calculated that one million sheep are annually lost in the United Kingdom from what the farmers call " liver rot " alone.

The TREMATODA are divided into the **Monogenea** and the **Digenea.**

A. *The* **Monogenea** *develop directly, without the intervention of asexual forms. They inhabit therefore one host only, and are with few exceptions ectoparasitic.*

Amongst the exceptions to the last statement, is the species *Aspidogaster conchicola*, which inhabits the pericardial cavity and the nephridia of the freshwater mussel (Anodon) (Fig. **74**).

A very curious Trematode inhabits the gills of the minnow. Its embryo is ciliated and free-swimming, and is termed a *Diporpa*; it, however, soon loses its cilia and settles down on the gills of its host. At first it lives singly, but after a time two individuals come in contact, and one seizes the dorsal papilla of the other by its ventral sucker; they then twist round so that the ventral sucker of the second is

able to attach itself to the dorsal papilla of the first, and in
this condition there is an actual fusion of the tissues of the

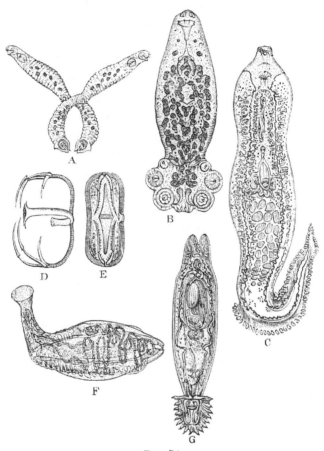

FIG. 74.

A. *Diplozoon paradoxum*, two united individuals.
B. *Polystoma intergerrimum.* × about 100.   After Zeller.
C. *Microcotyle mormyri.* × 7.
D, E. Two views of the chitinous framework of a sucker of *Axine belones*, highly
     magnified.   After Lorenz.
  F. *Aspidogaster conchicola.* × about 25.   After Aubert.
  G. *Gyrodactylus elegans.* × about 80.   After Wagner.

suckers and papillae.   This double organism is known as the
Diplozoon paradoxum (Fig. 74).

Polystoma intergerrimum, another of the Monogenea, is
common in the bladder of the frog (Fig. 74).   It lays its eggs
in the spring, and by protruding its body from the anus of the
frog, manages to deposit them in the water.   The young larvae

after a short free life attach themselves to the gills of a tad-pole. When the gills atrophy, the larvae proceed down the alimentary canal and eventually reach the bladder of the young frog. Here they take five or six years to reach maturity. This Trematode has at its hinder end a disc, round which are grouped numerous hooks and suckers.

Gyrodactylus elegans has a similarly-situated triangular plate which bears two large hooks in the centre, and sixteen smaller ones round the edge, by means of which it attaches itself to the fins of sticklebacks and other freshwater fish (Fig. 74). Its most remarkable feature is that it is viviparous, and its embryos before they leave the body of their mother have already developed their embryos inside them; and the latter may contain their embryos, so that four generations may be included under the cuticle of the sexually mature animal.

B. *The* **Digenea** *have always one, and usually several asexual intermediate generations intercalated between the sexual, and their life-history usually involves residence in two distinct hosts.*

The asexual generations usually inhabit some Mollusc, more rarely they attack fish. The sexual forms are found in all classes of the Vertebrata. The genus *Distoma* includes more than three hundred species, eight of which infest the human race.

One of the most dangerous human parasites is *Bilharzia haematobia*; it is remarkable amongst Trematodes for its sexes being separate. The mature worms are found in couples, the female partly enclosed in a gynaecophoric canal or groove on the under side of the thicker male. They inhabit the blood-vessels of the bladder and give rise to considerable disturbance in the system. Their eggs escape with the urine, but their future fate is not known.

Leucochloridium paradoxum is parasitic in the body of a snail, *Succinea putris*; it develops two sacs which grow into the tentacles of its host, which may ultimately be ruptured by the increase of these structures.

Both the CESTODA and the TREMATODA have been consider-ably modified by leading an endoparasitic life. They have lost their locomotor organs, and are dependent on cilia or on

the boring hooks of their larvae for a change of host; as some
compensation they are amply endowed with organs of adhesion
in the form of either hooks or suckers. They have further

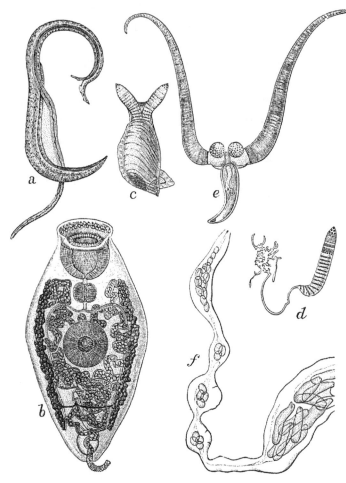

Fig. 75.—*a, Bilharzia haematobia*, the thin female in the gynoecophoric canal of the
stouter male, × 15 (after Leuckart). *b, Distoma macrostomum*, showing the
digestive, and the greater part of the genital apparatus, with the cirrus protruded, × 30. *c*, Snail (*Succinea*), the tentacles deformed by Leucochloridium,
nat. size. *d*, Leucochloridium removed from the tentacle, nat. size (after Zeller).
e, Bucephalus polymorphus, highly magnified (after Ziegler). *f*, Portion of a
sporocyst containing *Bucephali* in process of developement, × about 50 (after
Lacaze Duthiers).

acquired the power of producing an enormous number of eggs,
a necessary provision when the remoteness of their chance of
hitting on the right host is taken into account; this power of
propagation is further increased in many cases by the asexual

budding of the embryos. In the CESTODA, the group in which
parasitism has left the deepest mark, the whole alimentary system
has disappeared leaving no trace, and the food is absorbed
through the skin.   In both groups the sense organs, for which
they can have but little use, are very poorly or not at all
developed.

There can be little doubt that the CESTODA and TREMA-
TODA are connected, or that the former have become more
modified and departed farther from their common ancestor,
than the latter.   Lang's researches lead him to look for the
ancestors of the TREMATODA amongst the TURBELLARIA, and
especially amongst the **Triclades**, which group contains some
species not easily distinguished from the ectoparasitic Trematodes, but for their ciliation.

CHAPTER VII

NEMERTEA

NEMERTEA $\left\{\begin{array}{l}\text{Palaeonemertea—}Carinella,\ Polia.\\ \text{Schizonemertea—}Lineus,\ Cerebratulus.\\ \text{Hoplonemertea—}Tetrastemma,\ Geonemertes,\ Malacobdella.\end{array}\right.$

CHARACTERISTICS. — *This class is characterised by a ciliated ectoderm, which at the anterior end of the body is sunk in, and forms a pair of ciliated grooves or pits. There is a protrusible introvert, which may be armed with hooks and spines, opening above the mouth. A nerve commissure surrounds it. The nervous system consists of two cerebral ganglia, giving off two lateral nerves which extend throughout the body, and may unite above the anus. The alimentary canal is not branched, but may bear lateral caeca; it terminates in an anus. The generative organs are simple and paired; the sexes are usually distinct. With few exceptions, the Nemertea are marine.*

The Nemertines are mostly found amongst seaweed and coral rock, and they are frequently of the most brilliant colour. *Pelagonemertes* is, however, pelagic, and like most pelagic organisms is transparent, and *Malacobdella* is parasitic, living in certain Lamellibranchs. A few inhabit fresh water, such as *Tetrastemma aquarum dulcium* from North America, and an unknown species of the same genus recently found in England. Two species of *Tetrastemma* and two species of *Geonemertes* are terrestrial.

Probably the members of no class vary so much in size as do the Nemertines; many of them are quite small, whilst others attain the length of many feet, and Professor M'Intosh records finding a *Lineus marinus* which measured thirty yards, and even then was only half uncoiled. They possess extreme

powers of contractility, but are very easily broken into fragments.

Tetrastemma flavida, which may serve as a type of the
class, is a pinkish Nemertine an inch or more in length, which
is found under stones or in fissures of the rock when the tide
is down. It occurs in all European waters from Scotland to
the Red Sea; it is extremely delicate, and the slightest touch
may serve to break it.

The ectoderm of this animal is columnar and ciliated.
Beneath the ectoderm, and separated from it by a basement
membrane, is the thin layer of circular muscles, and this layer
in its turn surrounds a thickish layer of longitudinal muscles

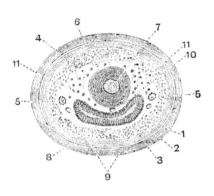

FIG. 76.—Diagrammatic section to
show disposition of parts in a
Hoplonemertean (such as Tetrastemma). After Hubrecht.

1. Cellular portion of integument.
2. Basement membrane.
3. Circular muscular layer.
4. Longitudinal muscular layer.
5. Lateral nerve.
6. Cavity of proboscis sheath.
7. Proboscis.
8. Intestine.
9. Lateral blood-vessels.
10. Dorsal blood-vessel.
11. Connective tissue.

(Fig. 76). From the latter, fibres arise and pass in various
directions through the body, certain of these are especially
conspicuous, running in a dorso-ventral direction between the
diverticula of the alimentary canal, and these give the body
the appearance of being segmented, but the bundles of muscles
are not very regular, nor are they always opposite each other.
A layer of diagonal muscle-fibres usually occurs within the
longitudinal.

The integument of Nemertines encloses a more or less
solid mass of parenchyma, in which the various organs of the
body are embedded. In this there are two kinds of spaces.
Firstly, there is the space into which the proboscis is withdrawn, and the blood-vessels. These contain a fluid in which
corpuscles float, and they represent that kind of coelom known
as an archicoel. And then there are the spaces in which the

generative cells arise, and certain spaces found in one or two species between the diverticula of the alimentary canal. The morphological value of these spaces is still under discussion. They do not contain a corpusculated fluid.

The mouth is situated on the ventral side, and is sur-

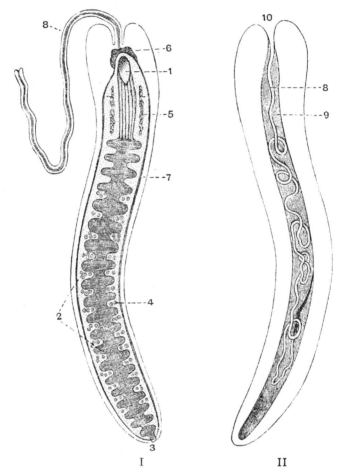

Fig. 77.—Diagram of the organs of a Nemertine. I. from below. II. from above.
After Hubrecht.

1. Mouth.	6. Brain lobes.
2. Intestinal diverticula.	7. Longitudinal nerves.
3. Anus.	8. Proboscis.
4. Ovaries.	9. Proboscis sheath.
5. Nephridia.	10. Proboscis pore.

rounded by swollen lips, it leads into a muscular oesophagus which soon opens into a spacious stomach. The stomach has a wide lumen, and it occupies a large part of the animal; it is produced into a certain number of lateral diverticula, but these do not seem to be very definite in number or size (Fig. 77). The stomach is lined by a layer of cells which are capable of assuming very different outlines at different times; they often break away from the wall and are seen floating in the lumen of the digestive canal. There is no special muscular coat, but some of the muscle-fibres running through the parenchyma are attached to the walls of the stomach. In most Nemertines the alimentary canal is ciliated. The anus is terminal.

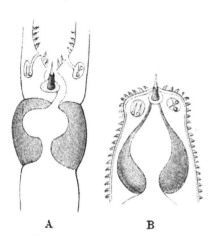

A   B

FIG. 78.—Introvert of a Hoplonemertine, with stylet, "reserve" sacs, and muscular bulb. After Hubrecht.

A. Retracted.
B. Everted.

The most characteristic organ in the Nemertines is the introvert or proboscis, which consists of the hollow eversible anterior end of the animal. In its retracted condition this is invaginated into a cavity, the lumen of the proboscis sheath, just as the finger of a glove may be inverted into the glove. The cavity of the proboscis sheath is a closed one, and full of a corpusculated fluid; the walls of this cavity, i.e. the pro- boscis sheath, are extremely muscular, when they contract the pressure of the fluid drives the introvert forward and it is everted. In many Nemertines the proboscis sheath extends the whole length of the animal, and only ends just above the anus. The method of the eversion of this proboscis is interest- ing; when it begins to protrude, it is the walls of the organ which first grow forward, and the extreme end of the proboscis —often armed with a spine—is the last part to appear, and is therefore only to be seen when the proboscis is fully ex- tended. It is retracted by a special muscle inserted into the tip of the proboscis behind the spine, and arising from the base of the proboscis sheath; when this contracts, the first portion to disappear is the tip. The aperture through which the proboscis appears is either terminal or ventral, but almost invariably in front of the mouth ; more rarely it opens with it (*Malacobdella*).

Tetrastemma has a well-developed spine at the end of its proboscis, and on each side a couple of small secondary ones. Certain glandular structures open by a duct near the base of the spine, and possibly secrete a poison. In those Nemertines in which the introvert is constantly in use, the walls of the proboscis sheath are extremely muscular ; and this defensive organ can be shot out with the greatest velocity, and at times with such force as to break off. When this is the case, it retains its vitality for some time, and crawls about independently. This may be accounted for by the enormous developement of nervous tissue found in its walls. The animal is capable of reproducing its lost introvert.

A closed system of blood-vessels lined with an epithelium is present (Figs. 76 and 80). It comprises a median dorsal vessel

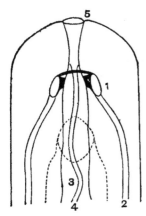

FIG. 79.—Anterior portion of the body of a Nemertine. After Hubrecht.

1. Brain.
2. Lateral nerves.
3. Proboscis sheath.
4. Proboscis.
5. External aperture through which the introvert is everted. The mouth and oesophagus are indicated by dotted lines.

which runs along the body just below the proboscis sheath, at the posterior end this divides into two branches above the anus, and the branches run forward as two lateral trunks situated in the longitudinal muscle layer. At the anterior end the three trunks again unite, and from their point of union give off a loop which in many species encircles the proboscis sheath. The dorsal and lateral trunks in most forms communicate by transverse vessels which lie between the diverticula of the alimentary canal. The blood is colourless in *Tetrastemma*, but in some other species it contains haemoglobin.

It is stated to flow forwards in the lateral, and backwards in the dorsal vessel.

The nephridia of Nemertines are paired and situated anteriorly (Fig. 77). Their ducts open to the exterior by one or more openings on each side of the body, and they always lie above the nerve trunk. These ducts are lined by a single layer of ciliated epithelial cells, and are sometimes much branched ; their inner ends vary a good deal in different genera, and there is considerable discrepancy in the accounts of different observers. In the freshwater *Tetrastemma* the ends of the branched ducts are said to terminate in flame cells, and

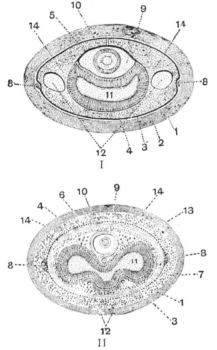

FIG. 80. — Diagrammatic sections to show disposition of internal organs in (I.) *Carinella*, a Palaeonemertine, and (II.) a Schizonemertine.

1. Cellular portion of integument.
2. Basement membrane.
3. Circular muscular layer.
4. Longitudinal muscular layer.
5. Second circular muscular layer in I.
6. Second longitudinal muscular layer in II.
7. Nervous layer.
8. Lateral nerves.
9. Cavity of proboscis sheath.
10. Proboscis or introvert.
11. Intestine.
12. Lateral blood-vessels.
13. Dorsal blood-vessel.
14. Connective tissue.

the whole system is compared to that of Turbellarians. The system is embedded in the parenchyma. In *Carinella* a portion of the wall of the lateral blood-vessel is modified to form the internal end of the nephridium. This forms a spongy gland which is continuous with the duct, the latter is also stated to open at two points into the blood-vessel.

The central nervous system consists of two pairs of ganglia in the head united by two commissures. One of these, the smaller, lies dorsal to the proboscis sheath, the other between the proboscis sheath and the oesophagus. It will thus be seen that the proboscis is surrounded by a nerve ring, a relationship to the nervous system which is usually confined to the alimentary canal of Invertebrates.

The dorsal and ventral ganglion are separated by a deep groove. The dorsal half gives off nerves to the eyes and fore part of the head. The ventral half is continuous with the main lateral nerve trunks. These two trunks run back on each side of the body, embedded in the layer of longitudinal muscle fibres. In *Schizonemertea* there is a third lobe borne on the dorsal aspect of the brain; and in *Hoplonemertea* this lobe may be distinct and connected with the brain only by a nerve. In some species it is hollow, and its walls ciliated. In the last-named subdivision the longitudinal nerves give off numerous segmentally-arranged nerves, but in the Palaeo- and Schizo-nemertines these are replaced by a delicate plexus which lies between the external longitudinal and the circular muscles (Fig. 80). The main trunks may unite above the anus in the Hoplonemertines, as in *Peripatus* and *Chaetoderma*. A median nerve runs back from the supra-proboscidian commissure and supplies the proboscis sheath and proboscis.

The sense organs in *Tetrastemma* consist of four eyes which seem to be little more than pigment spots devoid of lens or other accessories.

A ciliated groove exists on each side of the head; each of these leads into an oval sac which comes into close relationship with the cerebral ganglia. These are the lateral organs, and their nature is the cause of much discussion. They appear to arise in the Schizo- and Hoplo-nemertines partly from the epiblast of the skin and partly from the oesophagus. In the Schizo-nemertines, where the nervous system is coloured red with haemoglobin, they have been regarded as respiratory organs; but this does not explain their use in the other two subdivisions, and they have been variously regarded as sense pits and as excretory organs. The arrangement of their external ciliated openings affords a useful basis for classification.

Tetrastemma, like most members of the class NEMERTEA, is dioecious. The ovaries and testes are arranged along each side of the body, alternating with the diverticula of the alimentary canal (Fig. 77). They consist of sacs which arise in the dorso-ventral muscles (dissepiments), and are at first closed. The ova and mother-cells of the spermatozoa are probably derived from the cells lining the walls of these sacs. When the reproductive cells are ripe, each sac opens to the exterior by a dorsally-placed pore. The eggs are often deposited in mucous tubes secreted by the skin. *Geonemertes palaensis*, *Tetrastemma hermaphroditica*, and *T. Kefersteinii*, are hermaphrodite; and *Prosorhochmus Claparedii* and *Monopora vivipara* bring forth their young alive.

Among the ciliated ectodermal cells of many Nemertines, a number of unicellular glands occur; their secretion forms a copious mucus, which usually takes the form of a tube, in which the animal lives for a time, and which may be strengthened by grains of sand, etc.

The opening of the mouth is beneath or behind the cerebral ganglia, and in *Akrostoma, Malacobdella,* and some others, the proboscis opens into the dorsal side of the mouth. The proboscis may be armed with rhabdites, and some observers have described nematocysts in *Cerebratulus,* etc.; these observations, however, have not been confirmed. The morphological nature of the proboscis and its sheath affords matter for much divergence of opinion. It is usually regarded as a developement of the anterior protrusible and retractile part of the body which occurs in the Turbellarian *Proboscidea.* Hubrecht, who has advocated a relationship between the NEMERTINES and the CHORDATA, regards the hypophysis cerebri of the latter as representing the proboscis, whilst the notochord represents its sheath. The latest writer on the subject, Bürger, lays stress upon the fact that the opening of the proboscis is never quite terminal, and on the relationship it holds to the mouth. He is inclined to regard the organ as a great developement of the Turbellarian pharynx, which has ceased to open into the alimentary canal, and has acquired a hollow sheath into which it can be withdrawn.

In the Palaeo- and Schizo-nemertines the blood-vessels break up into a series of lacunar cavities in the head. In *Drepanophorus* the corpuscles are red with haemoglobin.

In *Langia* the lateral cords approach each other dorsally : an arrangement which, according to Hubrecht, might result in the formation of a dorsal cord such as is found in the Chordata.

The group as a whole is carnivorous, the larger species feeding on the tubicolous Chaetopods. They have the power of breaking up into pieces when irritated, and it is said the Schizonemertines can reproduce a head in connection with the various fragments.

The NEMERTEA are classified as follows :

(i.) **Palaeonemertea.**—*No deep lateral slit on the side of the head. No stylet in the proboscis. Mouth behind the level of the cerebral ganglia.* Carinella, Polia.

(ii.) **Schizonemertea.**—*A deep longitudinal slit on the side of the head, which leads to a ciliated duct which passes down to the cerebral ganglion. Lateral nerve trunks between the longitudinal and inner circular muscle layer. Haemoglobin in the nervous system. Mouth behind the level of cerebral ganglia.* Lineus, Cerebratulus, Langia.

(iii.) **Hoplonemertea.**—*One or more stylets in the proboscis. Mouth generally in front of cerebral ganglia. Lateral nerves internal to the muscular layers. No deep longitudinal slits on the side of the head.* Akrostoma, Drepanophorus, Tetrastemma, Geonemertes, Malacobdella.

The older classifications divided the group into two subdivisions : the Anopla and the Enopla. The former corresponded with the Palaeo- and Schizo-nemertines, the latter with the Hoplonemertines.

CHAPTER VIII

NEMATODA

CHARACTERISTICS.—*Animals with an elongated unsegmented body, tapering at each end. A well-developed cuticle is secreted by the epidermis. A digestive system is present, and the excretory system takes the form of lateral ducts which open anteriorly by a median ventral pore. As a rule, they are dioecious, and many are endoparasitic; among the parasitic forms an alternation of hermaphrodite and bisexual generations may occur. A ciliated epithelium is universally absent.*

The Nematoda are colloquially known as thread-worms. The order contains a great number of species, many of them parasitic; in fact, there are said to be as many species of parasitic Nematodes as all the other endoparasites together. About twenty different species attack man, and they occur in almost every organ of the body, often inducing sufficient trouble to cause death. The free species are usually small, often microscopic. The parasitic forms are as a rule larger, the Guinea worm, *Filaria medinensis*, which lives in the subcutaneous tissues of men and horses in the tropics, attains a length of 6 feet, and the female *Eustrongylus gigas*, which lives in the kidneys of mammals, may be 3 feet or more long.

Ascaris lumbricoides inhabits the human intestine and stomach, and is not uncommon in children. It is a white cylindrical animal pointed at each end. The female measures from 9 to 14 inches in length. The male is about half as long; it is rarer than the female, and may be distinguished by its curved hinder end, and the presence of two bristles in the neighbourhood of the anus.

The mouth is terminal and surrounded by three lips, a

median dorsal one bearing two tactile papillae, and two lateral. The anus is a transverse opening close to the hinder end. The excretory opening is minute ; it is situated on the ventral surface, a little way behind the anterior end of the body. The female generative pore is also in the middle ventral line, about a third of the total length of the body from the head. The male generative organs open with the anus.

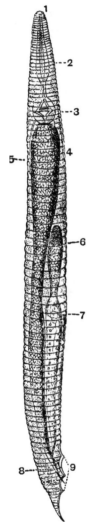

The integument consists of a well-developed cuticle, which is turned in at the various openings. The hypodermis which lies under this cuticle is a granular layer of protoplasm, with scattered nuclei. Within this layer are the longitudinally-arranged muscle - fibres (Fig. 82). These are very peculiar cells, that part of them next the hypodermis is transversely striated, and con-tractile ; the inner half, however, which is turned towards the coelom, is granular, and contains the nucleus of the cell. This mus-cular layer is broken up into quadrants by the presence of a dorsal, a ventral, and two lateral areas. The coelom, which is bounded by the granular portion of the muscle-cells, contains a fluid, but it is doubtful if this is corpusculated. There appears to be no splanchnic mesoderm surrounding the ali-mentary canal, and the coelom is probably not homologous with that of a Chaetopod, for instance.

The alimentary canal runs straight from the mouth to the anus. The mouth leads into an oesophagus lined by an infolding of chitin, whose walls are very thick and muscular, and its lumen is triradiate in transverse section. The oesophagus is

Fig. 81.—A lateral view of a Nematode, *Oxyuris*, to show the disposition of the various organs. After Galeb.

1. Mouth. 2. Oesophagus. 3. Enlargement of the oesophagus armed with chitinous teeth. 4. Intestine. 5. Opening of the segmental tubes (placed by mistake on the dorsal instead of the ventral surface). 6. Testes. 7. Vas deferens. 8. Cloaca. 9. Papillae.

separated from the intestine by a slight constriction. The latter is often flattened dorso-ventrally, it possesses no intrinsic muscles, and is lined by a layer of columnar cells (Fig. 82). On both the outside and the inside of the canal is a thin cuticle, apparently secreted by the columnar cells, and the internal cuticle is perforated by a number of minute pores. The rectum is of smaller diameter than the intestine, its walls contain muscle-fibres; in the male the vas deferens opens into it, and its posterior end thus forms a cloaca (Fig. 81).

There is no vascular system in Nematodes. The excretory system consists of two tubes, which run along the lateral line

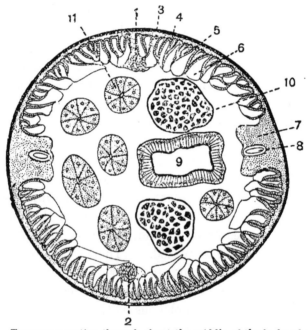

Fig. 82.—Transverse section through about the middle of the body of *Ascaris*. Slightly diagrammatic.

1. Dorsal median nerve.
2. Ventral median nerve.
3. Cuticle.
4. Hypodermis, a granular layer.
5. Muscle-cells, the striated outer border.
6. Muscle-cells, the granular nucleated inner part.
7. Accumulation of granular protoplasm round—
8. Lateral excretory canals.
9. Intestine.
10. Uterus with eggs.
11. Ovarian tubules.

of the animal, separating the longitudinal muscles of each side into a dorsal and a ventral portion (Fig. 82). Almost a third of the body-length from the anterior end these two canals give off transverse branches, which unite, and open by a minute pore in the ventral middle line. The canals end blindly, their walls consist of a granular protoplasm containing nuclei continuous with the subcuticular protoplasmic layer, and of an internal refractive layer. They contain a fluid.

The granular layer of protoplasm which lies between the cuticle and the longitudinal muscles is also heaped up in the middle dorsal and ventral lines, thus forming a ridge surrounding the dorsal and ventral nerves. This separates the dorsal and ventral longitudinal muscles into two lateral halves.

The nervous system consists of a ring surrounding the oesophagus, which may be swollen into an inconspicuous ganglion on the ventral side. The ring gives off anteriorly six short nerves which run towards the head; of these two are lateral and run in the lateral line, and the other four are arranged symmetrically, one each side of the dorsal, and one each side of the ventral middle line. The ring gives off posteriorly a dorsal and ventral median nerve, the chief nerves in the body. The ventral nerve stops in front of the anus, where it bears a ganglion. The dorsal and ventral nerves are connected at intervals by lateral commissures, which usually arise alternately. Probably four smaller nerves also pass backward from the oesophageal ring, lying in the same lines as the four small nerves which run to the head. The nerves all lie in the granular protoplasm surrounded by the longitudinal muscles.

The sexes are separate. The male reproductive organs lie in the hinder third of the body (Fig. 81). The testis is single, and consists of a long tube which winds about in the body-cavity, and at its lower end opens into the long vesicula seminalis. The testis is lined with a layer of nucleated protoplasm. The mother cells of the spermatozoa arise from a central rachis; when they break off from this they divide into two and then into four, each quarter then becomes a spermatozoon. Whilst in the body of the male, the spermatozoa have a rounded outline, but when introduced into the female they exhibit amoeboid movements. This peculiarity, together with the absence of any tail to the spermatozoa, is characteristic of the group of Nematodes.

The testis opens into a vesicula seminalis of a much greater diameter than the testis tube, its lining epithelium throws out processes into the lumen which resemble pseudopodia. These possibly perform the same function as the cilia which usually occur in the vesicula seminalis of other animals, cilia being entirely unknown in Nematodes.

The vesicula seminalis opens into a ductus ejaculatorius, with muscular walls. This ductus opens on the dorsal side of the rectum. On each side of it is a sac containing a chitinous spicule which is protrusible, and is doubtless used in copulation.

The female organs are double, and consist of ovaries, oviducts, uteri, and a vagina, the latter opening on the ventral middle line. The ovary consists of a tube in which the egg cells are formed in enormous numbers as stalked structures borne on a rachis. The oviduct differs from the ovary only in containing free ova, it leads on each side into a uterus in which numerous spermatozoa are found, and where the fertilisation of the ovum takes place. The two uteri unite, and open by means of a short vagina to the exterior.

The eggs are laid in millions, each surrounded by a smooth shell. The embryos develope in water or in damp earth, and are probably introduced into their human host by the drinking of dirty water.

The majority of Nematodes are parasitic, at any rate during a portion of their life, but a good many lead a free existence in damp earth, moss, and decaying matter, or in salt or fresh water. These are mostly minute forms, and are capable of withstanding a considerable amount of desiccation. The free-living forms, with certain others that inhabit plants, are included in the family Anguillulidae.

Tylenchus tritici does great damage to corn crops, its presence leads to the grains of corn being replaced by a dark brown gall. Inside this gall a small cluster of these minute worms are found. When one of the galls is sown with the seed, and rain follows, the parasites leave the gall and infest the young plants. The parasites pair within the gall on the corn ear, and eggs are laid within the gall.

Amongst the Nematodes which are parasitic in animals,

Ascaris nigrovenosa has a curious history. It is a common parasite in the lungs of frogs and toads, and in these hosts the parasites are hermaphrodite. Their eggs pass into the alimentary canal of the Amphibian, and leave the body, the embryo then developes into a bisexual form known as the *Rhabditis* generation; in this form the ova develope in the uteri, and the young embryos, making their way through its walls, devour the whole interior of their mother until only the cuticle remains, they then emerge and live in mud or water until swallowed by a frog, when they resume the first form.

Oxyuris vermicularis inhabits the human intestine, and is particularly common in the caecum; its ova when laid contain embryos already mature, hence it spreads with great rapidity. The ova are swallowed, and the solvent action of the gastric juice sets free the young embryos in the stomach, whence they pass into the intestine.

Filaria sanguinis hominis passes its larval life in the body of mosquitos, but the sexual female inhabits the lymphatic glands of man in Australia, India, China, and Egypt, giving rise to elephantiasis, etc. The embryos circulate in the blood and give rise to further disease; they are readily sucked up by a biting mosquito, and in this way the parasites are doubtless disseminated.

Trichina spiralis (Fig. 83) is a very minute Nematode which encysts in or between the muscle fibres. The adult worm lives

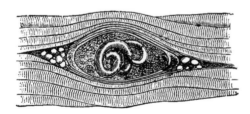

FIG. 83. — *Trichina* encysted amongst muscular fibres. After Leuckart.

in the alimentary canal of man and of other carnivorous mammals; it is viviparous. The young bore their way through the wall of the intestine of their host and encyst in the muscles. They do not become sexually mature unless eaten by some animal,—often a rat, and sometimes a pig,—in which case its flesh is liable to become "trichinised," and may carry the disease Trichinosis to man.

Nematodes are usually arranged in several families, as the Ascaridae, the Filariidae, the Anguillulidae, the Strongylidae. The grouping of these families into larger subdivisions is a matter of considerable difficulty, and no system which has as yet been proposed has met with general acceptance.

CHAPTER IX

HIRUDINEA

Hirudinea { Rhynchobdellidae—*Pontobdella, Clepsine.*
{ Gnathobdellidae—*Hirudo, Nephelis.*

CHARACTERISTICS.—*Animals with a ringed integument, a certain number of the annuli or rings corresponding with each true segment. A posterior ventral sucker formed by the fusion of some of the posterior somites is present. The coelom is much reduced by the ingrowth of connective tissue; it communicates with the vascular system, and contains the same fluid. The mouth is anterior, and usually surrounded by a sucker; the anus is dorsal to the posterior sucker. Hermaphrodite, with genital openings ventral and median, the male in front of the female. Mostly aquatic and blood sucking.*

Amongst the Triclade Turbellarians a certain segmentation begins to appear, and reaches its highest point in *Gunda segmentata*, where the alimentary canal has 25 lateral diverticula, there are 25 testes and 24 pairs of vitellaria, and the dorso-ventral muscles are arranged segmentally. In the Nemertines we also find every stage from entirely unsegmented animals to those in which the alimentary canal, the generative organs, the blood-vessels, the muscles, and even the proboscis sheath, present a certain repetition of parts which is called segmentation. It is usual, but by no means universal, for the segmentation of the various organs to agree, so that one segment of the body contains a segment or representative of each system of organs. Very often, however, some segments of one organ may be suppressed, or fail to develope; and again many segments of one system of organs, as, for example, the segmentally-arranged nerve ganglia, may fuse together, and thus the segmentation becomes irregular.

The Hirudinea are the first group in which segmentation forms a distinctive feature. The integument is ringed, and in the medicinal leech, *Hirudo medicinalis*, five annuli correspond with a segment, in *Pontobdella* four, and in *Branchellion* three. The limits of each true segment are, however, marked out by the arrangement of the colour bands.

A cuticle corresponding with the mucous tubes of Nemertines is formed from the secretion of unicellular glands; it is constantly being worn off and replaced.

The body-cavity is much reduced by the great developement of muscles and connective tissue, and in the medicinal leech its chief remains form the dorsal and ventral sinuses, it is, however, more conspicuous in the **Rhynchobdellidae.**

The bodies of those forms, such as *Clepsine* and *Nephelis*, where the muscles are strongly developed and the connective tissue is sparse, are peculiarly firm and rigid, but forms like *Aulostoma*, and to a less extent *Hirudo*, where the connective tissue predominates, are extremely limp and flabby. The cells of this connective tissue are embedded in a gelatinous matrix, and they may assume the following characters : (i.) fat cells, or cells crowded with fat globules, common in *Clepsine*; (ii.) elongated branched cells crowded with globules which are not fat, these pass into fibres at times ; (iii.) pigment cells ; (iv.) vaso-fibrous and botryoidal cells, forming a tissue which is composed of certain rounded cells crowded with brown pigment and arranged in rows ; by a change in their interior, channels arise which pierce the cells, and these ultimately open on the one side into the closed system of blood-vessels, and on the other into the coelomic sinuses. The botryoidal tissue appears to pass into the vaso-fibrous tissue by the cells dividing and becoming small, and the walls becoming very thin, and the nuclei of the cells dropping out. Many of the minute capillaries thus formed run between the columnar epithelial cells of the ectoderm, and here the oxygenation of the blood takes place.

In Nemertines the vascular system is one with the coelomic. The space which contained the corpusculated fluid was the space in which the inner ends of the nephridia lay, and was archicoelic in its origin. In Hirudinea this space is divided into two: a series of spaces lined with an endothelium, the true vascular system; and a series of sinuses with no special

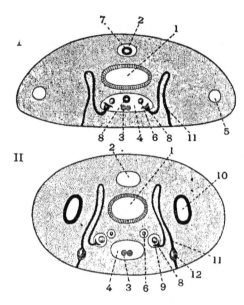

FIG. 84.—Diagrams of transverse sections: I., through *Clepsine*; II., through *Hirudo*. A. G. Bourne.

1. Alimentary canal.
2. Dorsal sinus.
3. Nerve cord.
4. Ventral sinus.
5. Lateral sinus.
6. Ovary.
7. Dorsal blood-vessel.
8. Inner end of nephridium.
9. Testis.
10. Lateral blood-vessel.
11. Nephridium.
12. Vesicle of nephridium.

cellular lining, and enclosing various organs of the body, the coelom. These two series of spaces open into one another directly in the **Rhynchobdellidae,** and through the botryoidal tissue in the **Gnathobdellidae.** In *Clepsine* (Fig. 84, I.) and *Pontobdella* the dorsal sinus contains a dorsal vessel, the ventral sinus contains the nervous system, a ventral vessel, the ovaries, and in *Clepsine* the inner ends of the nephridia. There are also two lateral sinuses in which in *Pontobdella* a lateral vessel is found. In *Hirudo* (Fig. 84, II.) there are a dorsal and ventral sinus, and two lateral vessels. The ventral sinus encloses the nervous system. The ovaries and testes lie in special sinuses. The internal end of the nephridium is placed in the testis sinus in those segments in which both occur.

The alimentary canal of Leeches falls into five sections: (i.) the muscular pharynx, which may possess two or more commonly three jaws, and numerous glands whose secretion in *Hirudo* serves to prevent the blood upon which the animal lives from coagulating; (ii.) the oesophagus and proventriculns, which in *Hirudo* is enormous, and produced into eleven pairs of lateral caeca—it serves as a storehouse for the blood; (iii.) the digestive stomach, usually very small; (iv.) the intestine; and (v.) the rectum. Sections (i.) and (v.) are formed by epiblastic invaginations. The alimentary canal of *Aulostoma* is ciliated; an interesting peculiarity, since this leech does not live on blood, but on small water-worms, etc. The **Rhynchobdellidae** have a protrusible proboscis.

The nephridium in *Pontobdella* (Fig. 85, I.) is a continuous

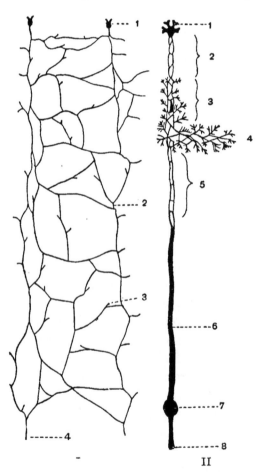

FIG. 85.—Diagrams of the nephridia in (I.) *Pontobdella* and (II.) *Hirudo*. A. G. Bourne.

I.
1. Funnels.
2. Branched network.
3. Caecal tubules.
4. External opening.

II.
1. Funnel.
2. Ducts in testis lobe.
3. Ducts in main lobe.
4. Ducts in caecal end of main lobe.
5. Ducts in apical lobe.
6. Unbranched tube passing to exterior.
7. Vesicle.
8. External opening.

II

network of fine tubules, which is spread through the greater part of the body. This network opens at intervals into a blood sinus by 10 pairs of internal funnels occurring in the segments 9-18. The lumen of the funnel is usually occluded, a condition which is even more pronounced in *Hirudo*. The tubules consist of simple or branched cells with an intracellular lumen, they unite at intervals, increase in size, and open by an intercellular duct at 10 external pores. *Branchellion* and *Piscicola* probably possess a similar nephridial network.

Clepsine and *Hirudo* have paired nephridia distinct from one another. The funnel in *Hirudo* lies in the same blood sinus as the testis (Figs. 84, II., and 86); it consists of lobed ciliated cells, and its lumen is always occluded. The ducts leading from the funnel are much branched and intracellular, they at length unite and open into a vesicle which leads to the exterior (Fig. 85, II.). In *Nephelis* and *Trocheta* the funnel lies in a hollow of the botryoidal tissue. It has recently been maintained that the structures, usually ciliated and often occluded, which are found at the inner ends of the nephridia have nothing to do with those organs, but may take some part in maintaining the circulation of the blood.

The nervous system consists of two cerebral ganglia and a ventral chain, which in *Hirudo* contains 23 ganglia (Fig. 86). The cerebral ganglia give off nerves to three minute ganglia which supply the jaws, and also nerves to the eyes and goblet-shaped sense organs. A number of simple eyes are found in the head in most forms; in *Piscicola* they are also found in the posterior sucker. The ventral nerve-chain lies in the ventral blood-sinus (Fig. 84, II.).

Leeches are hermaphrodite; the genital openings are ventral, median, and unpaired, the male being in front of the female. In *Hirudo* there are nine pairs of testes, arranged in segments 8 to 16; in *Nephelis* the testes are numerous, and scattered irregularly. They open into short transverse ducts, which unite into a longitudinal vas deferens. Each of the latter becomes coiled at its anterior end, forming an epididymis, and the two unite to form a single short duct. This opens to the exterior by a muscular protrusible penis, at the base of which prostatic glands are usually found (Fig. 86).

The true ovaries are filamentous bodies contained in capsules. These capsules, usually called the ovaries, occur in *Hirudo* in the seventh segment, one on each side of the nerve cord. The internal openings of the oviducts perforate the walls of the capsule, and lie in its lumen. The cavity of the capsule also contains certain amoeboid corpuscles, and probably repre-

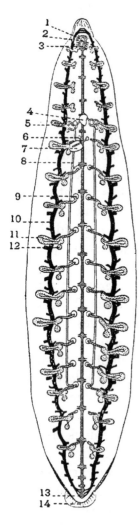

Fig. 86.—*Hirudo medicinalis,* opened along the median dorsal line, slightly diagrammatic. A. G. Bourne.

1. Cerebral ganglion.
2. Oesophagus.
3. 1st postoral ganglion.
4. Penis.
5. Epididymis.
6. Ovisacs.
7. Glandular enlargement of the oviducts.
8. Vas deferens.
9. 2nd pair of testes.
10. Lateral vessel.
11. 9th pair of nephridia.
12. Vesicle near external opening of nephridium.
13. 23rd ganglion.
14. Posterior sucker.

sents part of the coelom. The oviducts unite and pass into a muscular vagina, their walls are glandular, and secrete an albuminous fluid in which the eggs float in the cocoon.

The segments on which the reproductive organs open form the clitellum, which consists of certain segments with glandular walls, the secretion of which hardens and forms in most forms a cocoon in which the fertilised eggs mature. In *Clepsine* the eggs are attached to some foreign substance, and the female sits over them till they hatch, and then they attach themselves to her body. *Nephelis* deposits its cocoons on water-plants, *Aulostoma* and *Hirudo* in damp earth.

Leeches are usually inhabitants of fresh water; sometimes they live in salt water, and more rarely on land. They usually move in loops by the aid of their anterior and posterior suckers, but they can swim well. The land forms are most common in Asia south of the Himalayas, and in the East Indies and Australia. A gigantic form, *Macrobdella Valdiviana*, lives underground in Chili, and is said to reach the length of $2\frac{1}{2}$ feet. They are, with few exceptions, parasitic, living on the blood of Vertebrates.

The HIRUDINEA are divided into two groups:

(i.) **Rhynchobdellidae.**—*Cylindrical or flat, elongated body with both suckers well marked, fore part of the body retractile, forming a proboscis. The vascular and the coelomic spaces are in direct continuity; the blood does not contain haemoglobin.* Pontobdella, Clepsine, Piscicola.

(ii.) **Gnathobdellidae.**—*Mouth sucker-like, pharynx armed with three jaws. No proboscis. The vascular and the coelomic spaces are in indirect continuity. The blood contains haemoglobin.* Hirudo, Aulostoma, Nephelis.

CHAPTER X

CHAETOPODA

Chaetopoda
- Archiannelida.
- Oligochaeta
  - Naidomorpha—*Nais, Chaetogaster.*
  - Lumbricomorpha — *Lumbricus, Megascolides, Eudrilus, Perichaeta.*
- Polychaeta
  - Errantia—*Nereis, Aphrodite, Eunice, Tomopteris.*
  - Sedentaria—*Arenicola, Sabella, Capitella.*

CHARACTERISTICS.—*Segmented animals, with a more or less prominent prostomium or region in front of the mouth. Locomotion effected by cilia, or by setae implanted in the body wall or borne by lateral processes of the body termed parapodia. Each seta is the product of a single cell. The segments are divided externally by grooves, internally by septa. A pair of nephridia are typically found in each segment.*

The Chaetopoda are divided into three sub-classes: the ARCHIANNELIDA, the OLIGOCHAETA, and the POLYCHAETA.

ARCHIANNELIDA.

CHARACTERISTICS.—*Marine worms with small prostomium. The segmentation of the body is externally marked by rings of ciliated cells, and by slight grooves. There are no setae or parapodia or branchiae, but the head bears one or more pairs of tentacles. The longitudinal muscles are in four bands. The nervous system retains its connection with the hypodermis throughout life. The head bears a pair of ciliated grooves.*

The Archiannelida comprise a group of minute marine animals, which are to some extent intermediate between the Turbellarians and the Chaetopoda. The group includes four genera: *Polygordius, Protodrilus, Histriodrilus,* formerly known as *Histriobdella* and classified with the leeches, and *Dinophilus.*

Protodrilus Leuckartii is a small worm-like animal found in the sand at Pantano, an inland arm of the sea in the neighbourhood of Messina (Fig. 87). It creeps about in a Nemertine-like manner by means of the cilia which clothe the body.

The segmentation is shown externally by two rows of cilia on each segment, and by slight grooves which separate neighbouring segments from one another. The number of the segments increases with the age of the animal. The head bears a pair of hollow ciliated tentacles, into which a section of the coelom extends (Fig. 87).

The ectoderm consists of cubical epithelial cells, amongst which the ducts of many unicellular glands open. The cells lining a shallow groove which runs all along the ventral aspect of the worm bear specially long cilia. There is a double row of cilia on the head in front of the mouth, and an anterior and a posterior circlet upon each segment. *Protodrilus*, like *Polygordius*, has no circular muscles; the longitudinal fibres are arranged in four bands, two dorso-lateral and two ventro-lateral. An oblique longitudinal muscular septum running from each side of the body to near the ventral median line divides the body-cavity into a median and two lateral portions (Fig. 90).

The alimentary canal consists of a ciliated oesophagus

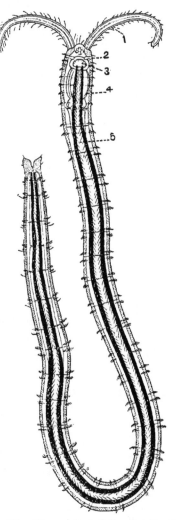

FIG. 87.—View of *Protodrilus Leuckartii.* After Hatschek.

1. Tentacle.
2. Ciliated Pit.
3. Oral cavity.
4. Muscular appendage of oral cavity.
5. Alimentary canal.

and an intestine. On the ventral wall of the oesophagus a U-shaped tube opens; one limb of this tube is enormously enlarged, and is very muscular. A somewhat similar stomodaeal musculature occurs in the pharynx of most Chaetopods, and would seem to be comparable to the odontophore of Molluscs, and with the ventral part of the muscular pharynx in Turbellarians. The intestine is moniliform, there being a constriction between each segment. It is supported by the transverse septa and by median dorsal and ventral mesenteries (Fig. 90). Its lumen is ciliated. The anus is terminal.

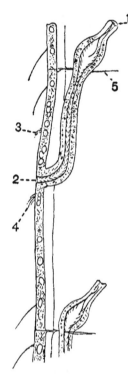

FIG. 88.—View of a nephridium of *Protodrilus Leuckartii.* After Hatschek.

1. Internal opening with flagellum.
2. External opening.
3. Cilia of anterior ring of the segment.
4. Cilia of posterior ring of the segment.
5. Septum.

The vascular system consists of a median dorsal vessel, which sends a branch along each tentacle. Other branches bring the colourless blood back from the tentacles, and these then fuse to form the median ventral vessel. Certain lacunae between the epithelial cells of the intestine and its musculature appear to supply the dorsal vessel with the fluid it contains.

The nephridia consist of an internal funnel (Fig. 88), from the edge of which a large cilium depends into the lumen of the tube which is ciliated; this passes through a septum, as in *Lumbricus*, and finally opens to the exterior on the lateral line.

The nervous system remains in the skin (Fig. 90); it consists of a cerebral mass, with circum-oesophageal commissures, which pass into two ventral cords. The latter are separated from one another by the ventral groove, but are connected by transverse commissures.

Protodrilus is hermaphrodite, *Polygordius* dioecious; the ovaries lie in the first seven segments. The ova are derived from some of the cells lining the coelom in the neighbourhood of the ventral middle line (Fig. 90). The testes are similarly built up from peritoneal cells, situated on both sides of the oblique septa, in the segments which succeed the seventh.

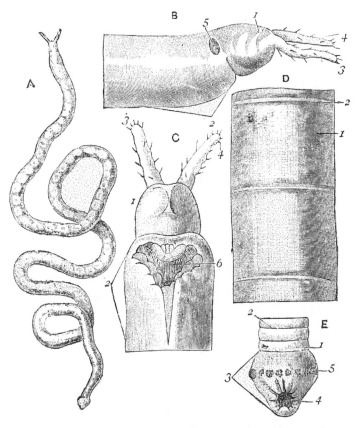

FIG. 89.—*Polygordius neapolitanus.* (After Fraipont.)

A. The living animal. × about 5.
B. Anterior end of the worm seen from the right side, more highly magnified.
  1. Prostomium.
  2. Peristomium.
  3. Tentacle.
  4. Setae on tentacle.
  5. Ciliated pit.
C. Ventral view of the same. Numbers as in B.
  6. Mouth.

D. Portion of body, showing
  1. Segments separated by grooves.
  2. Grooves.
E. Ventral view of posterior end, showing the last three segments.
  1. Segment.
  2. Groove.
  3. Anal segment.
  4. Anus.
  5. Ring of papillae.

Histriodrilus, which lives parasitically upon lobsters' eggs, is somewhat more highly organised than *Protodrilus.* The body is differentiated into regions, the nervous system has distinct ganglia in each somite, the stomodaeal muscular bulb is armed with three teeth, and the sexes are separate.

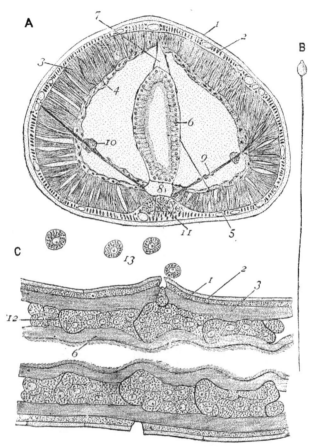

FIG. 90.—*Polygordius neapolitanus.*

A. Transverse section of a male Polygordius.
1. Cuticle.
2. Ectodermic epithelium.
3. Muscle plates.
4. Parietal coelomic epithelium.
5. Visceral coelomic epithelium.
6. Ciliated endodermal epithelium.
7. Dorsal blood-vessel in dorsal mesentery.
8. Ventral blood-vessel in ventral
   mesentery.

9. Oblique muscular septum covered
   with coelomic epithelium.
10. The testes.
11. Ventral nerve cord, continuous
    with 2.
B. A spermatozoon.
C. Horizontal section of a mature female
   Polygordius.
1, 2, 3, and 6 as in A.
12. The septa.
13. Ova.

The body-wall has undergone partial histological degeneration, and is ruptured in two places to allow the escape of the ova (13) which crowd the coelomic space.

Polygordius is the largest of the Archiannelids. *P. lacteus* is 40 mm. long (Fig. 89). The longitudinal ventral groove of *Protodrilus* has in this genus closed in, and forms a canal within the nerve cord. Red blood occurs in this genus, and the dorsal vessel gives off lateral branches, which, however, end caecally. The sexes are distinct, and the ovaries or testes arise in the posterior segments (Fig. 90).

The last member of the group, *Dinophilus*, is a minute marine animal. Two species, *D. gigas* and *D. taeniatus*, have recently been described by Weldon and Harmer from the coast of Devonshire and Cornwall. The body consists of a head or prostomium, which bears two eye-spots, and whose cilia are uniform or arranged in two preoral circlets. The mouth opens on the second segment or peristomium, and then follow five or six segments, and finally a postanal unsegmented tail. In both the above-mentioned specimens the entire ventral surface of the animal is uniformly ciliated, and each segment has one or two bands of cilia. *D. vorticoides* is uniformly ciliated all over. The nervous system is in contact with the skin. The coelom is traversed by strands of connective tissue in *D. gigas*; and in *D. taeniatus* there are more definite spaces connected with the inner ends of the nephridia. In *D. gigas* an excretory system of the Platyhelminthine type, with flame cells, has been described, but in *D. taeniatus* and *D. gyrociliatus*, 5 pairs of nephridia are found, each with a triangular appendage hanging into the lumen of their ciliated duct. The sexes are separate, and in the male *D. taeniatus* the fifth pair of nephridia appear to have become modified and form vesiculae seminales. A penis is present, and seems to be inserted indifferently into any part of the skin of the female.

There is little doubt that *Dinophilus* should be classified with the other members of the group Archiannelids; on the other hand, its median genital pore, the presence in some species of the Platyhelminthine excretory system, and the method of fertilisation adopted by *D. taeniatus* which is paralleled in the *Polycladida*, support the view of the Platyhelminthine origin of these worms.

THE OLIGOCHAETA.

CHARACTERISTICS.—*The Oligochaeta are characterised by the absence of antennae, parapodia, branchiae, and cirrhi. Their pharynx is devoid of armature. They are hermaphrodite, and their reproductive organs are confined to a few segments. Their ova are laid in cocoons. Their developement is direct.*

The Oligochaeta were at one time divided into two groups: the *Terricolae*, which live chiefly on land, and the *Limicolae*, which are mostly aquatic. Recent research has, however, broken down the structural barriers which were believed to exist between the members of these two groups. The oligochaet worms are now arranged by Benham in a number of families, which allow themselves to be grouped in two divisions: (i.) the **Naidomorpha**, *in which asexual repro-duction takes place*; *and* (ii.) the **Lumbricomorpha**, *in which it does not.* The earthworm is the most familiar example of the latter subdivision.

One of the most curious features found in many of the Oligochaets is the dorsal pore. In *Lumbricus* this pierces the skin on each segment in the middle dorsal line, and places the coelom directly in communication with the exterior. The pores occur in this genus on all the segments except the first six or seven. They are closed by a sphincter muscle, and opened by an anterior and posterior longitudinal band of muscles. They are found in several species of the Oligochaets,—*Lumbricus*, etc., —but do not occur in Polychaets. *Megascolides*, a gigantic Australian worm, measuring from 4 to 6 feet in length, ejects through its dorsal pores the milky coelomic fluid with which it coats the walls of its burrows.

The function of the modified skin of certain segments which constitutes the *clitellum* is to form the cocoons in which the eggs are deposited. It may completely enclose the body, and is then known as a *cingulum*, or it may be incom-plete. In the aquatic forms it only includes one segment: that on which the vas deferens opens. The capsulogenous glands also found in the skin give rise to the albuminous fluid found in the cocoon in which the ova and spermatozoa are deposited. This secretion serves to nourish the developing embryos.

The septa which divide the body internally into segments are almost absent in *Aeolosoma*; only one, dividing the head from the body, is present.

The setae vary a good deal in number and shape in different species (Fig. 91), but each is the product of a single cell which lies at the base of the sac from which the seta protrudes.

The alimentary canal of many of the lower Oligochaets is ciliated; in *Lumbricus* the lining epithelium from the mouth to the gizzard secretes a cuticle, but the intestine is lined by

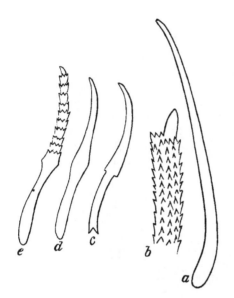

Fig. 91.

a. Penial seta of *Perichaeta ceylonica.*

b. Extremity of penial seta of *Acanthodrilus.* After Horst.

c. Seta of *Urochaeta.* After Perrier.

d. Seta of *Lumbricus.*

e. Seta of *Criodrilus.*

modified retractile cilia. *Criodrilus,* which inhabits the mud, and *Pontodrilus,* which lives on the sea-shore, have no gizzard; both these genera are also without nephridia in the anterior 10 or 15 segments. The typhlosole which is so characteristic in the intestine of *Lumbricus* (Fig. 92) is also absent in the latter genera as well as in *Megascolides.* In *Rhinodrilus* it forms a spiral fold running round the intestine.

The blood is contained in a series of closed vessels. The plasma of the blood is usually coloured red by haemoglobin which is dissolved in it, and not confined to the corpuscles. Numerous flattened corpuscles float in it. The coelomic fluid found in the body-cavity contains colourless amoeboid corpuscles.

The nephridial system of leeches shows how a single pair of nephridia in each somite, distinct from all the others, may arise from a scattered network. In Oligochaets a similar series of stages in the developement of a single pair of nephridia in each segment is found, and has been described by Beddard, Spencer, and others. *Perichaeta aspergillum* has a nephridial network of fine tubules which permeates the body. It is doubtful if any internal funnels opening into the coelom exist in the anterior segments, but they do in the posterior half of the body, and here they are very numerous. On the other

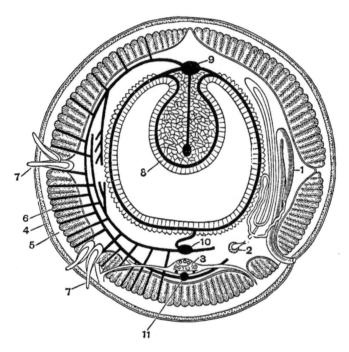

Fig. 92.—Diagrammatic transverse section through one of the posterior segments of *Lumbricus*. Partly after Marshall and Hurst.

1. Nephridium.
2. Funnel of nephridium.
3. Nerve cord.
4. Epidermis.
5. Circular muscles.
6. Longitudinal muscles.
7. Dorsal and ventral pairs of setae.
8. Typhlosole.
9. Dorsal blood-vessel connected by a vertical branch with typhlosole, and by branches with intestinal blood plexus.
10. Supraneural vessel.
11. Infraneural vessel.

On the left side are indicated the chief vessels given off from the main trunk to the body-wall and nephridium. After Beddard.

hand there are numerous openings to the exterior, both anteriorly and posteriorly. This network is continuous, and shows no trace of segmentation. In *P. armata* there is a similar network of fine tubules, but in addition there is in each segment a pair of large nephridia which pierce the septum in front and open into the preceding segment by a well-developed funnel. In *Megascolides* there are a great number of minute nephridial tubules, consisting of a short straight tube and a longer coiled tube, scattered all over the inner surface of the skin. These small nephridia have an intracellular duct, and are well supplied with blood-vessels. They open to the exterior, but no internal opening has been found. In addition to these smaller nephridia, the posterior half of the body has in each segment a pair of large nephridia, with an internal funnel-shaped opening. When · these large nephridia are traced forward through the region of the middle of the body, it is seen that they first lose their internal funnel, and then gradually decrease in size, and ultimately merge into the smaller nephridia. Thus the specialisation of the nephridia appears to commence posteriorly. The small and large nephridia are connected by a longitudinal duct.

The next stage towards the condition found in *Lumbricus* is when the network becomes discontinuous at the septa, and does not spread from segment to segment. This stage is almost reached by *Deinodrilus*, and quite by *Acanthodrilus* and *Dichogaster*. Then, as is shown in the case of *P. armata* and *Megascolides*, certain of the tubules of the network enlarge, and form large nephridia, and the network gradually ceases to be formed. Two pairs of such large nephridia exist in each segment in *Brachydrilus*; one pair then disappears, and the condition of *Lumbricus* (Fig. 92) is attained.

The aquatic Oligochaets have one pair of nephridia in each somite; the funnel is absent in *Chaetogaster*.

In certain land worms the nephridia of the anterior segments become modified, and undergo a very remarkable change of function. In *Acanthodrilus dissimilis*, in *Dichogaster*, and in *Digaster*, all three possessing a nephridial network, some of the tubules on each side of the pharynx become connected with a duct which opens into the buccal cavity. In *Megascolides* we have a similar change of function. The walls of the pharynx are pierced by a number of tubules, with an intracellular lumen, which opens into the cavity of the alimentary canal, and whose secretions pass into the pharynx. In every respect these tubules resemble the tubules

of the nephridial network.   This extraordinary change of a
nephridium into a salivary gland is paralleled in the Arthropod *Peripatus*, in which developement shows that the salivary
glands are modified nephridia.

Three giant fibres, consisting of a sheath with a clear contents, occur dorsal to the ventral nerve cord in nearly all
Oligochaets.   Connections have been recently traced between
them and the nerve fibres. Their function was formerly
thought to be solely for the purpose of support; hence they
have been termed the *neurochord,* and have been compared
with the notochord of the Chordata in their physiological
action.

The Oligochaets are hermaphrodite: in the CHAETOGASTRIDAE
the spermatozoa develope in the coelom, in *Lumbricus* the

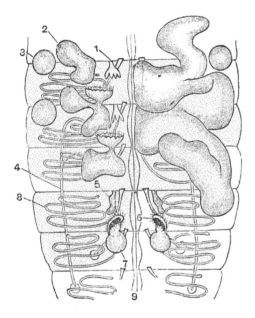

FIG. 93.—Genital segments of
Lumbricus (slightly altered
from Howes' *Biological Atlas*).
The left side represents the
immature, the right, the
mature condition, so far as the
male reproductive organs are
concerned.   After Beddard.

1. Anterior pair of testes ; the
   second pair are in the
   next segment.

2. Seminal vesicles.

3. Spermathecae.

4. Vas deferens.

   Ovary.

   Oviduct.

   Receptacula ovorum.

   Nephridia.

9: Nerve cord.

testes become enclosed in special vesiculae seminales which
are outgrowths from three of the septa (Fig. 93).   In these
vesiculae the spermatozoa mature.

In the aquatic Oligochaets the ova ripen in the coelom
or in an egg sac similar to the vesiculae seminales of Lum-
bricns.

The testes are usually four in number, but there may be
only one pair, as in *Geoscolex*.   There is a single pair of ovaries,

which are very constant in position, being with hardly an exception in the 13th segment. In *Eudrilus* (Fig. 94) the ovary is enclosed in a muscular sheath. The ciliated oviduct passes through the sheath, and ends in a funnel-shaped mouth in the ovary. The muscular sheaths of the oviduct and ovary are

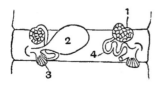

FIG. 94.—Female reproductive apparatus of *Eudrilus*. On the right side the spermatheca has been cut away to show the contorted oviduct, 4.
1. Ovary. 2. Spermatheca.
3. Gland opening into conjoined duct of spermatheca and oviduct. After Beddard.
4. Oviduct.

continuous. The oviduct is convoluted, and opens to the exterior on the 14th segment, together with a spermatheca and a small glandular body. The opening of the oviduct in *Peri-*

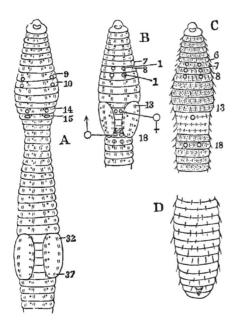

FIG. 95.—Diagrams of various earthworms to illustrate external characters. A, B, C, anterior segments from the ventral surface. D, hinder end of body of *Urochaeta*.

A. *Lumbricus*; 9 and 10, segments contain spermathecae, the orifices of which are indicated; 14, segment bears oviducal pores; 15, segment bears male pores; 32, 37, first and last segments of the clitellum.

B. *Acanthodrilus*; 1, orifice of spermathecae; ♀, oviducal pores; ♂, male pores.

C. *Perichaeta*; the spermathecal pores are between segments 6 and 7, 7 and 8, and 8 and 9, the oviducal pore on the 14th, the male pores on the 18th segment.

In all the figures the nephridial pores are indicated by dots, the setae by strokes.

chaeta is single and median. The various positions of the genital apertures, and their relations to the nephridia and setae in various genera, are shown in Fig. 95. In the aquatic Oligochaets the nephridia are not found in those segments which lodge the reproductive organs and their ducts; in the terrestrial forms they coexist. Some of the setae in the neighbourhood of the reproductive segments are modified and assist in copulation (Fig. 91).

Certain of the aquatic Oligochaets multiply asexually by fission. In *Aeolosoma*, in many respects the most primitive of the Oligochaets, one of the segments enlarges, forms a prostomium, and then breaks off from the anterior half. In the NAIDIDAE and the CHAETOGASTRIDAE, a "zone of fission" is formed between two segments when the worm has reached a certain size. This zone divides into two halves; the posterior of these forms a head for the posterior set of segments, the anterior gives rise to a series of new segments forming the tail of the anterior animal. In this way chains of zooids are formed. These at length are set free, and differ from the mature worm only in the absence of the reproductive organs, clitellum, and genital setae, which they acquire later.

No asexual reproduction is known amongst the Lumbricomorpha, but they possess a considerable power of reproducing lost parts.

THE POLYCHAETA.

CHARACTERISTICS.—*Marine worms, with numerous setae in bundles borne on parapodia. The head is distinct, and usually bears tentacles and palps; the somites of the trunk carry cirrhi and sometimes branchiae. As a rule the Polychaeta are dioecious and have an indirect metamorphosis.*

The Polychaeta are divided in (i.) the **Errantia** and (ii.) the **Sedentaria** or **Tubicola**; these subdivisions are characterised as follows :

(i.) *The Errantia are free and carnivorous, with a large prostomium, which usually bears tentacles and eyes. The body is rarely divided into regions; the parapodia are large; the pharynx is protrusible and provided with chitinous jaws or with papillae.*

APHRODITIDAE, EUNICIDAE, NEREIDAE, SYLLIDAE, ALCIOPIDAE, TOMOPTERIDAE.

(ii.) *The Sedentaria are tube-building worms, whose tube may be fixed. Body often divided into regions. Prostomium and parapodia small; pharynx never armed with teeth; vegetable feeders.*

ARENICOLIDAE, CAPITELLIDAE, CHAETOPTERIDAE, TERE-
BELLIDAE, SERPULIDAE.

The Polychaeta include a vast variety of worms, which
either swim about freely in the sea or inhabit tubes, from the
open mouth of which they often protrude the anterior end of
their bodies. They are very generally brightly coloured, and
many of them, especially the fixed forms, with their feathery
tentacles and branchiae, are objects of great beauty. With
three exceptions, they are exclusively marine; a few are pelagic,
and, as is usual with such a habit of life, their body is transparent. One or two only are parasitic, one
living in the coelom of the Gephyrean *Bonellia*,
another in the branchial cavity of a barnacle,
Lepas.

Arenicola piscatorum, the common lugworm,
is a member of the sub-division Sedentaria,
which tunnels out tubular passages in the
sand, boring down into it with its head, and
then turning the anterior end of the body
up again, thus assuming the shape of a **U**. It
can be dug up in considerable quantities in
sandy places round our coasts when the tide
is low; its presence being indicated by numerous little heaps of cylindrical sand castings, the
undigestible remnants of its food.

The worm may attain the length of ten or
more inches, and is of a blackish-brown colour
with a tinge of green.

The body of the animal is divisible into
three regions (Fig. 96): an anterior or neck of
6 segments, a middle or gill-bearing region of
13 segments, and a tail region of variable
length, in which the segments are not well
marked.

FIG. 96. — *Areni-
cola piscatorum.*

The chief characteristic which separates Polychaetous
from Oligochaetous worms is the presence of *parapodia*.
These, when typically developed, are lateral outgrowths of
the body-wall of each segment, into which the coelom is con-
tinned. The parapodium is usually divided into a dorsal and a

ventral half; the *notopodium* and the *neuropodium* respectively. Each of these may bear (i.) a bundle of bristles, the *setae*; (ii.) in the midst of the setae, a single large bristle, the *aciculum*; (iii.) solid fleshy prolongations of the body-wall, containing a nerve, and probably tactile in function, the *cirrhi*; (iv.) respiratory organs, processes of the body-wall well supplied with blood-vessels, and sometimes containing a prolongation of the body-cavity, the *branchiae*. These last are borne as a rule only by the notopodium.

The parapodia are not well developed in *Arenicola*. The first nineteen segments bear each a small notopodium in which a bundle of setae spring, and on the ventral surface a small neuropodium, which bears a row of hooked bristles.

The gills borne on the 7th to the 19th segments are feathery branched structures, through whose thin walls the red blood is visible. The tail bears no parapodia.

Each of the segments of the body is divided into five small rings: an unusual feature in Chaetopods, recalling the annulation of the Hirudinea.

The integument consists of the same elements as are found in Lumbricus: (i.) a cuticle, (ii.) an epidermis of columnar cells crowded with pigment cells, (iii.) a continuous sheath of circular muscles, (iv.) a layer of longitudinal muscles, much broken up by the presence of the bundles of setae, etc., and (v.) a lining of peritoneal epithelium.

The coelom is very spacious; at the anterior end of the body it is traversed by three septa, which mark the limits of the first three segments. There are no other septa in the first nineteen segments, but the tail is divided into as many chambers as there are rings by vertical septa. The body is further partially divided into three divisions by two longitudinal incomplete mesenteries, which run obliquely from the side of the body to near the middle ventral line. The central division lodges the alimentary canal, the lateral contain the nephridia. This dividing up of the body-cavity recalls the arrangement in the Archiannelids. The coelom is full of a corpusculated fluid, in which, during the breeding season, the ova and spermatozoa are found in great quantities.

The alimentary canal runs in a straight line from the

CHAETOPODA 153

mouth to the terminal anus. The pharynx can be protruded, and is then seen to be covered with papillae. The mouth opens into the oesophagus, which bears at its hinder end a pair of long glandular bodies, possibly homologous with the calciferous

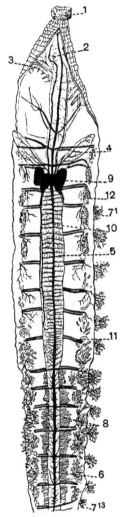

Fig. 97.—Anterior end of *Arenicola piscatorum* laid open by a median dorsal incision to show the internal organs. After Vogt and Yung.

1. Proboscis beset with papillae.
2. Oesophagus.
3. Muscles which retract the oesophagus.
4. Diverticula which open into the hinder end of the oesophagus.
5. Stomach.
6. Intestine.
7^1 and 7^{13}. 1st and 13th, or last, branchia.
8. Masses of chloragogenous cells.
9. Heart.
10. Dorsal blood-vessel.
11. Vessels to and from the branchiae.
12. Nephridia.

glands of *Lumbricus* (Fig. 97). The intestine, which traverses the gill-bearing region, is coloured yellow by the presence of yellow chloragogen cells in its walls, resembling the similar cells on the walls of the intestine in *Lumbricus*. These cells contain concretions, which seem to be set free in the coelomic fluid, and are possibly excreted by the nephridia. The walls of the intestine are somewhat wrinkled, and are rather thin. The lumen of the alimentary canal is usually distended with sand, which is eaten in large quantities by the worm for the sake of the small amount of vegetable debris which may be mixed with it. At the commencement of the tail the intestine passes into the rectum, which is supported by the numerous septa of this region, and ends in the terminal anus.

The blood-vessels consist of (i.) a dorsal vessel (Fig. 97), which at the anterior end anastomoses with the ventral vessel —the blood flows forward in this; (ii.) a ventral vessel underneath the alimentary canal, in which the blood flows backward; (iii.) a subintestinal vessel which lies in the wall of the intestine parallel and dorsal to (ii.). In the first six of the gill-bearing segments this vessel receives the efferent vessels from the gills. There are also a pair of small lateral vessels which end anteriorly in the heart.

The heart consists of a pair of enlarged, muscular, contractile transverse vessels, which lie in the sixth segment. They receive blood from the dorsal, subintestinal, and lateral vessels, and by their contraction force it into the ventral vessel. There are numerous capillaries given off from the chief vessels to supply the various organs of the body; the blood is red.

The blood in the ventral vessel is mainly venous, in each of the thirteen segments which carry gills this vessel gives off a pair of afferent branchial vessels, one of which passes to each gill. The gill consists of a number of branching filaments, into each of which the body-cavity is prolonged. Up one side of the filament runs the afferent vessel; down the other side courses the efferent vessel to open in the seventh to the twelfth segments into the subintestinal vessel, and in the thirteenth to the nineteenth segments into the dorsal vessel.

The nephridia are twelve in number, a pair being found in each of the last four segments of the neck and the first two of the gill-bearing region. They consist of the usual funnel-shaped opening into the body-cavity, of a large vesicle which opens to the exterior, and of a glandular swelling which opens into the vesicle, and is probably the secreting portion of the apparatus. In the breeding season the whole organ is crowded with ova or spermatozoa.

The nervous system consists of two small cerebral ganglia, which are connected by circum-oesophageal commissures with a ventral cord which is embedded amongst the longitudinal muscles. This gives off a number of lateral nerves, and is supported by two giant fibres, but does not exhibit any division into ganglia and inter-ganglionic connectives.

Arenicola has no eyes, but it possesses what are not common in Chaetopods, namely otocysts. On each of the cerebral ganglia a small hollow vesicle is found. The walls of this consist of connective tissue with a lining of very columnar cells, probably ciliated. The vesicle contains a fluid in which a number of concretions—otoliths—float; a special nerve passes to its walls. The whole structure is strikingly like the otocyst of many Lamellibranchs.

Arenicola is dioecious, and the ovaries and testes occupy similar positions in the male and female. The ova and spermatozoa are formed from certain of the peritoneal epithelial cells, which become in the breeding season heaped up round the bases of the nephridia. They break off and float in the coelomic fluid, and leave the body through the nephridia.

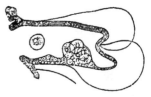

Fig. 98.—Ova originating from the lining epithelium of a parapodium of *Tomopteris*. After Gegenbaur.

The head of *Arenicola* is not provided with any special appendage, but in those worms which live permanently in fixed tubes, the anterior end of the body often bears the branchiae, and is usually provided with tentacles. *Terebella* is provided with numerous tentacles, into which the coelom is prolonged; they are exceedingly extensile, and stretch out in the form of a network all round the worm. Behind the tentacles are situated the branchiae. The appendages of the prostomium are sometimes distinguished by the name *antennae* from those of the peristomium, on which the mouth opens, which are termed the *tentacles*. Ventrally-situated *palps*, probably tactile organs, are also common on the head. In some of the *Serpulidae* a modified tentacle on the head forms an *operculum*, which closes the tube when the worm is retracted.

The division into different regions, which is well marked

in *Arenicola*, is even more conspicuous in some worms, *e.g.* *Chaetopterus*; but it is not a general feature of the group.

Aphrodite, the sea-mouse, is a Polychaet of oval outline, its notopodia bear a number of hairs, some iridescent, and others which are matted together into a feltwork covering the whole animal. This worm is further protected by a number of plate-like *elytra*, also borne by the notopodia, but situated beneath the feltwork; they may be modified cirrhi, but the two structures exist in some of the segments. Elytra are also found on *Polynoe*.

The nature of the tubes of the Sedentaria is very various. It may be soft, or of a parchment-like consistency, and it may be strengthened by a deposit of grains of sand or shell, or it may consist entirely of the latter, very skilfully agglutinated together.

The SABELLIDAE (Fig. 99) and SERPULIDAE, which live in fixed tubes closed at the lower end, have a ventral ciliated band, which is grooved in the former family, whose function is to carry up the undigested matter extruded from the alimentary canal, and pass it out of the tube.

In both subdivisions of the Polychaeta the pharynx is often protrusible; and in many Errantia it is armed with stout teeth, which in some species of SYLLIDAE are said to be traversed by the duct of a poison gland.

FIG. 99.—Sabella vesiculosa, Mont. After Montagu.

In the HESIONIDAE (Fig. 100) and a few others a pair of diverticula from the oesophagus, resembling in position the glandular appendages of *Arenicola*, contain air, probably secreted from the blood. The resemblance of these structures to lungs has been noticed by many observers. Those families provided with such structures have as a rule no branchiae. Another family of worms, the CAPITELLIDAE, are provided with a "siphon," that is, a tube which opens at both ends into the alimentary canal. The siphon never contains food, and its function is probably respiratory. A similar structure runs from one part of the alimentary canal to another in the Echiuridae, and in Echinoids. The CAPITELLIDAE and some other families are without any blood system. In other worms the principal vessels are similar to those described in *Arenicola*; the red blood of some forms is due to haemoglobin dissolved in the plasma, in others the blood is green or almost colourless.

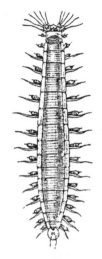

FIG. 100. — *Hesione splendida*, Sav. After Savigny.

The typical arrangement of the nephridia, one pair in each segment, is often interfered with. They usually fail in the anterior segments when there is a large pharynx, and in the tubicolous forms their number is usually much reduced: *e.g.* eight pairs in *Terebella* and six pairs in *Arenicola*. The genus *Capitella* are remarkable for having several pairs of nephridia in each segment, the number increasing in the posterior end up to six or seven pairs in this genus. The nephridia are themselves subject to much variation; one nephridium may have several funnels, and may be connected by a tube with another, and sometimes the organ breaks up into small tubules. The whole arrangement recalls the excretory system of some of the earthworms described above.

The ventral nerve cord of some of the tubicolous Polychaeta has its right and left half divaricated, and connected by numerous transverse commissures. This is well shown in *Serpula*, and in a less degree in *Sabella*. Eyes are very generally present, and are usually confined to the prostomium. *Polyophthalmus*, however, has a number of lateral eyes, a pair to each somite; whilst *Branchiomma* bears them on its branchial filaments. Otocysts, such as those of *Arenicola*, are rare.

The Polychaets, with some exceptions, are dioecious. The generative organs are usually developed in relation to a blood-vessel, which no doubt serves to nourish them; in the Sedentaria they often correspond in number with the nephridia. Their products ripen in the coelomic fluid (Fig. 98), and usually escape through the nephridia, they may however escape by rupturing the body-wall. Impregnation takes place externally. The eggs are sometimes laid in small masses of jelly; sometimes they remain under the care of the parent, under the elytra in *Polynoe,* in a cavity in the operculum in some SER-PULINAE, and attached to the tube amongst the TEREBELLIDAE.

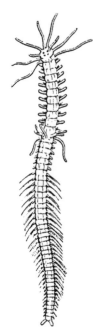

FIG. 101. — Parent stock of *Autolytus cornutus.* After A. Agassiz.

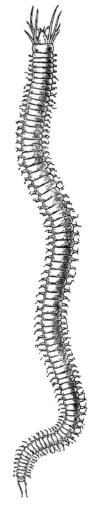

FIG. 102.—*Nereis pelagica,* L. After Oersted.

Asexual reproduction is not common; it occurs, however, in the SERPULIDAE and SYLLIDAE. In the former family a head is formed by one of the segments in the middle of the body, and the animal then divides just in front of this. In some of the SYLLIDAE, *Autolytus,* for example, one of the posterior segments, usually the last, gives rise to a new individual; this may be repeated, and chains of zooids are formed (Fig. 101). These zooids break off, develope generative organs, and reproduce sexually. As the original worm was without sexual organs, this genus exhibits an alternation of generations; a very uncommon phenomenon in Chaetopods. It is further complicated by sexual dimorphism, the male worm being in many respects different in appearance from the female. A somewhat similar phenomenon occurs in *Nereis* (Fig. 102), one form being known as *Heteronereis:* this genus is polymorphic, for in addition to the male and female forms, hermaphrodite individuals also occur.

CHAPTER XI

GEPHYREA

Gephyrea $\left\{\begin{array}{l}\text{Achaeta—}\textit{Phymosoma, Sipunculus, Phascolion.}\\ \text{Chaetifera—}\textit{Bonellia, Echiurus, Thalassema.}\end{array}\right.$

CHARACTERISTICS.—*Subcylindrical marine animals, with very slight indications of segmentation. The anterior part of the body is either retractile, forming an Introvert, or it bears an extensile Prostomium. Setae or chitinoid hooks usually present. The nervous system consists of a circum-oesophageal ring and a ventral non-ganglionated cord. No special respiratory or locomotor organs exist. A closed vascular system is present, and the coelom is spacious. Nephridia are present, and serve as a rule as ducts for the exit of the reproductive cells, as well as functioning as excretory organs.*
The Gephyrea are divided into two sub-classes :

I. The Achaeta, *characterised by the anterior end of the body being retractile. The introvert is withdrawn into the body by special retractor muscles in the same way as the proboscis of a Nemertine. No setae are found, but the introvert is usually armed by rows of chitinoid hooks. The mouth is anterior and terminal.*

II. The Chaetifera *have a segmented larva, but the segmentation is lost in the adult. There is a long extensile prostomium which is easily broken off. A pair of ventral setae are found. Anus terminal ; special branching organs open on the one hand into the coelom, and on the other into the rectum, they possibly function as nephridia. One to eight anterior nephridia serve as genital ducts.*

The Gephyrea Achaeta include ten genera; of these *Phascolion* and *Phascolosoma* occur in our seas, the former usually making its home in the shells of dead molluscs, the latter living at the bottom of the sea half buried in sand.

The largest genus of the class is *Phymosoma*; it

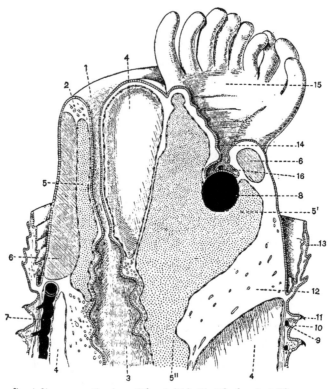

FIG. 103.—Semi-diagrammatic view of the right half of the head of *Phymosoma varians*, seen from the inner surface.

1. Mouth.
2. Lower ventral lip.
3. Oesophagus.
4. Portions of the coelom, seen in three places.
5. Blood sinus in lower lip corresponding with 4 in Fig. 105.
5'. Blood sinus surrounding brain, and opening into 5″, the dorsal vessel.
6. Skeletal tissue surrounding mouth.
7. Ventral nerve cord.
8. Brain.
9. Circular nerves at base of ridges bearing hooks.
10. Sense organ at base of ridges bearing hooks.
11. Rings of hooks.
12. Retractor muscle.
13. Extensile collar.
14. Pit leading to brain.
15. Lophophore.
16. Eye-spot.

contains twenty-eight species, which, with very few exceptions, are confined to the tropics, in many respects it resembles *Phascolosoma*, and, as its structure has been lately worked out, it will form the most convenient type for description.

Phymosoma varians is a West Indian species found embedded in the soft coral rock, in which it bores tubular
passages, probably dissolving the soft rock by some chemical
excretion. Its colour is brownish-yellow. In its extended
condition it is about 5 cm. long and about $\frac{1}{2}$ cm. broad,
tapering at each end. The anterior half of the body, the
introvert, can be withdrawn into the posterior half, just as
the finger of a glove can be invaginated into the hand.

The mouth is terminal, and is at the end of the introvert.
Dorsal to the mouth is a crown of eighteen or twenty short
tentacles arranged in a horse-shoe, the *lophophore* (Figs. 103 and
105). The dorsal ends of this horse-shoe are continuous with
the dorsal ends of a thickened lower lip, between which and
the crown of tentacles or lophophore the mouth opens. The
mouth has therefore the form of a crescentiform slit. In
the hollow of the horseshoe-shaped lophophore the skin is
wrinkled and pigmented; close beneath it, and in direct continuity with it, lies the bilobed supra-oesophageal ganglion.
About 2 mm. behind the mouth, a very extensile fold of tissue
forms a ring-like collar round the base of the head. This
collar can be produced so as to cover in the whole head.

The introvert is distinguished from the rest of the body
by the presence of numerous rows of minute chitinoid hooks
(Fig. 103), which alternate irregularly with certain papillae
to be described below.

The integument consists of the following layers : (i.) the
ectoderm, (ii.) circular muscles, (iii.) longitudinal muscles, and
(iv.) peritoneal epithelium. The ectoderm is a single layer of
cubical cells. Those covering the lower lip, and that side of
the tentacles turned towards the mouth, bear cilia. The
ectoderm of the concave side of the lophophore and its hollow
is crowded with black pigment, and at two places it is continuous with the substance of the brain. Over the rest of the
body the ectoderm secretes a thick cuticle, which is only
broken by the presence of the skin papillae.

These papillae are very characteristic of the Gephyrea;
they are formed by the ectoderm becoming folded into the
shape of a double narrow-mouthed conical cup. The outer
layer of cells resembles the ordinary ectoderm; the inner, how-

ever, are enlarged wedge-shaped cells which almost fill up
the cavity of the cup. They form a secretion which passes
out through the narrow opening. The mouth of this is
protected by special horny plates, modifications of the
cuticle. The papillae are scattered all over the body, and
as they stand out from the surface, they give the animal a
rough appearance.

Within the ectoderm is a layer of circularly-arranged
muscle fibres broken up into circular bands in the introvert,

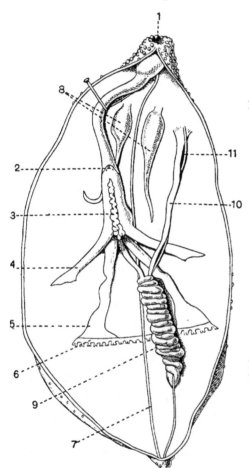

FIG. 104.—The body of *Phymosoma
varians* laid open by an incision
a little to the left of the median
dorsal line, so as to show the
internal organs. The introvert
is retracted.

1. Opening of introvert.
2. Position of brain, the two eye-
   spots are shown. This marks
   the level of the head.
3. Dorsal vessel.
4. Left dorsal retractor muscle.
5. Left ventral retractor muscle.
6. Generative ridge.
7. Ventral nerve cord.
8. Nephridia.
9. Coiled intestine.
10. Rectum.
11. Spindle muscle.

and forming a continuous sheath in the trunk; internal to
this are the longitudinal muscles, continuous in the introvert,
but arranged in about twenty anastomosing bundles in the
trunk. Within this layer, the coelom is lined by flat epithelial

cells. From the longitudinal bundles four stout muscles arise, two dorsal and two ventral. These pass to form a muscular ring ensheathing the oesophagus, just behind the head. They are termed the four *retractors*, and their function is to draw in the introvert.

The alimentary canal consists of a straight oesophagus, into which the mouth passes without any armature, and which in its turn passes into a coiled intestine. Both these parts are ciliated, the cilia of the oesophagus being continuous with those of the lower lip and tentacles. The intestine is coiled round a special " spindle " muscle, which arises from the extreme posterior end of the body, passes up the axis of the coil, and joins the longitudinal muscles of the body-wall near the anus (Fig. 104). A short rectum passes to the anus which terminates the alimentary canal, the anus pierces the body-wall just behind the line of division between the introvert and the trunk.

The vascular system is closed and is confined to the anterior end of the animal. Its most conspicuous part is a vessel which lies on the dorsal side of the oesophagus between the retractor muscles. The vessel is closed behind, and gives off no capillaries. At the anterior end it opens into a large sinus into which the brain protrudes ; from this sinus a circular vessel is given off which runs round the lower lip, and when full of blood, it serves to distend the latter. Another part of the vessel runs along the base of the lophophore, giving off branches into each tentacle. It is possible that the blood may become oxygenated in the tentacles, but the chief function of the whole system is to distend the tentacular crown and lower lip. The fluid in this system is corpusculated.

The coelom is very spacious, and contains a corpusculated fluid which bathes all the internal organs. The corpuscles are larger than those of the vascular system. The contraction of the circular muscles of the skin forces this fluid forward, and in this way the introvert is everted.

The nephridia, or excretory organs, of the Gephyrea are often termed " brown tubes." In *Phymosoma* they are two in number, one on each side of the ventral nerve cord (Fig. 104). They have the form of elongated sacs, which hang down into the body-cavity. At their upper ends the sacs are attached to the body-wall, and open to the exterior a little in front of the level of the anus. Each sac consists of two portions: a posterior glandular part lined by large glandular cells, which give off vesicles containing their excretion, and a muscular non-glandular anterior half, which opens both on to the

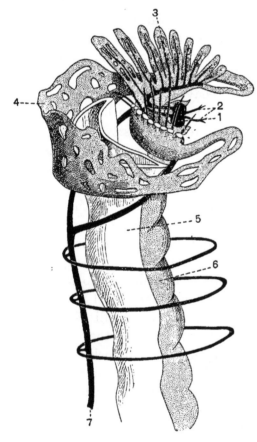

Fig. 105. — Diagram showing relation of nervous system, vasonlar system, and oesophagus in *Phymosoma varians*. Partly after Selenka.

1. The brain, represented relatively too small.

2. Nerves to skin of preoral lobe.

3. Lophophore ; each tentacle is represented by its blood sinuses and its nerve.

4. Blood sinuses of lower lip.

5. Oesophagus.

6. Dorsal blood-vessel.

7. Ventral nerve cord.

exterior and into the coelom. The opening into the latter space is situated close to the external opening, and is guarded by a frilled, funnel-shaped lip, thickly ciliated. The wall of the organ contains many muscle fibres, and it is capable of considerable change of form.

The nervous system consists of a bilobed brain in continuity with the epidermis of the concavity of the lophophore (Fig. 103). It gives off a pair of lophophoral nerves, which run along the base of the tentacles, sending off a nerve into

each.   Laterally the two lobes are continued into stout
nerves which embrace the oesophagus, and fuse to form a
ventral cord.   On each side the ventral cord is supported by
two longitudinal muscles, and the whole is loosely attached to
the ventral surface of the body-wall by muscular strands.
The cord shows but slight traces of double origin, it bears no
ganglia, but ganglion cells are uniformly distributed on its
ventral surface.   It gives off a series of lateral nerves, which
form complete rings round the body, situated in the skin
(Figs. 103 and 105).

Two pits of large ectodermal cells, crowded with dense
black pigment, have sunk on each side into the brain.   They
are hollow, and contain a coagulum in dead specimens.   They
are usually spoken of as eyes.

Phymosoma is dioecious.   Both the ovary and testis are
formed of a ridge of the peritoneal epithelium which runs
across the body at the base of the ventral retractor muscles.
Certain of the cells of this ridge break off and float in the
coelomic fluid.   In the female they become ova, in the male
they are the mother cells of the spermatozoa.   The ova grow
a good deal whilst in the body-cavity, and secrete a thick egg
shell; ultimately they leave the body through the nephridia.
The spermatozoa derived from one mother cell always remain
connected as long as they are in the body-cavity, and in this
condition are taken up by the funnel-shaped internal openings
of the nephridia.   The ova are fertilised externally in the
water.

Certain of the **Gephyrea achaeta** differ in many points
from *Phymosoma*.   *Sipunculus* has no lophophore, and the
mouth is surrounded by a frayed fringe, which, like the
tentacles of other forms, is well supplied with nerves and
blood-vessels.

Many species are without the hooks on the introvert.
A layer of oblique muscles lies very commonly between the circular and longitudinal fibres.   The capacity
of the dorsal vessel, which acts as a reservoir for receiving
the blood when the tentacles and head are retracted, is increased in some species of *Phymosoma* by a number of lateral
diverticula, and in some Sipunculids by the addition of a

ventral vessel. *Sipunculus* and *Phascolosoma* have remarkable bodies known as "urns" floating in their coelomic fluid. They are bell-shaped structures, with a ring of cilia round the mouth, and a nucleus. These remarkable corpuscles are formed by the division of certain large cells on the wall of the dorsal blood-vessel, they were formerly thought to be parasitic Infusoria.

The **Achaeta** have no special organs of locomotion, and probably do not move about much. *Sipunculus* and *Phascolosoma* usually live half embedded in the sand, which they swallow in large quantities. *Phascolion* lives in empty worms' tubes or in mollusc shells, and its body is often permanently twisted, accommodating its shape to that of its home. *Phymosoma* lives in holes or passages in coral rock, or in holes between stones. As a rule the members of this subdivision occur only in comparatively shallow water.

The **Gephyrea chaetifera** are provided with a prostomium, which may acquire enormous proportions. In *Bonellia* it may, when fully extended, attain a length of 2 or 3 feet, whilst the body is only $1\frac{1}{2}$ to 2 inches long. In this genus it is bifid at the end. In *Echiurus*, *Bonellia*, and *Thalassema* there are a pair of large chitinoid hooks placed anteriorly on the ventral side of the body, and in some species of *Echiurus* there is one, sometimes two, posterior circlets of setae, each seta originating from a single cell, like those of the Chaetopods.

Bonellia viridis is coloured a bright green by a pigment termed "bonellein," which is not identical with chlorophyll. The mouth in the **Chaetifera** lies at the base of the prostomium, which is ciliated and grooved, and is doubtless used to catch minute organisms for food; the intestine is looped and the anus terminal. In *Bonellia*, *Echiurus*, and *Thalassema* a "siphon" or collateral intestine, such as is found in the CAPITELLIDAE and ECHINIDS, is present.

Branched organs open into the rectum in most of the **Chaetifera**. At the end of each branch is a small funnel-shaped ciliated opening leading into the coelom. The cells lining the tubes of these branches have been seen crowded with excretory granules, and they may possibly function as nephridia as well as serve to regulate the amount of fluid in the coelom.

The vascular system is more complex in the **Chaetifera** than in the **Achaeta**. The dorsal vessel in *Echiurus* opens behind into a circular blood-vessel which surrounds the oesophagus. At its anterior end it enters the prostomium and runs to the tip of this organ, here it splits, and the two branches return, one down each side of the prostomium, till they have passed the mouth, when they unite to form a median supraneural blood - vessel. This is connected with the perioesophageal circular ring by a transverse vessel. Haemoglobin has been detected in the coelomic corpuscles of *Thalassema*.

The last-named genus may have from one to four pairs of nephridia, according to the species, *Echiurus* has usually two pairs.

The nervous system, like the vascular system, is continued into the prostomium, running all round the edge, and finally uniting below the oesophagus, thus forming a circum-oesophageal ring, which gives off the ventral cord. In no place is the nerve ring or cord thickened to form anything like a ganglion.

In *Echiurus* and the female *Bonellia* the coloemic epithelial cells which surround the ventral vessel enlarge and form the reproductive cells, which are thus favourably situated for receiving nourishment.

There is a very remarkable dimorphism in the genus *Bonellia*. The female is a fair-sized animal, with a body 2 inches long, but the male is a microscopic planarian-like organism which lives in a recess of the nephridium of the female. It is from 1 to 5 mm. long, and is ciliated all over. Its intestine is not functional, and it ends blindly both in front and behind. The spermatozoa arise from the coelomic epithelium, and escape by a modified nephridium. A nervous system; but no vascular system, is present.

The male larva is said to cling to the prostomium of the female, and thence to pass into the mouth, where it undergoes its final changes, then it creeps out from the mouth and into the nephridium, where it spends the rest of its life. Another genus, *Hamingia*, has a similarly degenerate male, which also lives in the nephridia of the female.

Those members of the armed Gephyrea whose developement has been investigated show unmistakable affinities to the Chaetopods. Their larvae exhibit a metameric segmentation, but the somites disappear early. Traces of segmentation are retained in the adult in a few cases, such as the four pairs of nephridia in one species of *Thalassema*, the double ring of setae in *Echiurus Pallasii*, and possibly in the rings of hooks and circular nerves of many forms. A connecting link between the Gephyrea armata and the Chaetopoda may exist in the curious worm *Sternaspis*. This animal, usually classed with the Chaetopoda, retains a well-marked segmentation; and its blood - vessels, whilst resembling in their disposition the more important vessels of the Gephyrea, open into a well-developed system of capillaries. On the other hand the looped intestine, one pair of brown tubes, retractile anterior end of its body, and—in *Sternaspis spinosa*—a long bifid prostomium, described by Sluiter, are all features shared in common with the Gephyrea.

The unarmed Gephyrea have an abbreviated developement which shows no traces of metameric segmentation, but this hardly seems a sufficiently important difference to warrant the breaking up of the group.

CHAPTER XII

BRACHIOPODA

Brachiopoda $\left\{\begin{array}{l}\text{Ecardines—}\textit{Lingula, Crania, Discina.}\\ \text{Testicardines—}\textit{Argiope, Terebratula, Waldheimia.}\end{array}\right.$

CHARACTERISTICS.——*Coelomata devoid of organs of locomotion, and usually fixed in the sand on to some Voreign body, by a peduncle. A bivalved shell encloses the body. The valves are dorsal and ventral, and in one subdivision are hinged to one another. They are lined by dorsal and ventral extensions of the body-wall, termed the mantles; these often bear chitinoid setae round their edges. A lophophore surrounds the mouth, bearing ciliated tentacles. The alimentary canal is ciliated, and receives the secretion of two branched glands, the liver; it is in one sub-division aproctous. One, rarely two, pair of nephridia exist. Exclusively marine.*

The existing Brachiopoda are interesting as the survival of what in early geological time was a very widely distributed and very numerous group of animals. The two genera *Lingula* and *Discina* extend from the Cambrian, the oldest group of the Silurian rocks, to the present day; and, judging by their shells, they appear to have undergone but little change during the vast period of time which must have elapsed since they lived. They are found in great numbers, both of individuals and of species, in these older

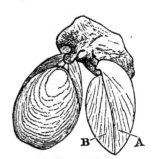

FIG. 106.— *Waldheimia cranium.*
A. Ventral,
B. Dorsal valve.

Paleozoic formations; but the group seems to have been most flourishing in the Devonian seas, for upwards of 60 genera and

1100 species have been described from Devonian rocks. Since this epoch they have dwindled, and at the present day not more than about 100 species exist.

Argiope (*Cistella*) *neapolitana* is a small Brachiopod found attached by a peduncle to pieces of rock at a depth of about 70 metres in the Mediterranean. The dorsal and ventral shells entirely cover the body except the peduncle, which projects through a " beak " formed by the ventral or larger shell. The

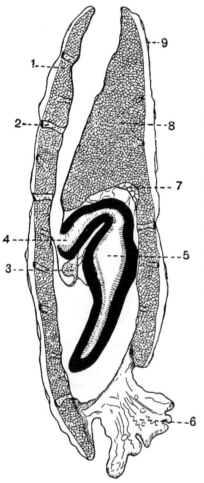

Fig. 107.—A longitudinal vertical median section through *Argiope neapolitana*.

1. Ventral shell.

2. Canal containing blood-vessel.

3. Sub-oesophageal nerve ganglion.

4. Mouth.

5. Stomach.

6. Peduncle.

7. Plexus of blood-vessels.

8. Median crest on dorsal shell.

9. Organic membrane which has separated from shell during the process of decalcification.

dorsal shell is rather the smaller ; both are of a brownish hue with small white spots. The body of the *Argiope* lies almost entirely in the dorsal shell (Fig. 107), and is supported by certain ridges which this shell bears on its inner surface. The whole animal is about 2·5 mm. long, and about the same in breadth.

The shells are secreted by the body-wall or by the mantle. Since the body lies chiefly in the dorsal shell, the larger part of the latter is secreted by the body-wall, and the dorsal mantle is of small extent; on the other hand, the greater part of the ventral shell is lined by a fold of integument, the ventral mantle.

The substance of the shell is composed of minute calcareous spicules kept together by a network of organic fibrils (Fig. 107). The shell is pierced by numerous canals, whose outer ends are somewhat enlarged and covered with a cuticle.

Since the mantle is formed of a duplicature of the body-wall, it is necessarily double, and the body-cavity extends into it, in some places this space lodges the reproductive organs. The mantle sends a prolongation into each of the canals in the shell, which is continuous with some of its blood-vessels. These prolongations contain blood-corpuscles, and doubtless serve to nourish the organic fibrils which keep together the calcareous spicules of the shell. The lophophore occupies a considerable part of the dorsal shell, and forms a large part of the body-wall. Its shape is oval, its border running parallel to the edge of the shell, except at the anterior median line, where a narrow deep indentation almost divides it into two, and thus gives it a somewhat horse-shoe shape. The indentation is occupied by a median ridge

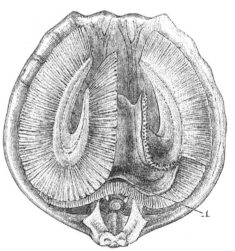

FIG. 108.—*Waldheimia flavescens.* Interior of dorsal valve, to show the position of the lophophore. A portion of the fringe of cirrhi has been removed to show the brachial membrane, and a portion of the spiral extremities of the arms.
A. Position of mouth.

in the dorsal shell (Fig. 107). The lophophore carries round its edge, on the dorsal side of the mouth, from 70 to 100 tentacles; at the base of the tentacles is a ciliated groove, whose other side is formed by a lip which also runs round the edge of the lophophore (Fig. 109). The tentacles are partially ciliated as well as grooved, and any particles of food they come in contact with are carried down the groove to the mouth, which opens in its posterior median line. In other genera the lophophore stands out from the surface of the body and becomes curiously coiled and rolled up, as in *Waldheimia* (Fig. 111), in which animal it is supported by a calcareous loop.

The mouth is a transverse slit leading into a short oesophagus; this is attached by mesenteric strands to the

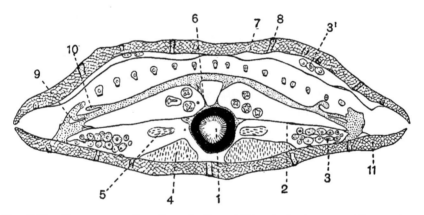

FIG. 109.—Transverse section through the middle of *Argiope neapolitana*. The section includes the posterior limit of the lophophore, but is anterior to the brood pouches.

1. Stomach.
2. Gastroparietal bands.
3. Ovary in dorsal shell.
3'. Ovary in ventral shell.
4. Dorsal adjustor muscle.
5. Occlusor muscle.
6. Left mesentery; posteriorly this fuses with the right to form a single mesentery.
7. Ventral shell.
8. Vascular canal in shell.
9. Canal at base of lophophore, which sends a branch into each tentacle.
10. Lip forming with the tentacles a groove.
11. Dorsal shell.

end of the median projection of the dorsal shell, and it opens directly into the globular stomach. On each side of the alimentary canal is the liver, composed of six or seven thick tubules, which unite and open into the stomach by a broad mouth. The lumen of the liver is often full of secretion, it is lined by vacuolated cells. The stomach opens behind into a short intestine which has no anus, and which, like the rest of the alimentary canal, is ciliated. The alimentary canal is supported by a median sheet of connective tissue, the mesentery, which passes from it to the ventral shell' (Fig. 109), and by two lateral sheets, termed the gastroparietal bands, which pass out from the stomach to the sides of the body-wall.

Owing to the peculiar relations of the animal to its shell, the body-cavity becomes very complicated, it is partly produced into the mantles which line the shells, and here the reproductive organs partially lie. At the posterior and lateral regions the body-wall is pushed in, in such a way as to form two lateral brood pouches, which lie behind the level of the lophophore, and are enclosed by the shell. The embryos undergo the early stages of their developement in these pouches.

The coelom is traversed by four bundles of muscle fibres, two of which open and close the shell, the other two move the shell

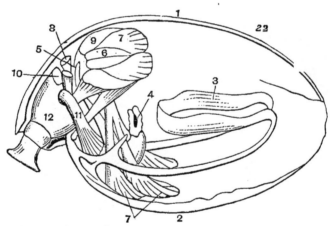

FIG. 110.—*Waldheimia flavescens.* Diagram showing the muscular system. After Hancock.

1. Ventral valve.	7. Divaricators.
2. Dorsal valve.	8. Accessory divaricators.
3. Calcareous loop.	9. Ventral adjustors.
4. Mouth.	10. Peduncular muscles.
5. Extremity of intestine	11. Dorsal adjustors.
6. Adductor.	12. Peduncle.

on its peduncle. The latter are termed adjustors, and a pair arise from each valve of the shell and are inserted into the peduncle. By their contraction they raise or depress the shell, and by contracting alternately they may also serve to rotate it. The occlusor muscles have a double origin from the dorsal shell, but the two parts unite to form a single tendon, which is inserted into the ventral shell. The divaricators are very small. They arise from the ventral shell, and are inserted into the dorsal valve in such a relation to the hinge as to cause the shell to open when they contract. Additional muscles are found in other members of the group, those of *Waldheimia* and *Lingula* are shown in Figs. 110 and 113.

Running round the edge of the lophophore, at the base of the tentacles, is a canal which is probably continuous with the general body-cavity. It gives off a branch into each tentacle, and the latter are probably extended by the entrance of the coelomic fluid into them (Fig. 109).

There is a closed vascular system containing a corpus-

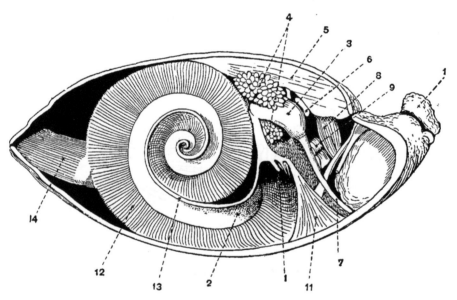

FIG. 111.—View of the inner side of the right half of *Waldheimia australis*.
From a dissection by J. J. Lister.

1. Mouth.
2. Lophophore.
3. Stomach.
4. Liver tubules.
5. Median ridge on dorsal shell.
6. Heart.
7. Intestine ending blindly.
8. Peduncular muscle.
9. Internal funnel-shaped opening of nephridium.
10. Peduncle.
11. Body-wall.
12. Tentacles.
13. Coil of lip.
14. Terminal tentacles.

culated fluid. The vessels composing it are irregularly scattered through the tissue of the body. They are especially developed in the mantle and in those parts of the body-wall next to the shell, and send off numerous caecal processes into the canals permeating the substance of the latter. The blood is probably aerated as it comes through the vessels of the mantle. The presence of a central heart in *Argiope* is a matter of dispute if it exists, it is a contractile vesicle situated dorsal to the stomach; a spherical vesicle is found in this position in *Waldheimia*, but its relation to the blood-vessels is not very definitely known.

The nephridia, which also function as genital ducts, open internally, with large funnel-shaped mouths, which are directed towards the posterior end of the dorsal shell. These openings lead to short tubes which run in that part of the body-wall which is pushed in to form the brood pouch; into this they

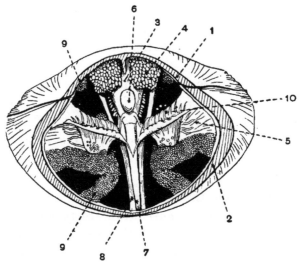

FIG. 112.—The posterior quarter of a *Waldheimia australis* has been removed, in order to show the relations of alimentary canal, nephridia, etc. The coelom is represented black. From a dissection by J. J. Lister.

1. Dorsal shell.
2. Ventral shell.
3. Stomach.
4. Liver.
5. Gastroparietal bands.
6. Heart.
7. Intestine.
8. Tendons of adductor muscles.
9. Ovary.
10. Internal funnel-shaped opening of nephridia with three ova dropping in.

eventually open. The walls of the tube consist of a connective tissue basement membrane lined by glandular cells crowded with brown concretions. Some of these cells are also ciliated.

The ova are modifications of the cells lining the coelom. There are four ovaries, one on each side in the dorsal and in the ventral shell. The cells are borne on an axis, and all stages from the ripe ovum to the unmodified peritoneal cell may be seen on the same axis. The ripe ova fall into the coelom and leave the body through the nephridia. They undergo the early stages of their developement in the brood pouches. No testes have been described, and it is uncertain whether this species is hermaphrodite or not.

The nervous system consists of a circum-oesophageal nerve ring, which is enlarged into a well-marked sub-oesophageal ganglion lying in the epidermis. This lies in that part of the body-wall which overhangs the mouth, just behind the base of the tentacles. The nerve ring swells into a small supra-oesophageal ganglion, which is not so large or so well marked as the sub-oesophageal. The latter gives off a nerve which runs round the edge of the lophophore, and nerves to the dorsal and ventral mantles.

The Brachiopoda are divided into two orders:

(i.) *The Ecardines, whose shell is chitinous and but slightly strengthened by a deposit of calcareous spicules. The shell has no hinge and no internal skeleton to support the arms. The alimentary canal terminates in an anus, median and ventral in Crania and lateral in Lingula.*

Lingula, Glottidia, Crania, and Discina.

(ii.) *The Testicardines have shells composed of calcareous spicules, the valves are hinged together, and there is usually an internal skeleton supporting the arms of the lophophore. There is no anus.*

Argiope, Terebratula, Terebratulina, Rhynchonella, Thecidium, Waldheimia.

In the Lingulidae the dorsal and ventral valves are about the same size; in all other Brachiopods the ventral is much the larger, and except in *Crania* always lies uppermost.

In some genera, as *Argiope* and *Lingula*, the body occupies most of the space enclosed by the valves of the shell; in *Terebratula* and some others the body takes up but a small portion of this space, the remainder being occupied by the arms of the lophophore, which stand out from the surface of the body, and may be coiled in a very complicated manner. In *Rhynchonella* the lophophore is protrusible, but this is exceptional.

The edges of the mantle usually carry a row of setae, which arise from ectodermal pits, as in Chaetopods.

The intestine in *Lingula* is of some length, it takes one or two twists, and terminates in an anus which opens at the right side into the mantle cavity between the shells. In *Discina* also the anus is lateral, but in *Crania* it opens in the median line into a cavity which lies between the posterior ends of the valves where the peduncle would normally be found.

The food of Brachiopods consists chiefly of Diatoms and minute unicellular Algæ, which are brought to the mouth by the action of the cilia on the lophophore.

The only case of serial repetition of parts presented by the Brachiopoda is the two pairs of nephridia found in the genus *Rhynchonella*.

The sexes are separate in *Crania*, but *Lingula*, and probably some others, are hermaphrodite.

The recent Brachiopods are found in all seas, usually at moderate depths, within 100 fathoms. *Lingula* and *Glottidia* sometimes live between tide marks, but may extend to a

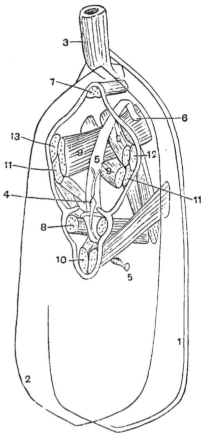

Fig. 113.—*Lingula anatina.* Diagram showing the muscular system : after Hancock.

1. Dorsal valve.
2. Ventral valve.
3. Peduncle.
4. Heart.
5. Alimentary canal.
6. Anal aperture.
7. Umbonal muscle.
8. Central muscle.
9. Transmedial muscle or sliding muscle.
10. Anterior muscle.
11. Middle muscle.
12. Adjustors, enabling valves to move forward and backward on each other.

12

depth of 70 metres.   They anchor themselves in the mud, and their long, hollow peduncle forms a sand-tube around itself. Some Testicardines have been found at great depths ; they usually attach themselves to the rocks, and, when they are found at all, occur in considerable numbers.

POLYZOA

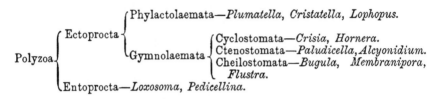

Polyzoa
- Ectoprocta
  - Phylactolaemata—*Plumatella, Cristatella, Lophopus.*
  - Gymnolaemata
    - Cyclostomata—*Crisia, Hornera.*
    - Ctenostomata—*Paludicella, Alcyonidium.*
    - Cheilostomata—*Bugula, Membranipora, Flustra.*
- Entoprocta—*Loxosoma, Pedicellina.*

CHARACTERISTICS.—*Small coelomate animals, invariably possessing the faculty of budding. The colonies nearly always fixed. The ectoderm secretes, as a rule, a cuticle, which may be horny or calcareous. The intestine is bent in the form of a* U, *the anus and mouth being approximated; between them is situated the nerve ganglion. The mouth is surrounded by a series of ciliated tentacles. The individuals of the colonies may be hermaphrodite, but the generative cells may ripen at different times.*

The Polyzoa comprise a very great number of species, which can be grouped into two main subdivisions :

(i.) *The Ectoprocta, in which the anus lies outside the circlet of tentacles which surrounds the mouth.*

(ii.) *The Entoprocta, or those forms in which the circlet of tentacles or lophophore embraces both ends of the alimentary canal, the anus as well as the mouth.*

The latter subdivision contains very few forms, but the former includes a great number of species, mostly marine ; a few, however, inhabit fresh water.

Plumatella fungosa is a fairly common representative of the freshwater Ectoprocta. It occurs all over Europe, on pieces of submerged trees, etc., living by preference in stagnant or slowly-flowing water. Each individual of the colony lives in a thin chitinous tube, from the mouth of which the animal protrudes under favourable circumstances, and into which it withdraws in time of danger. The size of the colony varies a good deal, it may, however, attain a diameter of some inches, and may weigh a pound or more.

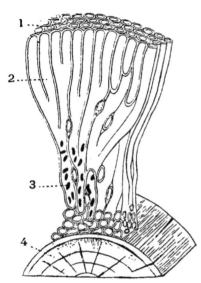

The chitinous tube is a modification of the cuticle, and is secreted by the epidermis of the body-wall. The cuticle is usually spoken of as the *ectocyst*, the body-wall underlying it being called the *endocyst*. The anterior portion of the body is capable of being extended beyond the mouth of the tube, and bears a series of tentacles arranged in a horseshoe-shaped lophophore. This portion can also be retracted, and in this condition the space in which the tentacles lie is termed the tentacle-sheath.

Fig. 114.—A portion of *Plumatella fungosa* seen in vertical section. Allman.

1. Mouth of tubes.
2. Cavity of tubes.
3. Statoblasts.
4. Piece of wood in which the colony is growing.

The body-wall, with its cuticle, is horny in *Plumatella*, gelatinous in *Lophopus*, and calcareous in most marine Polyzoa. It is often known as the *zooecium*, whilst the extrusible part of the organism, with the tentacular crown and the alimentary canal, etc., is distinguished as the *polypide*.

The body-wall of *Plumatella* contains a layer of external circular, and of internal longitudinal muscles, and is lined by a ciliated epithelium; at its lower end it becomes free from the ectocyst, and by the contraction of the muscle fibres in this unattached region the coelomic fluid is forced forward, and serves to extrude the polypide. Certain muscle fibres stretch from the endocyst to the wall of the extrusible portion of the body. These have been termed the parieto-vaginal muscles, and serve to prevent the full extrusion of the polypide. The zooecia at their basal ends open into one another, and the body-cavities of the various members of the colony are in free communication.

The lophophore is in the form of a double horse-shoe, it bears from forty to fifty ciliated tentacles, which, by the current they set up, assist in bringing minute algae, etc., as food to the mouth. The bases of the tentacles are connected together by a fine membrane or web, sometimes called the calyx. On the side of the horse-shoe nearest the mouth is situated a ciliated extension of the body-wall. This lobe, which more or less overhangs the mouth, is the epistome.

The alimentary canal is U-shaped, the mouth opens into an oesophagus, which is ciliated in some species. This leads into a stomach. From the stomach a rectum turns forward again and opens to the exterior outside the ring of tentacles, but not very far from their base. The walls of the intestine contain muscle fibres; the single layer of cells lining the stomach enclose brown granules, which apparently increase in number with the age of the polypide. These are probably excreta, which for some reason or another do not find their way out of the body.

A strand of tissue of considerable importance in the life-history of the animal passes from the posterior end of the stomach, and is attached to the body-wall of the animal, near the posterior end. It is termed the *funiculus*, and doubtless serves to prevent the polypide from being extruded too far. The coelom is spacious, and contains a corpusculated fluid which is kept in motion by the ciliated cells lining the body-wall. It is continued into the lophophore, and into each tentacle, but is partly divided into two by an incomplete septum which stretches across the body below the level of the base of the lophophore.

The nervous system consists of a bilobed ganglion lying on the oesophagus, between it and the anus. It is situated just in front of the imperfect septum which stretches across the animal in this region. At the sides the two lobes extend round the oesophagus, forming a complete circum-oesophageal nerve ring, each lobe also gives off a nerve which runs along the base of the lophophore, and which furnishes a small nerve to each tentacle.

No nephridium has been described in *Plumatella*; in an allied form, *Cristatella*, a pair of ciliated tubes are, however, said to lead from the body-cavity and to open by a common pore between the anus and brain.   The interpretation of these

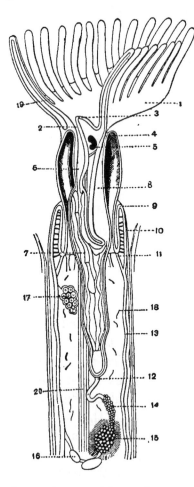

FIG. 115.—View of right half of *Plumatella fungosa*, slightly diagrammatic. After Allman and Nitsche.

1. Lophophore.
2. Mouth.
3. Epistome.
4. Anus.
5. Nerve ganglion.
6. Oesophagus.
7. Stomach.
8. Rectum.
9. Edge of fold of body-wall.
10. Wall of tube.   Ectocyst.
11. Parieto-vaginal muscles.
12. Funiculus.
13. Body-wall.   Endocyst.
14. Testis.
15. Testis, more mature.
16. Statoblast.
17. Ovary.
18. Spermatozoa free in body-cavity.
19. Calyx.
20. Retractor muscle.

ducts is a subject of uncertainty.   The functions of a renal apparatus may possibly be delegated to those large cells in the stomach in which brown granules accumulate during the life of the polypide.

Plumatella is hermaphrodite.   The ova develope from the cells which line the body-wall, near the anterior end.   The testes are formed from cells covering the upper end of the funiculus; these cells multiply, and become spermatozoa.

The cells of the funiculus at its lower end give rise

to some very remarkable structures known as *statoblasts* These have the morphological significance of buds; they consist of a little heap of cells which secrete around them a chitinous shell.

The freshwater Polyzoa usually die down at the approach of winter, and the continuance of the race is provided for by the persistence of the statoblasts. These structures are usually formed during the autumn, and escape after the decay of the parent organism. Their chitinous shell in most cases is a complicated structure; part of it contains air vesicles, the air being secreted from the protoplasm of the cells which have formed the shell. Thus the latter forms a float; in some cases, however, the statoblasts are attached to submerged stones, and the floating ring is then rudimentary or absent. The formation of these structures, capable of resisting the winter frosts, in freshwater Polyzoa acquires an additional interest when they are compared with similar contrivances found in other members of the freshwater fauna. *Spongilla fluviatilis*, the freshwater sponge, also dies down at the approach of winter; it also forms remarkable bodies, termed gemmules, which consist of a collection of cells protected by curious spicules termed amphidiscs, and from which a new sponge arises in the spring. In the same connection the ephippian eggs of *Daphnia*, which are also supported by an air float, and in other respects have a striking resemblance to some statoblasts, and the winter eggs of Rotifers and Planarians, may be mentioned. The low temperature of winter, which affects the comparatively small bulk of fresh water much more than the ocean, where beyond a depth of a few fathoms the cold hardly affects the temperature of the water, has apparently called forth these modifications.

A. ECTOPROCTA.

The Polyzoa are divided into two groups,—(*a*) the ENTOPROCTA and (*b*) the ECTOPROCTA,—according as to whether the anus is included or not in the circlet of tentacles which surrounds the mouth. The Ectoprocta are further characterised by the possession of a well-developed coelom. This group includes an immense number of species, and has been divided into two subdivisions : (i.) the PHYLACTOLAEMATA, which inhabit fresh water; and (ii.) the GYMNOLAEMATA, which are almost invariably marine.

The Phylactolaemata are further distinguished from the Gymnolaemata by the presence of an epistome and the shape of their lophophore, which is that of a horse-shoe, and by the formation of statoblasts. The structure of a member of this subdivision has been illustrated by the description of *Plumatella* ; in this genus the body-cavities of the various polypides are in communication, though some of the polypides are partially separated by an imperfect septum. In *Lophopus* and *Cristatella* the coelom in each polypide is in free and open communication with that of all the others. The last-mentioned genus forms colonies, which may attain the length of over two inches. The colony is oval in outline, and the polypides project from its upper convex surface; the lower surface is flat, and on this the whole colony creeps slowly along on submerged stems or stones. This mode of progression of the colony is one of the very few instances of any co-ordination of function which exists between the various individuals which compose a Polyzoan colony. The Phylactolaemata are all hermaphrodite.

The Gymnolaemata have a circular lophophore, and are devoid of an epistome. With the exception of a few genera, they are marine. In *Paludicella* the funicular tissue of the various individuals communicates by means of certain perforations known as rosette plates, but in the marine forms the zooecia are more independent. The ectocyst may be calcareous, horny, or gelatinous, and the various zooecia may be aggregated together in an almost infinite variety of ways. From time to time the polypide dies down, the tentacular crown and alimentary canal degenerating and forming what is known as the " brown body," which is coloured by the concretions which have accumulated within the wall of the stomach. This brown body lies in the zooecium until the endocyst produces a bud, and then it may become included in the alimentary canal of the bud or young polypide ; here the nutritive matter which it may contain is doubtless absorbed, and the undigestible matter passes out of the intestine of the young polypide.

In some of the Gymnolaemata there is a pore by means of which the sea-water can be admitted into the coelom when it is desirable to expand the polypide. This pore may be guarded by a circlet of bristles, which tend to prevent the entrance of grains of sand or other foreign bodies. In certain individuals of some species of *Alcyonidium*, etc., there is a ciliated canal which leads from the coelom to the exterior ; this is known as the " inter-tentacular organ," it probably serves as an exit for the generative cells, at any rate spermatozoa have been observed to leave the body by this channel. In most species, however, the generative products escape only by the dying down of the polypide.

The Gymnolaemata are divided into three classes, according to the character of their zooecia and the nature of their cell mouths when the polypides are retracted.

(i.) The CYCLOSTOMATA.—*These have tubular zooecia, always calcareous. The cell mouth is circular, and with no apparatus for closing it. Many of them are found fossilised, probably because their calcareous skeleton is easily preserved.* Crisia, Hornera, etc.

(ii.) The CTENOSTOMATA.—*The zooecia are soft, and their apertures guarded by a folded frill.* Bowerbankia, one of this sub-order, has a muscular gizzard armed with teeth, situated between the oesophagus and the digestive stomach. *Paludicella,* one of the few freshwater members of the Gymnolaemata, belongs here, and *Alcyonidium.* There is some reason to suppose that the Phylactolaemata are derived from this group of the Gymnolaemata.

(iii.) The CHEILOSTOMATA.—*This is the largest subdivision, and its members are clearly characterised by the possession of a lid or operculum, which closes the mouth of the zooecium when the polypide is retracted. The zooecia are calcareous.*

This subdivision exhibits a considerable degree of poly-morphism. Some of the individuals of the colonies are modified to form structures known as *avicularia,* resembling in shape a parrot's beak. The two halves are constantly opening and closing, and by their action in catching small worms, etc., they probably serve as defensive organs, as well as assist in keeping the colony clean. The smaller beak is believed to be a modified operculum, whilst the larger corresponds with a much modified zooecium. These structures exist in very various degrees of perfection, those of the genus *Bugula* being amongst the most specialised.

The *vibracula* are long stiff processes which move up and down, and are possibly tactile in function, they are believed to be homologous with the lower beak of an avicularium. In one genus they move in unison, and thus the colony exhibits some degree of co-ordination.

B. ENTOPROCTA.

This group contains but few genera. The mouth and the anus are both surrounded by the lophophore, which is circular. The tentacles can be bent over the mouth, but the anterior end of the body cannot be retracted into the posterior half. The coelom is almost completely obliterated. A pair of nephridia are present.

This group is chiefly founded on the structure of two comparatively well-known marine genera, *Loxosoma* and *Pedicellina*. *Loxosoma* is unique amongst Polyzoa, inasmuch as it is not colonial; like most Polyzoa, it increases by budding, but the buds separate from the parent organism. Both the genera are stalked, in *Pedicellina* the stalk arises from a creeping stolon, and the calyces or the bodies of the individuals often drop off, and are replaced by the regeneration of new ones at the end of the stalks, a process apparently analogous to the formation of brown bodies in ectoproctous forms. In *Loxosoma* the stalk, which carries the calyx, is at least in the young condition provided at its lower end with a foot-gland, by means of which it is usually attached to some marine animal. The paired nephridia of *Loxosoma* consist of ciliated intra-cellular ducts piercing a few large cells, and probably each beginning with a flame cell. The ducts open to the exterior between the ganglion and the oesophagus.

Most species of the Entoprocta are dioecious, and the generative organs have special ducts. Besides the sexual reproduction, they multiply also by the formation of buds. This process amongst the Polyzoa is remarkable, inasmuch as the endoderm is not represented in the tissues of the bud; in most other animals which reproduce by budding, all the three embryonic layers occur in the bud.

The affinities of the Polyzoa are somewhat obscure. The developement of the Entoprocta and their adult structures point to the origin of the group from some ancestor which is represented in the ontogeny of the Mollusca, Chaetopoda, and armed Gephyrea by the Trochosphere larva. On the other hand, the Phylactolaemata have a certain marked resemblance to the unarmed Gephyrea, and if this resemblance be not homoplastic, we must regard the Gymnolaemata, and still more the Entoprocta, as degenerate forms.

MOLLUSCA
{
LIPOCEPHALA—LAMELLIBRANCHIATA—*Anodonta, Arca, Mytilus, Solen, Pecten.*

GLOSSOPHORA
{
GASTEROPODA
{
Isopleura—*Chiton, Neomenia, Chaetoderma.*

Anisopleura
{
Streptoneura (Prosobranchiata)
{
Zygobranchiata—*Haliotis, Fissurella, Patella.*
Azygobranchiata
{
Reptantia—*Buccinum, Paludina.*
Natantia (Heteropoda)—*Carinaria, Pterotrachea.*

Euthyneura
{
Pulmonata—*Helix, Onchidium.*
Opisthobranchiata
{
Palliata—*Bulla, Aplysia, Pleurobranchus.*
Non-Palliata—*Doris, Eolis, Phyllirhoe.*
Pteropoda.
}

SCAPHOPODA—*Dentalium.*

CEPHALOPODA
{
Tetrabranchiata—*Nautilus.*
Dibranchiata
{
Decapoda—*Sepia, Loligo, Spirula.*
Octopoda—*Octopus, Argonauta.*
}

CHAPTER XIV

MOLLUSCA

CHARACTERISTICS.—*Unsegmented Coelomata, with a primitive bilateral symmetry. Their body is soft, and is dorsally produced into a fold, the mantle, which usually secretes a shell. The ventral part of the body forms, as a rule, a muscular process, the foot, which may be modified in various ways, but whose function is usually to assist in locomotion. Respiration is typically carried on by a pair of vascular processes, which project from the body-wall, and are termed the ctenidia. Near the base of these organs is a modified patch of epithelium, whose function is olfactory, and this has been termed the osphradium. The portion of the body-cavity in which the heart lies, the pericardium, communicates directly with the exterior by means of the nephridia. The heart is systemic, and the circulation partly lacunar. The nervous system typically consists of a pair of cerebral ganglia in the head, a pair of pedal ganglia in the foot, and a pair of pleural ganglia in the body. The last pair are united by a long commissure, the visceral nerve cord, which may become twisted. The sense organs comprise the osphradia, otocysts in connection with the pedal ganglia, tactile tentacles on the head, and in many cases eyes. The developement includes a characteristic larva, the Veliger.*

The phylum Mollusca includes a large number of animals which exhibit the greatest variety of structure and habit. The majority of them are marine, some inhabit fresh water, and many are terrestrial. The group includes the class Cephalopoda, the members of which are the largest, and at the same time the most ferocious of invertebrates. Some members of the

phylum are pelagic, and consist of the most transparent and
delicate tissues, others are sessile, being fixed either by cords
secreted by a gland in the foot (*Mytilus*) or by the surface of
the shell (*Ostrea*), whilst, again, others bore long funnel-shaped
passages in the rocks or in submerged pieces of wood, etc.

The very various animals which compose this phylum
may be separated into two main divisions, according as to
whether they retain a well-marked prostomium or not.
Those which have lost a definite cephalic region have probably
done so in correlation with a sessile, inactive life.   They form
the division **Lipocephala.**   The other division comprises those
Mollusca which possess a well-developed head, associated with
a toothed lingual ribbon, capable of a biting or rasping action,
borne on a cushion and moved by certain muscles, the whole
apparatus constituting the odontophore.   This organ has given
a name to the division, the **Glossophora.**

Division I. **LIPOCEPHALA.**

Characteristics.—*Mollusca   with   rudimentary   prostomium,
no odontophore, and no eyes.   Either sessile, or with very
feeble powers of locomotion.*

This division contains but one class, the Lamellibranchiata.

Class **Lamellibranchiata.**

Lipocephala which have retained the primitive molluscan
bilateral symmetry.   The body is laterally compressed, and
the mantle is bilobed, each lobe secreting one valve of the
bivalved shell.   The two valves, right and left, are united by
a dorsal elastic ligament.   The *ctenidia* or gills are largely developed, and by the currents their cilia create, assist in bringing
food to the mouth.   The foot is usually plough-shaped, and
contains part of the viscera.   It may be used for boring in
sand or rock, more rarely for crawling.   The pericardium, part
of the coelom, is in communication with the exterior by means
of a pair of nephridia.   The generative glands are simple, and
have no accessory organs connected with them.

In the common freshwater mussel, *Anodonta cygnea*, the
shells are equivalve.   Each valve is composed of three layers:

(i.) the periostracum, or outermost layer—this is thin and horny, and not calcified, and is formed by the thickened free edge of the mantle; (ii.) the prismatic or middle layer, consisting of closely-packed calcareous polygonal prisms—this is also deposited by the edge of the mantle; (iii.) the nacreous or mother-of-pearl layer, which lines the inside of the shell—it is composed of laminae of calcareous matter, and is deposited by the whole of the surface of the mantle and body in contact with the shell. It is this last layer which, when deposited in concentric layers round foreign particles, such as grains of sand, etc., produces pearls.

The shells of some Lamellibranchs are not equivalve, *e.g.* the oyster, *Ostrea*, which is attached to rocks by means of its larger valve. In *Pholas* there are additional calcareous plates inserted dorsally between the two valves; and in *Teredo*, the mollusc which does so much damage by boring into wood, the valves fail to completely cover the body, which secretes a calcareous lining to the tube in which it lives.

The valves of the shell are kept in apposition by adductor

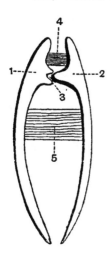

FIG. 116.—Section through *Anodonta*, to show mechanism of opening and of closing the valves. After Lankester— Zoological Articles reprinted from the *Encyclopædia Britannica*.

1 Right valve of shell.

Left valve of shell.

Hinge.

4 Elastic ligament.

5. Adductor muscles.

muscles. These may be two in number, an anterior and a posterior, or the posterior may alone persist (*Monomyaria*).

The edge of the mantle is thickened, and in some genera it bears tentacles and eyes. Posteriorly it is notched in such a way as to form two apertures, which remain open when the edges of the remainder of the mantle are in contact. These openings form the dorsal and ventral siphons. In some Lamellibranchs, *e.g.* *Mactra*, *Cyclas*, etc., these notches, by the fusion of their edges, are converted into tubes, which in some genera attain a length of several inches. The ventral siphon serves to admit fresh water, bringing with it oxygen and food, and the dorsal siphon gives exit to a stream of water which carries away the waste products and generative cells.

The foot is not developed in the OSTREIDAE, and is small in *Mytilus*, the marine mussel. In the cockle, *Cardium*, and in *Trigonia*, it can be suddenly bent, and by this means the animal jumps along. In *Solen* the foot is suddenly retracted, and in this way water is violently forced out of the siphons, and the animal is propelled forwards. *Pecten* flies through the water, with its dorsal surface downward, by the flapping of the valves of its shell. The foot often bears a special gland, which secretes a number of horny filaments known as the *byssus*, which serve to anchor the animal to the ground. This structure is well seen in *Arca* and in *Mytilus*.

The mouth, which is median, and ventral to the anterior adductor muscle when the latter is present, lies in a groove formed by the anterior and posterior labial palps. These are ciliated structures, which resemble to some extent the gills, and doubtless serve to convey minute organisms to the mouth as food. The alimentary canal is ciliated. The stomach gives off a caecum, which in many genera lodges a crystalline style. The function of the style is obscure, but it appears to consist of an albuminoid material. The intestine is coiled, and leads to a straight rectum, around which the ventricle of the heart is often folded. A fold of the intestine, or typhlosole, increases its surface. A paired gland, the so-called liver, pours its secretion into the stomach.

Two auricles return the arterialised blood to the ventricle, which in *Arca* is double ; the ventricle gives off an anterior and posterior aorta, which distributes the blood all over the body. The blood from the mantle is in *Anodonta* returned directly to the auricles ; the rest of the blood is collected into a vena cava in the floor of the pericardium, and is thence sent through the nephridia to the gills and returned to the auricles. The circulation is partly lacunar, the blood being contained in irregular splits in the tissues and not in distinct vessels. The blood contains amoeboid corpuscles, and is usually colourless; two species, however, *Solen legumen* and *Arca Noe*, contain haemoglobin in their corpuscles.

The gills consist primitively of an axis, which is fused to the body for the greater part of its course; this contains an efferent and an afferent blood-vessel. The axis gives off two series of filaments, which hang down parallel to one another, thus forming two lamellae. The filaments of both series may be bent up, forming V-shaped structures, those of the outer series having their free ends external and next to the mantle, whilst those of the inner series have their free ends internal and next to the foot, so that each series forms a gill with an outer and an inner lamella. In *Mytilus* and some others the outer and inner limbs of each filament are connected by certain pieces of tissue termed interlamellar concrescences. Neighbouring filaments are kept parallel to one another by an arrangement unique in the animal kingdom. Each filament bears certain patches of ciliated cells, and the cilia of two opposite patches are interlocked, in the same way as a couple of brushes when put together. In more complex genera these ciliary junctions are replaced by interfilamentous concrescences, and in *Anodonta* the interlamellar and interfilamentous concrescences are developed to such an extent as to leave but narrow passages through which the water circulates. The free ends of the filaments of the outer lamella of the external gill, and of the inner lamella of the internal gill, very frequently fuse with the contiguous organs, the mantle, or the foot.

Between the lamellae of each gill a certain space is developed which is more or less continuous with that of the other gills. This *epibranchial* space often serves to lodge the developing ova, it communicates with the dorsal siphon, through which the waste products leave the animal.

Each gill filament contains a blood-vessel, and it is often stiffened by two rods of a chitinous material. Its outer epithelium bears cilia, which serve to create a current of water, which enters the pallial chamber by the ventral siphon.

The nephridia of Lamellibranchs are usually known as the Organs of Bojanus. There is a single pair, and each consists of a glandular or secretory portion which has an opening

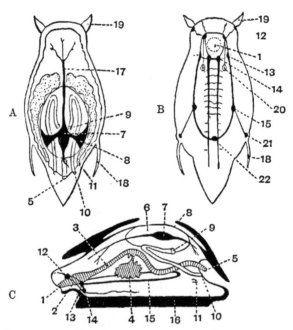

Fig. 117.—Diagrams of a Schematic Mollusc. After Lankester—Zoological Articles reprinted from the *Encyclopædia Britannica.*

A. Dorsal view, showing the heart, pericardium, generative organs, and nephridia.
B. Ventral view, showing the nervous system.
C. Lateral view.
    1. Mouth.
    2. Nerve ring.
    3. Oesophagus.
    4. Liver.
    5. Anus.
    6. Pericardium.
    7. Heart. Ventricle.
8 in A. Auricle.
8 in C. Wall of pericardium.

9. Internal opening of nephridium.
10. External opening of nephridium.
11. Opening of genital gland.
12. Cerebral ganglion.
13. Pleural ganglion.
14. Pedal ganglion.
15. Visceral ganglion.
16. Foot.
17. Anterior aorta.
18. Gill.
19. Tentacle.
20. Auditory vesicle.
21. Olfactory ganglion.
22. Abdominal ganglion developed on the visceral loop.

into the pericardium, and of a ureter which opens to the exterior in the neighbourhood of the orifice of the generative glands. In the oyster the kidney is much less compact, and its secretory part is scattered through the body, even reaching the mantle.

The nerve ganglia are usually rendered conspicuous by their bright orange colour. The cerebral ganglia, which lie one on each side of the mouth, probably represent the cerebral and pleural ganglia of other molluscs; they are united both with the pedal ganglia in the foot and with the olfactory (parieto-splanchnic) situated on the ventral face of the posterior adductor muscle. A pair of auditory vesicles, lined with ciliated cells and containing a single otolith, are usually present close to the pedal ganglia, and are innervated by a nerve from the cerebro-pedal commissure, which probably comes from the cerebral ganglia. Tactile papillae or tentacles are common round the edge of the mantle. In some cases the tentacles have been modified and form eyes, which attain a great degree of complexity. In *Pecten, Spondylus*, etc., these eyes have a remarkable resemblance to the vertebrate type of eye, inasmuch as the optic nerve passes in front of the retina, and the retinal elements are thus turned away from the light. The epithelium in the neighbourhood of the olfactory ganglion is modified to form an organ of smell, by means of which the quality of the water flowing in through the ventral siphon may be tested.

The Lamellibranchs with few exceptions are dioecious. The generative organs are branched glands usually situated in the foot, though in *Mytilus* they occur in the mantle. The generative cells are formed in the caecal processes of the gland, and they leave the body by a right and left simple duct which is continuous with the walls of the gland, and in some cases opens into the duct of the kidney (*Spondylus, Lima, and Pecten*).

Division II. GLOSSOPHORA.

CHARACTERISTICS.—*Mollusca with a prostomium more or less developed and a buccal cavity armed with a rasping tongue, the radula, which together with its accessory parts constitutes the odontophore.*

The Glossophora comprise three classes:

> (i.) Gasteropoda.
> (ii.) Scaphopoda.
> (iii.) Cephalopoda.

Class **Gasteropoda.**

CHARACTERISTICS.—*The* Gasteropoda *have a foot which is in the main a crawling organ, it is simple, median, and has a broad flat surface. The foot is often divisible into three divisions, termed the pro-, meso-, and meta-podium.*
The Gasteropoda are divided into two sub-classes:

i. **Gasteropoda Isopleura.**

CHARACTERISTICS.—*The Gasteropoda Isopleura retain the primitive bilateral symmetry of the group. The body is elongated, the mouth anterior and the anus posterior. The viscera generally are paired and bilaterally symmetrical.*
This subclass includes six genera, which are distributed amongst three orders. The best-known genus is *Chiton*, in which the shell is metamerically divided into eight parts. The gills or ctenidia are also metamerically repeated to the number of sixteen or more, and at the base of each is a patch of olfactory epithelium, the *osphradium*. *Chiton*, like *Chaetoderma*, another member of the subclass, is dioecious, in the former the generative cells escape by special ducts. In *Neomenia* and *Chaetoderma*, however, they leave the body by means of the nephridia.
The nerve ganglia are not very markedly developed, but ganglion cells are scattered all along the well-defined nerve-trunks. In some *Chitons*, eyes furnished with a lens, retina, cornea, etc., have been described as existing on the shell plates.

ii. **Gasteropoda Anisopleura.**

CHARACTERISTICS.—*In the members of this subdivision the head and the foot have retained a bilateral symmetry, but the visceral hump with its included organs has undergone a twist which has resulted in rotating the anus and posterior part of the viscera to the right. The angle through which the anus has been twisted varies in different groups; it may be as much as 180°, and in this case the anus lies above the middle line of the neck. One of the ctenidia is usually atrophied, and one of the nephridia specialised as a generative duct. The mantle developes a shell, which often increases the asymmetry of the animal by being spirally coiled. This shell is often capacious enough to shelter the whole animal, thus forming a kind of house into which the animal can withdraw. The foot is usually provided with a mucous gland.*

The Gasteropoda Anisopleura are subdivided into two branches: Streptoneura (Prosobranchiata) and Euthyneura.

Branch A. STREPTONEURA.

CHARACTERISTICS.—*The first branch comprises those Molluscs in which the torsion has proceeded to such an extent that the anus has become anterior, and the right gill and osphradium have crossed anteriorly to the left, whilst the left gill and osphradium have come round posteriorly to the right. As a consequence one limb of the visceral nerve loop is pulled over the other and a figure of 8 is produced.*

This branch includes two orders: Zygobranchiata and Azygobranchiata.

Order 1. ZYGOBRANCHIATA.

CHARACTERISTICS.—*The first order includes all those forms in which, although the torsion is complete, so as to bring the anus near to the anterior median line, the atrophy of the ctenidium of one side has not usually taken place, and the generative cells leave the body through one of the nephridia which still retains its renal function. No accessory generative organs occur, and the visceral hump is coextensive with the foot.*

This group includes three families. The best-known genera are *Haliotis,* known as the Ormer in the Channel Islands, where it forms an article of diet, *Fissurella,* and *Patella* or the limpet.

Patella vulgata, the common limpet, is protected by a conical dome-shaped shell, whose average length is about two inches. The edges of the shell are not quite smooth, and their inequalities generally correspond closely with those of the rock upon which the animal is situated. Limpets are usually found between the tide-marks, and if they wander away from the spot on which they usually occur when covered by the

tide, they are stated always to return to it before the water
has again receded.

The visceral hump is covered by the conical shell.   The
body-wall at its edge is produced into a fold, the mantle.   The
ventral surface of the animal consists of the muscular oval

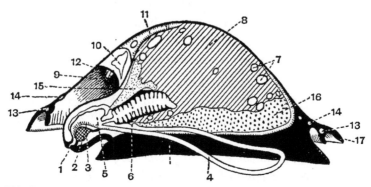

FIG. 118.   Diagram of a vertical median section of a Limpet, *Patella vulgata.*   After
    Lankester—Zoological Articles reprinted from the *Encyclopædia Britannica.*

1. Mouth.   10. Heart in pericardium.
2. Odontophore.   11. Nephridium.
3. Radula.   12. Opening of larger nephridium.
4. Radula sac.   13. Branchial efferent vessel (vein).
5. Buccal cavity.   14. Branchial afferent vessel (artery).
6. Laminated stomach.   15. Salivary gland.
7. Intestine cut across.   16. Generative gland.
8. Liver.   17. Edge of the mantle.
9. Anus.

foot, between which and the mantle a groove exists which
lodges the gills.   The foot is attached to the shell by a
circular muscle which is incomplete anteriorly.

A distinct head exists, and this carries a pair of tentacles
with a pair of eyes which appear as black specks near the base
of the tentacles.   Above the head the groove between the foot
and the mantle deepens into a large pallial cavity.   Into this, not
in the median line, but slightly to the left of it, the anus opens,
and on each side of the anus lie the openings of the renal
organs (Fig. 119).   On the neck are also situated two small
bodies representing the ctenidia, which are fully developed in
the allied forms *Haliotis* and *Fissurella*; in connection with
these a patch of olfactory epithelium, the osphradium, has also
been discovered.   The function of these ctenidia, the original
breathing organs, has been assumed by certain folds of the
mantle forming the actual gills.

The mouth leads into the cavity of the buccal mass, this
is partially obliterated by the developement of a large ventral

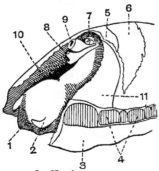

FIG. 119.—Side view of anterior end of Limpet,
Patella vulgata. Part of the mantle is cut
away to show the contents of the pallial chamber. After Lankester—Zoological Articles reprinted from the *Encyclopædia Britannica.*

5. Small nephridium.
6. Large nephridium.
7. External opening of small nephridium.
8. External opening of large nephridium.
9. Anus.
10. Rudimentary ctenidium.
11. Pericardium.

1. Head.
2. Tentacle.
3. Mantle skirt.
4. Muscles forming root of foot, and
   adherent to the shell.

mass, over which the tooth-ribbon or radula works. The
ventral mass contains certain cartilaginous nodules, and is very

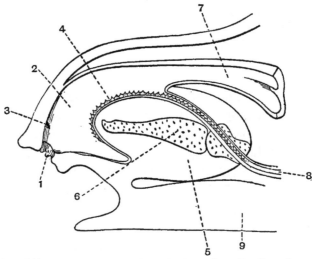

FIG. 120.—Vertical section through the neck of *Patella vulgata.*
After R. J. Harvey Gibson.

1. Mouth.
2. Buccal cavity.
3. Palatal tooth.
4. Radula.
5. Odontophore.

6. Anterior cartilage.
7. Oesophagus.
8. Radula sac.
9. Foot.

muscular. The radula, which runs over it, is continued into a
sac, from the blind end of which it grows (Fig. 120). The radula

and its sac attain an extraordinary length in the limpet, often twice the length of the animal; they lie between the viscera and the muscular foot.   Two pairs of yellowish salivary glands pour their secretion into the buccal cavity by two ducts on each side, and many mucous glands also open into it.

The oesophagus leads from the buccal mass into the stomach. The walls of this organ are much folded, it receives by numerous ducts the secretion of the liver.   The latter is a large organ occupying the greater portion of the space in the visceral hump, and enveloping a considerable proportion of the

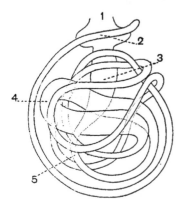

FIG.   121. — Semi - diagrammatic   view   of intestinal coils of *Patella vulgata*.   After R. J. Harvey Gibson.

1. Buccal mass.
2. Rectum.
3. Crop.
4. Stomach.
5. Coils of intestine.

alimentary tract.   The intestine which passes from the true stomach makes a loop and then again enlarges into a second stomach, which is bent upon itself; after this the intestine coils in a most complicated way and ultimately ends in a rectum, which opens to the exterior on the anal papilla in the anterior pallial chamber (Fig. 119).   The whole alimentary canal is lined throughout by ciliated cells ; the extent of its convolutions are shown by the fact that it may attain a length of over fourteen inches, in an animal a little more than an inch long.

The heart consists of a single auricle and ventricle, in the allied forms *Haliotis* and *Fissurella* two auricles exist.   It is enclosed in a pericardium situated in the posterior angle of the anterior pallial chamber.   A large vessel, the branchial vein, runs on each side round the edge of the mantle at the base of the gills ; anteriorly the two vessels unite and empty into the auricle.   A muscular valve separates the auricle from the ventricle.   The cavity of the latter is much broken up by strands of muscle fibres; it opens into the left and right aortae, the former supplying the circular muscle. Both aortae soon terminate in lacunar spaces, from whence the blood presumably passes to the gills. The blood is colourless, and contains amoeboid corpuscles.

The nephridia are paired, but the right is much larger than the left. They open to the exterior by small renal papillae, situated one on each side of the anal prominence, and also, according to some observers, internally by two minute pores into the pericardium. The existence of the reno-pericardial openings has recently been denied, both in *Patella* and in *Fissurella*. *Haliotis* and *Trochus* possess a left reno-pericardial duct only. The left kidney lies between the rectum and the pericardial chamber. The right kidney, which is aborted in other **Anisopleura,** occupies a large space in the visceral hump. In part of its course it is closely applied to the generative organs, and when the ova and spermatozoa are ripe they are stated to burst into the lumen of the kidney, and so to leave the body through the renal papilla on the right of the anus. The lumen of the kidney is much broken up by ridges which project into it from its walls. The ridges are covered with glandular epithelium, which is partly ciliated; in the substance of the ridges numerous blood-vessels ramify.

The nervous system is very complex, it comprises several pairs of ganglia, the most important of which are the cerebral, the pedal, and the pleural. The cerebral ganglia are situated at the base of the tentacles, they give off nerves to the eyes and to the tentacles. The two ganglia are united by a commissure above the pharynx; they also give off a commissure on each side which passes to an anterior superior buccal ganglion. From each buccal ganglion two commissures arise, one uniting it with the similar ganglion of the other side, the other passing posteriorly to a posterior superior buccal ganglion, which is in its turn united with the similar one on the other side. Thus the buccal nervous apparatus consists of a square of commissures with a ganglion at each angle. The cerebral ganglia are connected with one another by a commissure which runs underneath the buccal mass; this bears two small ganglia—the inferior buccal ganglia.

From the posterior end of each cerebral ganglion two commissures pass backward, the outer one passing into the pleural
ganglion, the inner to the pedal.   Each pleural ganglion is
connected with the pedal of its own side, and the two pedals
are united by a pedal commissure.   The pleural gives off two
stout nerves.   The outer of these soon splits, one branch going
to the gills and mantle, the other to the circular muscle which
attaches the animal to its shell.   The second nerve given
off from the pleural forms the origin of the visceral loop.   This
is a nervous loop, which, starting at each end from the pleural
ganglion, forms a figure of 8 twist.   In its course it gives off
a nerve to an olfactory ganglion lying at the base of each of
the rudimentary ctenidia.   The olfactory nerve going to the left
ctenidium arises from the loop near to the right pleural ganglion, that to the right ctenidium arises near the left ganglion.
This twisting of the visceral loop is characteristic of the
Streptoneura.

The pedal ganglia give off each two large nerves, which
supply the muscles of the foot.

The tentacles have a tactile function; at their base the
eyes are situated—they consist of a pair of pits sunk in the
surrounding tissue.   The epidermal cells lining these pits
become modified and deeply pigmented, and are connected by
an optic nerve with the cerebral ganglia.   A similar simple
eye, consisting of an open pit lined with pigmented cells, is
found in *Nautilus*.

Limpets are dioecious; the position of the generative glands
is similar in the two sexes (Fig. 118) between the muscular
foot and the digestive organs, rather near the posterior end.
Like other members of the Zygobranchiata, the generative glands
possess no ducts, and their contents leave the body through the
right nephridium.

Order 2. AZYGOBRANCHIATA.

CHARACTERISTICS.—*The Azygobranchiata, which constitute the
other order of the Streptoneura, have lost their original left
ctenidium and osphradium.   The right nephridium does not
exist as such, but is most probably represented by the generative*

duct. The left nephridium, which was in Patella smaller than the right, and the left ctenidium and osphradium, are retained. The anus lies on the right of the neck of the animal. The Azygobranchiata are dioecious, and the males are often furnished with a large penis.

The great majority of Azygobranchiata are adapted for creeping at the bottom of the sea (**Reptantia**), and for this purpose have a large muscular foot with a flat sole. They are often spoken of as sea-snails; the shell which encloses and protects the visceral hump is usually coiled. A large gland is not unfrequently found lying alongside the rectum; in the genera *Murex* and *Purpura* its secretion turns purple when exposed to the light, and it was in ancient times used as a dye. The posterior surface of the foot in some species bears a calcareous or horny plate, the operculum, which serves to close the mouth of the shell when the animal is retracted. The foot itself shows a tendency to break up into three portions: anteriorly the propodium, in the middle the mesopodium, and posteriorly the metapodium, which bears the operculum.

This group includes, amongst many others, *Buccinum*, the whelk, and *Littorina*, the periwinkle. *Paludina*, the river-snail, and *Valvata* are freshwater members of the group, and so is the terrestrial *Cyclostoma*, which has no gill, the mantle chamber having become a respiratory organ. It lives in damp places. *Entoconcha mirabilis* lives parasitically in the body of *Synapta digitata*; this is exceptional, as parasitism is very rare amongst the Mollusca.

A few Azygobranchiata have become modified in connection with a pelagic mode of life. These form the section **Natantia**, also known as the **Heteropoda**. As in other pelagic organisms, their tissues have become wonderfully transparent, and of a gelatinous consistency. The foot has become a swimming organ. Its division into pro-, meso-, and meta-podium is well marked. *Atlanta* has a coiled transparent shell, into which the body and foot may be withdrawn, and the metapodium carries an operculum. In *Carinaria* the foot is by far the largest part of the animal, and the relatively small visceral hump is covered by a small hyaline shell. In

Pterotrachea the visceral hump is still more reduced, and is devoid of a shell.

The sense organs and nervous system are unusually well developed in this subdivision, the otocysts being usually closely attached to the cerebral ganglia. The lingual ribbon has moveable lateral teeth, which divaricate when the tongue

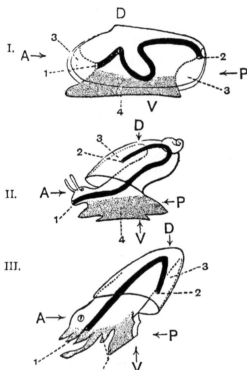

Fig. 122.—Diagrams of a series of Molluscs to show the form of the foot and its regions and the relations of the visceral hump to the antero-posterior and dorso-ventral axes. After Lankester—Zoological Articles reprinted from the *Encyclopædia Britannica.*

I. A Lamellibranch.
II. An Anisopleurous Gasteropod.
III. A Cephalopod.
A. Anterior surface.
P. Posterior surface.
D. Dorsal surface.
V. Ventral surface.
1. Mouth.
2. Anus.
3. Mantle cavity.
4. Foot.

is protruded, but come together when that organ is withdrawn, by means of these the Natantia, which are carnivorous, catch other pelagic organisms to feed upon.

Von Erlanger has shown in the developement of *Paludina* that the glandular portion of the nephridium arises as an evagination of the pericardium, and the mouth of the evagination remains as the reno-pericardial pore, when such an orifice exists. The genital gland arises as a proliferation from the pericardial wall at a spot where a rudimentary evagination has been formed which represents the missing nephridium. The duct of the nephridium and the duct of the gonad are symmetrically situated and homologous involutions of the epiblast.

Thus, in this animal at least, and probably in all Molluscs, the cavity of the generative organs and of the glandular part of the kidney is part of the coelom, and the generative cells are formed from coelomic epithelial cells.

Branch B. EUTHYNEURA.

CHARACTERISTICS.—*The second great branch into which the Gasteropoda Anisopleura is divided differs from the Strepto-neura in the fact that the torsion of the visceral hump has proceeded to a less extent, and that consequently the peri-cardium is placed obliquely in front of the mantle cavity and gill. For the same reason also the one side of the visceral loop has not become pulled over the other by the forward rotation of the gill and osphradium, and the loop is untwisted. The condition of the gill and kidneys is as in Azygobranchiata. The right auricle is wanting. With one exception, no operculum is found. All Euthyneura are hermaphrodite, a sharp distinction from the Streptoneura. The shell is often but slightly calcified, and it and the mantle tend to disappear.*

The Euthyneura comprise two orders : (i.) the Opistho-branchiata and (ii.) the Pulmonata.

The Opisthobranchiata are characterised, as their name implies, by the position of the ctenidium behind the heart. The branchial vein opens into the auricle, which is situated behind the ventricle. The shell is often absent, and when it is developed it is frequently covered by a fold of the mantle. Both the mantle and the ctenidium may, like the shell, be wanting, the function of the ctenidium may be carried on by processes of the body-wall.

In some Opisthobranchs the margin of the foot is produced into two broad flaps, which may stand out at right angles to the axis of the foot, or may in some cases be folded up against the sides of the body.

One of the most characteristic features of this order is the presence of processes of the body-wall termed *cerata* or dorsal papillae. The cerata vary very much in size, number, and arrangement. Processes of the liver are prolonged into some of them, these have been termed *hepatocerata* to distingusih them from those which are simply diverticula of the body-wall. These cerata often coexist with well-developed gills, and their minute structure does not point to any very definite respiratory function; it seems not improbable that their varied shape and colour may be in some cases protective and in others conspicuous and warning. In those Opisthobranchs which possess hepatocerata, such as *Doto* and *Eolis,* the liver is not a compact gland, but consists of a number of diverticula given off from the alimentary canal, each diverticulum passing into one of the cerata, and being large enough for food particles to pass into it and be there digested. In *Eolis* the liver diverticula do not end blindly, but are stated to open into an invagination of the ectoderm termed the *cnidophorous sac.* This opening is guarded by a minute sphincter muscle. The cnidophorous sac, which in its turn opens to the exterior, is lined by a number of large cells, cnidoblasts, which are crowded with nematocysts or thread-cells; these recall the stinging organs of the Coelenterata. The everted threads of these nematocysts are armed with both large and small spines.

In *Aplysia, Bulla,* and *Pleurobranchus* the original ctenidium has been retained, but is situated behind the heart, in *Doris* and its allies the ctenidium appears in a modified form as a circlet of feathered processes which surround the median dorsal anus; in *Eolis, Tethys,* etc., it has completely disappeared. In *Aplysia* there is a large gland, whose secretion is said to be poisonous, which opens just below the osphradium near the anterior end of the ctenidium; and numerous small cutaneous glands open on the under surface of the mantle, in *Aplysia* these produce a purple secretion.

The Opisthobranchs are, like the Pulmonata, hermaphrodite; and the generative organs consist of a hermaphrodite gland or ovo-testis. Some of the cells of this gland form ova, whilst others divide up and become spermatozoa. From the ovo-testis a hermaphrodite duct leads to an albuminiparous gland, in the substance of which the duct coils. Just where the duct leaves the gland it gives off a small diverticulum, the vesicula seminalis. The duct then passes on to the external opening, but just before it reaches that it receives the duct of a spherical spermatheca. When eggs leave the body by means of this external opening situated in front of the ctenidium, they are enveloped in a gelatinous coating deposited by the albuminiparous gland, and are impregnated by the spermatozoa of another individual, which has been stored up in the sperma- theca. The spermatozoa when they pass out of the genital pore pass along the spermatic groove, which runs along the right side of the head and terminates in a muscular penis, by the aid of which the spermatic fluid is introduced into the genital pore of another individual and finds its way to the spermatheca.

The Opisthobranchiata show a considerable tendency to lose some of their organs; this process of degeneration is carried farthest in one of the sub-orders, the Haplomorpha, in which neither mantle-fold, ctenidia, nor cerata are found. *Phyllirhoe* has little but its odontophore to show its relationship with the Mollusca. It is a flattened, Planarian-like, transparent pelagic organism, about half an inch long. Its skin contains numerous unicellular glands, which are said to secrete a phosphorescent slime. The ovo-testis in this animal is double. A small Hydromedusa, *Mnestra*, is frequently found attached by the aboral surface of its body to these animals. In *Rhodope* the degeneration has gone still farther, and the odontophore has disappeared.

Recent research has shown that the group of animals which formerly ranked as a class, the Pteropoda, are really allied to the Opisthobranch Gasteropoda. The Pteropoda are subdivided into two orders, the Thecosomata and the Gymnosomata. The former consists of three recent families, and is allied to the Bulloidea; the latter contains five families, and is allied to the Aplysioidea, members of the Opisthobranchiata. The failure to recognise the correct affinities of these animals was to some extent due to their external symmetry, but this is a secondary feature which does not affect their internal organs. The animals com- posing this class are all carnivorous and pelagic. In correspondence with their mode of life, they are delicate organisms with transparent tissues, those amongst them pro- vided with a shell—the Thecosomata—having a hyaline one. In both orders the margin of the foot is prolonged and modified into fins, which in the Thecosomata embrace the head. The anterior portion of the digestive tract is evaginable in *Clio*, and bears a number of unicellular glands opening upon cones. The secretion of these glands is adhesive, and they serve as organs for the capture of prey. *Pneumodermon* has a pair of appendages which bear suckers similar in structure to those found in the Cephalopoda.

Order 2. PULMONATA.

CHARACTERISTICS.—*This second order of the Euthyneura has possibly become modified from a palliate Opisthobranch ancestor, in correspondence with the altered conditions of life involved in the change from an aquatic to a terrestrial habitat. The ctenidium has atrophied, and the edge of the mantle has fused with the body-wall, leaving only one small opening which leads from the pallial chamber to the exterior—this is the respiratory pore. The walls of the pallial chamber are very vascular, and they function as lungs. An operculum is never present.*

This order includes the land-snails and slugs. As a rule, its members live on land, and they all breathe air when adult, even those like the pond-snail *Limnaea*, which lives in water, but whose mantle chamber contains air. The mantle cavity of the young freshwater Pulmonates is stated at first to contain water, this is afterwards replaced by air. The great extent to which the blood-vessels and capillaries are developed in the Pulmonata is possibly connected with this habit.

No true operculum is found in any Pulmonate, but many secrete a temporary lid to their shell when they withdraw into it for the winter. This temporary operculum is called a *hibernaculum*. In the slugs the shell may be small and even quite atrophied.

Onchidium and *Peronia* are members of a family which inhabit the sea-shore and brackish marshes. *Onchidium* is remarkable for possessing, in addition to the normal pair of cephalic eyes, a number of dorsal eyes scattered over the integument. These latter, like the eyes in the mantle of *Pecten* and *Spondylus*, are not constructed on the usual invertebrate type, but, as in the eyes of vertebrates, the nerve pierces the layer of sensory cells, and is distributed to that side of them which is nearest the lens.   In Onchidium, as in Pecten, etc., the eyes are probably modifications of certain simple tentacles which are borne on the mantle.

CLASS **Scaphopoda.**

CHARACTERISTICS.—*Elongated cylindrical Glossophora.   The mantle has extended on to the ventral surface, and has fused in the middle ventral line.   It has secreted around it a cylindrical shell.   The foot can be protruded through the anterior opening of the shell.   No heart is present.   The visceral loop of the nervous system is untwisted.   Dioecious, the generative products escape through the right nephridium.*

This class includes but three genera, of which the best-known is *Dentalium.*   The cylindrical shape of its body may be correlated with a burrowing habit of life, and it is interesting to note that a similar shape is found amongst

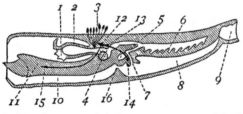

FIG. 123.—Diagram of the anatomy of *Dentalium.* Lateral view of organs, showing as though by transparency. After Lankester—Zoological Articles reprinted from the *Encyclopædia Britannica.*

1. Mouth surrounded by pinnate tentacles.
2. Oral process.
3. The ctenidial filaments.
4. Odontophore.
5. Oesophagus.
6. Left lobe of liver.
7. Anus.
8. Peri-anal part of mantle cavity.
9. Appendix of mantle skirt separated by a valve from 8.
10. Peri-oral part of mantle cavity.
11. Foot.
12. Cerebral ganglion.
13. Pleural ganglion.
14. Olfactory ganglion placed on visceral loop as in the Lamellibranchiata, according to Spengel.
15. Pedal ganglion.
16. Left nephridium.

the sand-boring Lamellibranchiata, such as *Solen.*   The mantle is ensheathed by a shell, which is at first incomplete ventrally.   The shell resembles a truncated elephant's tusk, it is open at both ends, the larger opening is the anterior one, and the concave surface is dorsal.

The foot is protrusible through the anterior opening of

the shell, it terminates in a trifid lobe. Dorsal to it is the oral cone or head, at the end of which the mouth opens, surrounded by short pinnate tentacles (Fig. 123). A buccal mass and a radula are present; two liver lobes symmetrically placed open into the stomach, from which the intestine passes to open by the anus in the ventral middle line.

A right and left nephridium are present, and open to the exterior on either side of the anus. There is no heart, but the coelom contains a colourless blood. At the base of the oral cone a number of ctenidial filaments have their origin. These are capable of very considerable extension. The nervous system consists of a pair of cerebral ganglia, close to which lie the pleural ganglia. Long commissures connect the cerebral with the pedal ganglia. The visceral commissure is also long, and bears the olfactory ganglia, situated in front of the anus, in the same position as in Lamellibranchs. The generative gland is alike in both sexes; it is situated dorsally, and its products make their exit through the right nephridium.

CLASS **Cephalopoda.**

CHARACTERISTICS.—*Bilaterally symmetrical Glossophora. The visceral hump is elongated, not twisted; the sub-pallial chamber is chiefly developed posteriorly, and contains the gills, anus, and excretory pores. The shell may be external or internal, in a few cases it is absent. The foot has grown round the head, and is broken up into the characteristic arms of the Cephalopoda, provided with suckers. Part of the foot forms a funnel-like siphon, which guides the water as it is expelled from the pallial cavity. The vascular system is well developed, in addition to the central heart consisting of a ventricle and two auricles, an accessory branchial heart exists at the base of each gill in all but Nautilus. Powerful beak-like horny or calcareous jaws guard the mouth, and the radula is well developed. Chromatophores are present in the integument. The Cephalopoda are dioecious.*

The Cephalopoda are divided into two orders: (i.) The TETRABRANCHIATA or the TENTACULIFERA and (ii.) the DI-BRANCHIATA or the ACETABULIFERA.

(i.) CHARACTERISTICS.—*The Tetrabranchiata are characterised as follows. The siphon is not complete ; the edges overlap one another, but are not fused ; the lobes of the circumoral foot carry tentacles, not suckers. There are two pairs of ctenidia and two pairs of nephridia. The coelom opens straight on to the exterior, and not into the nephridia. There are two oviducts and two vasa deferentia, but the left is in both cases rudimentary. The shell is large, external, and chambered. No ink sac, salivary glands, or branchial hearts exist.*

This order contains very many extinct forms, but only one living genus—*Nautilus.*

(ii.) CHARACTERISTICS.—*The Dibranchiata have but one pair of ctenidia and one pair of nephridia. The edges of the siphon have fused so as to form a complete funnel. The arms or processes of the foot which surround the head bear cup-like suckers. Branchial hearts exist at the base of the gills. The coelom communicates with the exterior through the nephridia, and not directly ; the oviducts may be paired or single ; the vas deferens, with one exception, is single. An ink sac and salivary glands exist. The sense organs are highly developed.*

The Dibranchiata comprise two sub-orders : the **Decapoda** and the **Octopoda.**

a. CHARACTERISTICS.—*The Decapoda have ten arms, two of which are very long, and differ in appearance from the others. The suckers are stalked, and provided with a horny ring. The body is elongated, and bears lateral fins. The shell is enclosed by an upgrowth of the mantle, and is therefore internal.*

β. CHARACTERISTICS.—*The Octopoda have eight similar arms, bearing sessile suckers, which are not strengthened by a horny ring. The body is short and globular. The oviducts are paired. There is no shell in or on the visceral hump.*

The Cuttle-fish or *Sepia officinalis* is common in most seas, and in the spring, when it approaches the shore to deposit its eggs amongst the rocks, it is easily caught. In considering the anatomy of this form, it is important to orientate the animal correctly; with this view it should be placed mouth downwards, then its foot will be ventral, its visceral hump dorsal, and its mantle cavity posterior. For the sake of convenience it is, however, better to twist the animal through a right angle, and describe it as it swims, with its foot and mouth anterior, the visceral hump posterior, the mantle cavity ventral,

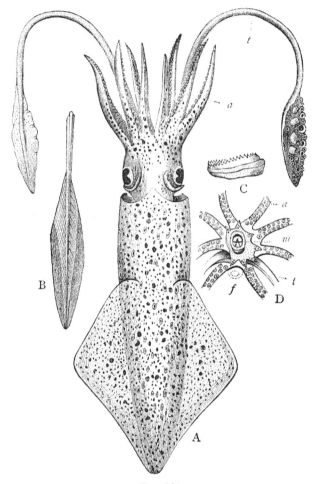

FIG. 124.

A. *Loligo vulgaris.*
　　a. Arms.
　　t. Tentacles.
C. Side view of one of the suckers, show-
　　ing the horny hooks surrounding the
　　margin.

B. Pen of the same, reduced in size.
D. View of the head from in front, show-
　　ing the arms (*a*), the tentacles (*t*), the
　　mouth (*m*), and the funnel (*f*).

and the shell or cuttle-bone, which may be felt through the skin, dorsal.　The most dorsally-placed pair of arms, really the most anterior, are termed the first pair.

Sepia, being a Decapod, has five pairs of arms, of these, the fourth pair are unlike the others.　They are much longer, and can be withdrawn into pouches at their base; they do not bear their suckers scattered uniformly over their inner surface, but the suckers are all aggregated in a swollen pad at the free end of the arm (Fig. 126). The suckers are very remarkable organs, they are cup-like structures whose rim is strengthened by a toothed horny ring. A retractor muscle can deepen the cavity of the cup, so that when the edge of the cup is applied to any object and the muscle contracts, the sucker adheres to the object by the pressure of the surrounding medium. In

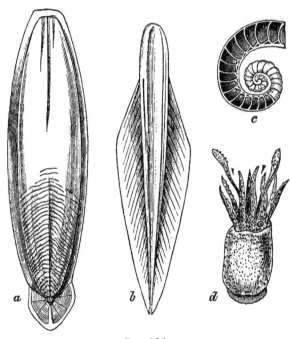

FIG. 125.

a. Internal skeleton of *Sepia ornata*.   Rang.
b. Internal skeleton of *Histioteuthis Bonelliana*.   D'Orb.
c. Internal skeleton of *Spirula fragilis*.   Lamarck.
d. Animal of *Spirula Peronii*.

Sepia, as in other Decapods, the suckers are stalked. In the male, the fifth arm on the left side has lost some of its suckers, and this is termed the hectocotylised arm, *vide* p. 222.

That portion of the foot which is modified to form the sucker-bearing arms is homologous with the fore-foot or propodium of other Mollusca (Fig. 122). The mid-foot or mesopodium has become converted into the siphon, a funnel-like structure open at both ends. The posterior aperture communicates with the mantle cavity. When the edges of the mantle are in close apposition to the body-wall, to aid which a pair of cartilaginous nodules exist on the mantle edge, which fit into corresponding depressions on the outside of the funnel, and the muscles of the mantle contract, the water in the mantle cavity is forced violently through the siphon. The result of this is, the *Sepia* darts backwards. In the lumen of the siphon is a small valve which only allows the water to pass one way; this possibly represents the hind-foot or metapodium.

The cuttle-bone or shell of *Sepia* is entirely internal. It lies along the dorsal surface in a sac formed by the concrescence of certain folds of the mantle in this region (Fig. 126). It consists of a posterior horny portion, which is continued forward by a series of calcareous plates deposited by the inner wall of the sac; between these plates air is found. This air must be secreted by the surrounding tissues, it probably assists the *Sepia* to balance itself.

If the mantle be divided and the mantle chamber exposed, the anus will be seen situated in the middle line near the posterior end of the siphon. Close to it the duct of the ink sac opens. A little way behind the anus, and on each side of it, is situated a nephridial opening, and at about the same level on the left side is the aperture of the genital duct. The large ctenidia lie one on either side, and in the female *Sepia* a pair of large nidamental glands are to be seen through the body-wall.

The mouth is surrounded by a circular lip and guarded by two strong horny beaks resembling those of a parrot, except for the fact that the under beak is the larger and more prominent (Fig. 126). The buccal mass is large and muscular, it contains a well-developed radula. The rows of teeth consist of five central conical teeth and one hook-shaped tooth on each side. The oesophagus, a narrow tube, passes from the buccal mass straight to the posterior end of the body, and opens into a thick-walled stomach. At this point the intestine bends forward ventrally, and gives off a curved caecum of considerable size. From this the rectum passes along the ventral body-wall to the anus. A pair of small salivary glands lie on either side of the oesophagus just behind the cartilaginous skeleton of the head; their two ducts unite into a single channel, which opens into the buccal mass in the neighbourhood of the radula.

The liver is of considerable size; it is bilobed, and the anterior end of each half reaches as far forward as the salivary glands; the duct of each lobe arises about its middle, and runs back parallel with the oesophagus, to open into that part

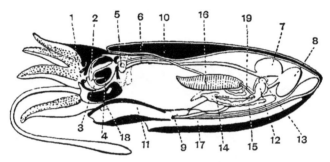

Fig. 126.—Diagram representing a vertical, approximately median antero-posterior section of *Sepia officinalis*. From a drawing by A. G. Bourne. After Lankester—Zoological Articles reprinted from *Encyclopædia Britannica*.

1. Mouth.
2. Upper beak.
3. Lower beak.
4. Odontophore.
5. Nerve ring.
6. Oesophagus.
7. Crop.
8. Gizzard.
9. Anus.
10. Cuttle bone enclosed by a growth of the mantle.
11. Lumen of siphon.
12. Branchial heart.
13. Appendage of branchial heart.
14. Viscero-pericardial aperture.
15. Renal glandular mass.
16. Left ctenidium.
17. Subpallial chamber.
18. Valve in siphon.
19. Afferent branchial vessel.

of the intestine just behind the stomach which gives off the caecum. The bile ducts are enwrapped by the unpaired diverticulum of the nephridia, and this, where in contact with them, developes spongy excretory tissue, the so-called pancreatic caeca. The ink bag lies to the right of the stomach, its duct runs parallel to the rectum, and opens to the exterior. Its secretion forms a pigment named after the animal, and by the expulsion of this the surrounding water is made inky, and thus serves as a screen to cover the escape of the squid.

The coelom of *Sepia* contains the stomach, the heart with

its chief vessels, the branchial hearts, and in its posterior half the genital gland. It is partially divided into two by an in-

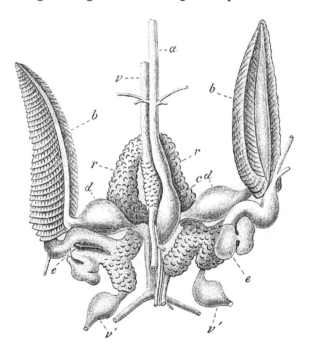

FIG. 127. — Central organs of the circulation, gills, and renal organs of *Sepia officinalis*. After John Hunter.

v' v'. Visceral veins.
   a. Aorta.
   v. Vena cava.
   c. Ventricle.
   d. Auricles.
   e. Branchial hearts.
   b. Branchiae.
   r. Renal organs.

complete septum. It opens into the nephridia by a minute pore on each side.

The heart consists of a ventricle with two auricles opening into it. The ventricle is continued in front into an anterior aorta, which gives off vessels to the mantle and liver and runs forward to the head and arms, and behind into the posterior aorta, which supplies the generative organs and the fins, etc. The blood passes largely by capillaries but partly by lacunar spaces, into the veins. Of these, the largest is the anterior vena cava, which has a ventral position and splits into two branchial veins. These latter are beset with diverticula of the nephridia, and they receive numerous veins which bring back the blood from the generative organs, the ink sac, etc. Just before they enter the branchial hearts they are joined by veins from the mantle and posterior end of the visceral hump. The branchial hearts are pulsating muscular enlargements at the base of the ctenidia, whose contractions force the blood through the gill. A mass of excretory tissue—the pericardial gland—

is developed on the wall of the coelom, beneath each branchial heart. The blood, after passing through the ctenidia, is returned to the auricles.

The blood contains colourless corpuscles: in the oxidised condition it is nearly colourless, but when venous it is bluish. The colour is due to haemocyanin, a substance containing oopper, which is diffused through the serum.

The ctenidia are organs of considerable size. In the normal state their long axis is parallel with the longitudinal axis of the body, and they are attached throughout their whole length to the body-wall. The axis bears a double row of plate-like lamellae, which decrease towards the anterior end, thus giving a pyramidal shape to the organ. An afferent vein from the branchial heart traverses the axis and gives off branches to the lamellae; here the blood is aerated, and is then returned by an efferent vein which runs parallel and close to the former, this leads to the auricle.

The nephridia are paired, right and left, but they are connected by two transverse portions, an anterior and a posterior. The former of these transverse communications gives off a diverticulum which stretches, as the unpaired nephridial sac, back as far as the genital gland. The ventral wall of the nephridia is smooth, but the dorsal wall, which is wrapped round the branchial veins, and into which numerous veinlets run, is spongy and glandular. At the anterior end of each kidney is a short ureter which opens to the exterior at the side of the anus. Near the inner end of the ureter there is a rosette-shaped opening covered with ciliated epithelium, which leads into the coelom.

Cephalopods, which are the largest and most ferocious of all the invertebrates, have developed an internal cartilaginous skeleton, a very unusual arrangement outside the phylum Vertebrata. The cartilage consists of a structureless matrix, through which numerous cells are scattered; the cells give off branching processes which permeate the substance in all direc-tions. Nodules of this cartilage exist in processes of the mantle edge, and fit into corresponding depressions on the edge of the siphon when the mantle is closed, and also along the base of the lateral fins, but the most considerable developement

of cartilage is in the head.   Here there is a cephalic cartilage
of complicated form, which is pierced by the oesophagus.   It
ensheaths the chief nerve ganglia, the ear is embedded in
it, and it forms two recesses which lodge the eyes.   Another
portion affords some support to the bases of the arms, and there
is also a flat piece situated in the neck known as the nuchal
plate.

The chief nerve ganglia are aggregated round the oesophagus, close behind the buccal mass, and are embedded in the

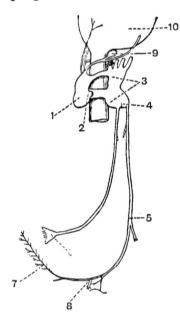

Fig. 128.—Lateral view of the nervous centres
and nerves of the right side of *Octopus vulgaris*.   From a drawing by A. G. Bourne.
After Lankester — Zoological Articles reprinted from the *Encyclopædia Britannica*.

1. Cerebral ganglion.
2. The optic nerve.
3. Pedal ganglion giving nerves to arms.
4. Pleuro-visceral ganglion.
5. Right visceral nerve.
6. Right stellate ganglion of the mantle, con-
    nected by a nerve to the pleural portion
    of 4.
7. Branchial branch of 5.
8. Olfactory branch of 5.
9. Buccal ganglion.
10. Buccal mass.

cartilaginous skeleton.   The cerebral ganglion on the dorsal
side of the oesophagus gives off a pair of nerves which end in
the superior buccal ganglion, from which a pair pass to
the inferior buccal ganglion, both lying on the surface of the
buccal mass.   In connection with these ganglia there is a well-
developed stomatogastric system.   Laterally each cerebral
ganglion is continued into two very stout optic nerves ;
these expand into the optic ganglia, situated at the back
of the eye.

The cerebral ganglion gives off two circum-oesophageal commissures, which pass down to the nervous mass on the ventral
surface of the oesophagus.   This mass is composed of three
ganglia very much fused together.   Anteriorly lie the pedal

ganglia, which give off ten large nerves, one to each arm; they also supply the siphon. The auditory nerves also arise from the pedal ganglia, although their fibres may be traced to the cerebral. The pedal ganglia are partially marked off from the fused pleural and visceral by the presence of a small foramen through which a blood-vessel passes. From the pleural portion of this compound nerve centre a stout nerve passes to the stellate ganglion, situated at the angle between the mantle and the head. It can be seen shining through the integument when the mantle cavity is exposed. From this ganglion nerves radiate to the muscles of the mantle. The visceral half of the fused ganglion gives off a pair of stout visceral nerves, which unite to form a loop. These visceral nerves supply the generative organs, the kidneys, and other viscera, and each sends a stout branch to a ctenidium.

The eyes of *Sepia* are of great complexity. They have a striking but superficial resemblance to the Vertebrate eye, and fundamental differences exist between these two types of visual organs. Anteriorly the eye is covered by a transparent cornea, which in *Sepia* is closed. The cornea is protected by certain folds of skin, which can cover it in by the contraction of a sphincter muscle, and there is also a horizontal lower eyelid. Within the cornea is the anterior chamber of the eye, into which the folds of the iris project; they partially cover the lens, which consists of an outer and an inner part separated by a membrane. The lens is supported by the ciliary body, which with the lens occupies the anterior half of the retinal chamber. The retina, which completes the wall of this chamber, is two-layered, and the nerves which pass to it from the optic ganglion enter the retina posteriorly.

The auditory apparatus consists of two otocysts sunk in the cephalic cartilage. Their cavities have an irregular shape, and are lined by an epithelium, which is ciliated in places, they contain an endolymph, in which a single spherical otolith floats.

An olfactory function is attributed to two small invaginations of the skin, situated one just behind each eye. The sacs open to the exterior by a small slit-like aperture; they are lined by a ciliated columnar epithelium, amongst which are certain special sense cells, each provided with a single sense hair. This organ is supplied by a nerve which arises from a special ganglion situated near the base of the optic ganglion. No osphradia corresponding with those of other Molluscs have yet been described in *Sepia*.

Certain large cells crowded with pigment, situated in the subepidermal connective tissue, play an important part in the life of a Cephalopod. Attached to these cells, which are called *chromatophores*, are a number of radiating muscle fibres; when these contract, the cavity of the cell enlarges, and the contained colour becomes diffuse; the chromatophores contract by their own elasticity, and when contracted the colour is concentrated. The whole system is under nervous control, and the colour of the animal may change with startling rapidity In the *Sepia* and other members of the group this faculty is used as a protection, the colour of the animal tending to assimilate itself to that of the surrounding rocks or sand. In addition to the chromatophores, the subepidermal tissues contain other modified connective tissue cells known as *iridocysts*; these cells are so modified as to produce iridescent colours by the diffraction of light.

Sepia is a dioecious animal which lays eggs. The male is usually somewhat smaller than the female, and its arms are relatively longer; the fifth arm on the left side is hectocotylised, that is, it is modified in connection with the process of depositing the spermatozoa. It is thickened at its base, and almost devoid of suckers. The testis lies at the extreme end of the visceral hump, in a capsule—part of the coelom—into which opens a more or less coiled vas deferens, the walls of which are much folded, and provided with numerous glandular diverticula. Whilst passing down this vas deferens the spermatozoa are divided up into packets, and the glandular walls secrete around each packet a cuticular spermatophore. Finally, the sperm duct opens into a large receptacle known as Needham's sac, in which the spermatophores are stored up; they pass to the exterior by the genital pore situated to the left of the anus, and they are deposited in the hectocotylised arm, and are possibly introduced by it into the mantle cavity of the female at the time of oviposition.

The spermatophores are complex structures about 2 cm. long, they have a receptacle in which the minute spermatozoa are stored up, and a long tightly-coiled spiral thread, the expansion of which explodes the capsule, and the spermatozoa rush out.

In the female the ovary occupies the same position as the testis in the male; the cavity of both these generative glands communicates with the pericardial portion of the coelom, though partly shut off from it by a septum. A cushion projects into the lumen of the ovary, which bears ova in various stages of developement; from the ovary the oviduct, which is ciliated, passes to its external opening to the left of the anus. Accessory glands are present; of these the most important are a large pair of nidamental glands, which deposit the substance of the egg capsules; in *Sepia* there is a second smaller pair of

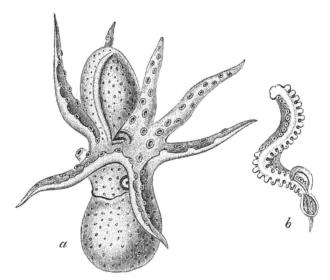

FIG. 129.—*a*, Male of *Argonauta argo*, with the hectocotylised arm still contained in its enveloping cyst, four times enlarged (after H. Muller). *b*, Hectocotylus of *Tremoctopus violaceus* (after Kölliker).

nidamental glands, as well as the large ones. The egg capsule is prolonged into a stalk, by means of which the eggs are kept together, and the collection of eggs somewhat resembles a cluster of grapes.

In *Sepia* one of the arms in the male is slightly modified, and probably assists in the deposition of the spermatozoa, but in certain Octopods this modification is carried much further. In *Argonauta argo* the third arm on the left, and in *Ocythoe tuberculata* the same arm on the right, becomes detached from the male, and is placed in the mantle cavity of the female. It carries a small sac charged with spermatophores, and was at one time looked upon as a parasite, and the name *Hectocotylus* was given it. The male, after losing its arm, always reproduces it again. In the female *Argonauta* the eggs are carried about in the shell; this is the only member of the Octopoda which has a shell, and it does not correspond with the shell of other Cephalopods, but is formed from the expanded ends of the two dorsal arms.

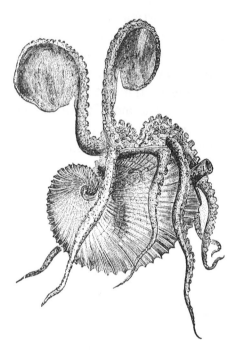

FIG. 130. — *Argonauta argo*, the Paper Nautilus, female. The animal is represented in its shell, but the webbed dorsal arms are separated from the shell which they ordinarily embrace.

In other Dibranchiata the shell varies from the external coiled chambered shell of *Spirula* to the horny pen of *Loligo*. Even in *Spirula* (Fig. 125) the shell is partially surrounded by folds of the mantle, and in other forms the folds have fused together so that the shell comes to lie in a closed sac. In the Tetrabranchiata, *Nautilus* and the extinct Ammonitidae, the shell is external, and chambered. The animal lies in the last-formed chamber, and closely fits it. The chambers are separated from one another by septa, and the whole is traversed by a membranous tube, the *siphuncle*, which is a continuation of the integument of the animal. The chambers are full of a gas probably secreted by the dorsal integument, and they doubtless serve as a float.

In *Nautilus* the fore-foot is broken up into certain flattened lobes, which differ in their arrangement in the two sexes. The lobes bear at their edges cylindrical tentacles, which can be retracted into muscular sheaths. Probably the tentacles correspond to the suckers in the Dibranchiata. In some species of cuttle-fish the suckers are replaced by hooks, or both may coexist; the arms in the Octopoda are usually connected by a fold of skin forming a web, which is no doubt of use in swimming. The arms of *Architeuthis*, a gigantic form, sometimes attain the length of 40 feet, and the total length of the body and arms may measure 60 feet.

The beak which guards the mouth is calcareous in *Nautilus*, and horny in other Cephalopods. The possession by *Nautilus* of two pairs of auricles which open into the single ventricle is correlated with the two pairs of ctenidia. There are in this same animal two pairs of nephridia; this repetition of parts is almost unknown in Mollusca, the only other case being the gills and shells of *Chiton*, and it is therefore particularly interesting.

The chief nerve ganglia in *Nautilus* are band-like, and hardly to be distinguished from the commissures which connect them. The nerves to the mantle are numerous, and are not aggregated into one stout cord as in the Dibranchiata. The same animal is provided with a pair of osphradia, situated at the base of the anterior ctenidia; these organs have not yet been discovered in other Cephalopods. The eye of *Nautilus* is one of the most remarkable organs found in the order. It has the shape of a kettledrum; the tense membrane, which is external, being pierced at its centre by a minute hole, which leads into a dark chamber lined by the retina. The latter is bathed by sea water, which enters through the minute pore. The mechanism by which images must be formed on the retina resembles that of a pin-hole camera.

CHAPTER XV

ECHINODERMATA

Echinodermata
{
Asteroidea—*Asterias, Solaster, Brisinga.*
Ophiuroidea—*Astrophyton, Ophiopholis.*
Crinoidea—*Comatula, Pentacrinus.*
Echinoidea .
{
Regulares—*Echinus, Toxopneustes.*
Clypeastroidea—*Clypeaster, Rotula.*
Spatangoidea—*Spatangus, Brissus.*
}
Holothuroidea
{
Actinopoda—*Holothuria, Cucumaria, Dcima.*
Paractinopoda—*Synapta, Chirodota.*
}
}

CHARACTERISTICS.—*Animals with a primitive bilateral symmetry, which is in the adult replaced by a more or less regular radial symmetry, usually pentamerous. The skin is hardened by calcareous deposits, which may take the form of scattered spicules or of plates which build up an almost complete shell, but in all cases they are mesodermic structures. A well-developed coelom is present, and part of it becomes cut off from the rest to form the water-vascular system, which is both locomotor and respiratory in function. The five radial vessels of this system correspond with five areas, the "ambulacra"; the angles between them form the "interambulacra." The alimentary canal usually opens to the exterior at both ends, but an anus may be absent. The sexes are usually distinct, and developement is nearly always associated with a metamorphosis. They are exclusively marine.*

The Echinodermata are divided into five classes:

I. ASTEROIDEA.

II. OPHIUROIDEA.

III. CRINOIDEA.

IV. ECHINOIDEA.

V. HOLOTHUROIDEA.

CLASS I. ASTEROIDEA (Starfishes).

CHARACTERISTICS.—*Echinodermata whose body is flattened dorso-ventrally, and is produced into arms or rays, which are usually five or more in number. These arms are longitudinally grooved on the ventral surface, and the tube-feet lie in this groove. The madreporic plate is dorsal and interradial in position. The alimentary canal sends caecal diverticula into the arms. The generative organs are interradial in position at the base of the arms. Pedicellariae usually present.*

Asterias rubens is one of the commonest of starfishes, and is constantly left stranded on our shores by the retreating tide. Its body consists of a central disk, from which five arms or radii project. The surface on which it habitually rests or moves, and on which the mouth opens, may be termed the ventral, the upper and more convex, where the anus is situated, may be called the dorsal.

From the mouth five grooves radiate along the arms, these are the *ambulacral* grooves, and they lodge the tube-feet; between each two grooves, and consequently interradial in position, are five sets of oral spines, which project over the mouth and perhaps assist in feeding. If the tube-feet be removed from each ray, it will be seen that the ambulacral groove is formed of two rows of ambulacral plates, situated right and left of the middle line of the radius (Fig. 131). Each right plate is so placed as to form an angle, open ventrally, with the corresponding left plate, and between the adjacent plates of each side certain pores exist which give exit to the tube-feet. The groove is covered in by the integument, and lodges two radial canals, of these the most ventral is divided by a vertical septum, and is called, for reasons mentioned below, the " peri-hæmal " space. The dorsal canal is the radial trunk of the water-vascular system. At the outer end of the ambulacral plate a series of *adambulacral* ossicles are situated, and these support three rows of moveable spines. Those spines which are nearest to the centre of the disk form the oral spines mentioned above ; these are borne by the first adambulacral ossicles, one set on each side of an inter-radius.

At the distal end of each arm the ambulacral plates end in

15

a single ossicle, which supports a terminal tentacle bearing a number of pits of pigmented cells, called collectively the eye-spot. Between this single ossicle and the other ambulacral

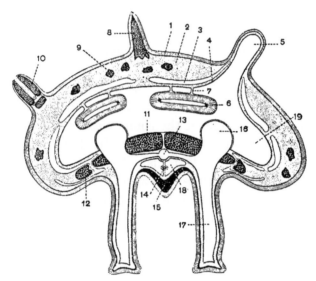

FIG. 131.—Diagram of a transverse section of the arm of a Starfish.

1. Epidermis.
2. Mesoderm.
3. Perihaemal space in the skin.
4. Peritoneal lining of body-cavity.
5. A branchia.
6. Paired caeca from intestine.
7. Mesentery supporting caeca.
8. Spine.
9. Ossicle in skin.
10. Pedicellaria.
11. Ambulacral ossicle.
12. Adambulacral ossicle.
13. Radial trunk of water-vascular system.
14. Radial trunk of blood vascular system of Ludwig.
15. Radial nerve connected with plexus under epidermis.
16. Ampulla of tube-foot.
17. Tube-foot.
18. Perihaemal space.
19. Coelom.

plates all the new plates appear. The tentacle at the tip of the arm, together with the eye-spot, is surrounded by a circlet of spines.

On the dorsal surface of both disk and arms numerous spines are scattered, and amongst them many pedicellariae (Fig. 131). These must be regarded as modified spines; they consist of a basilar plate and of two blades which snap against one another like the two limbs of a pair of forceps—in some of them the blades cross one another as they do in a pair of scissors. The function of these pedicellariae seems to be to catch hold of foreign bodies, and so keep parasites from settling on the skin or penetrating through the branchiae into the coelom.

On the dorsal surface of the disk, situated interradially, lies the madreporic plate, through which the water-vascular system communicates with the exterior. The two arms which lie right and left of this plate are termed the " bivium," and contrasted with the other three or " trivium "; in mapping out the various organs of the body these will be found to be convenient terms. The anus lies near the centre of the dorsal surface of the disk.

The skin is formed of (i.) an outer cylindrical epithelium with nerve fibrils at the base, (ii.) an intermediate connective tissue layer with some muscle fibres,—this is the matrix for the spines and plates,—and (iii.) an inner coelomic epithelium, which is ciliated; this last lines the true coelom or enterocoel, a spacious cavity containing the alimentary canal, the generative organs, etc. The coelom contains a fluid in which amoeboid corpuscles float.

The angle which the two series of ambulacral plates in each arm make with one another is floored in by the outer layer of the integument, the nerve plexus of which is thickened and forms the radial nerve (Fig. 131). The cavity thus formed is the radial perihaemal vessel or blood-vessel of French authors; it is divided into a right and a left portion by the presence of a median mesentery. This mesentery in this species, but not in others, has a certain amount of glandular tissue in it, which Ludwig describes as a blood-vessel.

On the dorsal surface of the starfish numerous delicate processes of the skin may be seen projecting above the general level of the body-wall. These thin-walled extensions of the integument are known as dermal branchiae; the coelomic fluid passes freely into them, and they doubtless serve as respiratory organs (Fig. 131). It has been recently shown that some of the amoeboid cells of the coelomic fluid (phagocytes), when they have eaten any particles which it is desirable should be ejected from the body, make their way to the walls of these dermal branchiae, and force a passage through them to the exterior, whence they are washed away.

Besides the enterocoelic ciliated body-cavity, there are a

number of vessels. They constitute the blood system according to French authors. The radial one has already been mentioned ; the five radials unite with a circum-oesophageal ring, which is stated to open into the body-cavity by five interradial pores. Inside this is another ring-shaped vessel, into which a large sinus surrounding the stone canal—the axial sinus—opens. Besides this there is an aboral pentagon which sends off interradially pairs of vessels which dilate and surround the genital organs.

The mouth is situated centrally on the ventral surface, surrounded by a ring of nervous matter. The mouth leads by a short oesophagus into a large stomach, the walls of which are folded in many sacculi. When the starfish attempts to devour young molluscs or shellfish which are too large to be taken in at the mouth, these sacculi are protruded and enclose the prey. They are retracted by special muscles. The walls of both the oesophagus and the stomach are ciliated, and the eversible portions contain many glands, the secretions of which possibly exercise a paralysing effect on the prey. The stomach is followed by a pentagonal pyloric portion with its angles situated radially. From each angle a short duct passes to the base of each arm, and here opens into two large hepatic caeca, which occupy a large portion of the space in each arm and extend to its tip. Each caecum is supported by two dorsal mesenteries. From the pyloric portion a short rectum passes to the anus, which is in the next interradius to that bearing the madreporic plate, and is almost central. The rectum gives off two short caeca, which lie in two neighbouring interradii —that between the left and central arm of the trivium, and between the left arm of the trivium and bivium.

The water-vascular system consists of a circumoral ring which gives off five radial vessels, one running along each arm, and a single interradial stone canal, which passes from the circumoral ring, and opens to the exterior at the madreporic plate, which is calcified.

The madreporic plate is marked externally by a number of radial grooves ; at the bottom of each of these is situated a row of pores ; these open into a series of tubules, which collect into an ampulla, and this in its turn opens into the lumen of the stone canal. The stone canal is lined by a ciliated epithelium, surrounded by calcified connective tissue, a ridge projects into its interior, and the free edge of the ridge may bifurcate, each half then folding back upon itself. The circumoral ring bears nine glandular bodies, composed of branching tubules lined with cubical cells, and opening into the ring. These bodies are known as Tiedemann's bodies. The stone canal opens in the position where the tenth of these bodies should be. It is possible that the corpuscles which float in the fluid of the water-vascular system are formed in these bodies.

The radial vessels which pass along the arms lie ventral to the ambulacral plates, between them and a transverse muscle which runs between each pair (Fig. 131). Opposite each tube-foot the radial vessel gives off a transverse branch. Each branch passes between the ambulacral ossicles, and opens into a vesicular expansion, the ampulla, situated in the coelom. From this another vessel passes to the tube-foot. The contraction of the ampulla forces fluid into the tube-foot, and so extends it. At the tip of the arm the radial tube ends in an unpaired terminal tentacle, at the base of which is a thickening beset with eyes. The tentacle has a very well-developed nervous layer.

The blood system described by German authors is founded on misinterpretation. They describe a radial vessel, an oral ring, and an aboral ring, and a connecting heart lying inside the corresponding organs described above. The radial and oral vessels are nothing but the thickened septa of the true vessels, the heart is a solid glandular organ, and the aboral vessel is the genital rhachis, partly degenerate. The rhachis is in connection with the so-called heart.

The nervous system is diffused all over the body, but better developed in some parts than in others. The epidermis contains numerous sense cells, prolonged at their bases into nerve fibrils; these are not very abundant on the dorsal surface, but along the ridge which lies between the tube-feet, and in a ring which surrounds the mouth, both sense cells and nerve fibres exist in great quantities. The triangular ridges which occupy the ventral surface of the arms unite in a ring round the mouth, and constitute the central nervous system

(Fig. 132). The outer cells of this ridge are mainly sense cells, and ganglion cells and nerve fibres occur at their bases. The nerve layer is also well developed on the tube-feet.

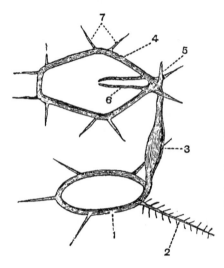

FIG. 132.—View of blood-vascular system of a Starfish as described by German writers. Modified from Ludwig.

1. Circumoral ring.
2. Radial vessel with branches to ampulla.
3. Heart.
4. Circumoral ring.
5. Dorsal end of heart passing into the skin.
6. Vessels to intestine.
7. Paired vessels passing to generative glands.

Asterias rubens is dioecious. The generative organs consist of five interradial pairs of glands, which are alike in both sexes, and when mature each extends into two neighbouring arms (Fig. 133). Except during the breeding season, the size of the glands is inconsiderable. Each gland opens to the exterior by a single duct, which terminates in a perforated plate situated dorsally and interradially. The various glands are connected together by a genital rhachis, and they are supplied by the above-mentioned genital vessels, which dilate to form a sinus round the glands. Fertilisation takes place externally.

The Asteroidea are mainly inhabitants of shallow water, though a considerable number of species from great depths have been described. The arms are usually five in number; one species of *Solaster* has, however, thirteen, and *Brisinga* has nine to twelve arms, which are more sharply marked off from the disk than is the case with other Asteroidea. The same genus is devoid of dermal branchiae, of eye-spots, and of ampullae at the base of the tube-feet.

The family ASTROPECTINIDAE is, with one exception, characterised by the anus being absent, and by the tube-feet being

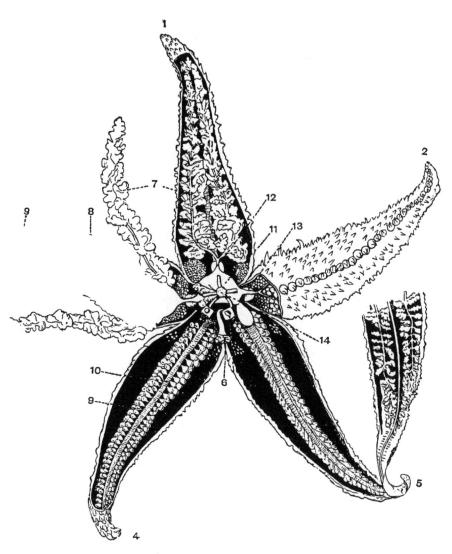

G. 133.—The common Starfish (*Asterias rubens*), dissected to show motor, digestive, and reproductive systems. After Rolleston and Jackson.

Central radius of trivium.
Right arm of trivium.
Left arm of trivium.
Left arm of bivium.
Right arm of bivium.
Madreporic plate and canal.
Arborescent "hepatic" caeca, two in each arm.
Generative glands.

9. Ampullae of tube-feet.
10. Ambulacral plates, inner surface.
11. Pyloric portion of stomach.
12. Duct leading from stomach to paired caeca.
13. Cardiac division of stomach bulging into arm.
14. Anus.

arranged in two rows on the ventral surface of each arm, and not in four, as appears to be the case in *Asterias*. Their tube-feet have pointed extremities, and not a sucking-disk.

Besides the ampullae on the radial vessels, additional

FIG. 134.—*Solaster papposus* (upper surface).

reservoirs for the water-vascular fluid usually occur on the circumoral ring; these are termed *Polian vesicles*, and are usually five or ten in number. It is doubtful whether the vesicles which occur near the right position in *Asterias rubens* are really Polian vesicles, that is, opening into the ring, or whether they are the first pair of ampullae of each radial vessel. In one species, *Cribella oculata*, some of the openings in the madreporic plate lead into that section of the body-cavity which surrounds the heart and stone canal, instead of into the latter canal.

The Asteroidea have great powers of regenerating lost parts. Arms broken off grow out again from the disk, and even the whole disk may be regenerated from a single separated arm.

CLASS II. OPHIUROIDEA (Brittle Stars).

CHARACTERISTICS.—*Echinodermata with a central disk bearing
long slender arms, into the cavity of which no part of the
alimentary canal is prolonged. There is no anus. The
madreporic plate is ventral, and usually is an oral plate.
There is no ambulacral groove, and the tube-feet are lateral in
position.*

This class is allied to the **Asteroidea,** and is sometimes
included with the latter in a single class. The Ophiuroids,

however, differ from the
Asteroids in the sharp distinction between disk and
arms, a condition approached
by *Brisinga*, in the absence
of any digestive diverticula
in the arms, in the ventral
position of the madreporic
plate, and in the almost
universal absence of pedicellariae. In the adult also
the ectoderm is absent except on the tube-feet.

The arms are long and
slender, in most cases they
are protected by four rows

FIG. 135.—*Ophiopholis bellis* (upper surface).

of plates, a ventral, a dorsal, and two lateral, the tube-feet
protrude between the ventral and lateral; they have no
ampullae. The nervous system has sunk under the skin,
and is protected by the ventral plates. Dorsal to it is the
radial blood-vessel, and dorsal to that the water-vascular
vessel. In a transverse section of the arm, the greater part
of the space is occupied by the ambulacral ossicles. Originally
paired, these have fused and become single; they are grooved
dorsally and ventrally. The dorsal groove lodges part of the
coelom, the ventral the above-mentioned vessels and nerve
cord.

The mouth is armed with certain modified ossicles; it is
central in position, and leads into a spacious stomach, which

is produced into five radial and five short interradial caeca. The walls of the stomach are lined by a ciliated epithelium,

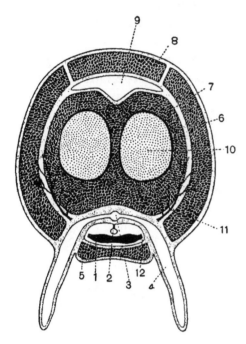

FIG. 136.—Diagram of a transverse section of an *Ophiuroid*.

1. Radial nerve, with lateral branches.
2. So-called radial blood-vessel.
3. Radial water-vascular trunk.
4. Tube-foot.
5. Ventral plate.
6. Lateral plate.
7. Ambulacral ossicles.
8. Dorsal plate.
9. Dorsal portion of coelom.
10. Muscles.
11. Lateral nerve.
12. Origin of lateral nerve.

and are supported by connective tissue strands, which traverse the coelom to the body-wall. There is no anus.

The water-vascular system consists of a circumoral ring, which bears four Polian vesicles; in the fifth interradius it gives off the ciliated stone canal, which is simple and un-calcified, this passes to the madreporic plate on the ventral surface. In *Astrophyton* there are five madreporic plates, one in each interradius, and five stone canals. The radial vessels which arise from the ring bear no ampullae, but give off branches which pass directly to the conical tube-feet. Cor-puscles tinged with haemoglobin occur in the water-vascular fluid of one species.

The true vascular system resembles that of Asterids. The aboral ring has, however, an undulatory curve, being ventral in the interradii. MacBride has recently proved that both the axial sinus and the aboral ring are involutions of the coelom. The so-called heart is nothing but a genital stolon, whence the genital rhachis grows out. The genital stolon in the earliest stage is a mere thickened ridge of peritoneum, so that here, as in other Coelomata, the generative cells are derived from the lining of the coelom.

The circumoral nerve ring, like the radial nerves, has lost its connection with the epidermis, and has sunk into the body.

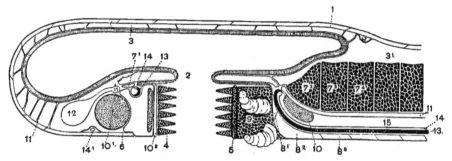

FIG. 137.—A diagrammatic vertical section of an *Ophiuroid*, after Ludwig. The circumoral systems of organs are seen to the left, cut across, their radial prolongations cut longitudinally, to the right.

1. Body-wall.
2. Mouth.
3. Body-cavity.
3[1]. Body-cavity of the arm.
4. Oral ossicles.
5. Torus angularis.
6. Oral plate.
7[1]. 1st ambulacral ossicle.
7[2], 7[3], 7[4]. 2nd to 4th ambulacral ossicle.
8[1], 8[2], 8[3]. 1st to 3rd ventral plate.
9. 1st oral foot.
10. Transverse muscle of the 2nd joint.
10[1]. External interradial muscle.
10[2]. Internal interradial muscle. (The line should point to the dotted tissue.)
11. Water-vascular system: to the left the circumoral ring, to the right the radial vessel.
12. Polian vesicle.
13. Nerve ring and radial nerve.
14. So-called blood-vessel.
14 (to the right). Genital rhachis enclosed in aboral sinus.
15. Radial perihaemal canal.

The radial nerves in the arms are frequently segmented, a ganglionated swelling occurring corresponding with each ossicle.

The generative organs consist of numerous caeca which open into a genital bursa. The bursae are ten in number, and lie one on each side of each arm; they open ventrally by a slit-like aperture at the base of each arm. A genital rhachis connects the generative organs, which are surrounded by a blood-sinus, as in Asterids. *Amphiura squamata* is hermaphrodite, and it is stated that when certain internal parasites, Orthonectidae, infest the coelom, it ceases to produce eggs, but produces a greater number of spermatozoa.

Some of the Ophiuroids give off a phosphorescent light from the back of their arms.

Ophiopholis bellis (Fig. 135) exists in great numbers in the northern European seas. Like many other members of the class, it is brilliantly coloured. The different specimens of the same species exhibit a surprising amount of variation both in their colour and markings.

Class III. CRINOIDEA (Sea Lilies).

CHARACTERISTICS.—*The dorsal or aboral surface usually prolonged into a jointed stalk, by which the animal is fixed. The calyx, consisting of the disk and arms, in some species breaks off from the stalk and leads a detached existence. The jointed arms bear lateral pinnules. The tube-feet take the form of tentacles arranged in groups on the disk, arms, and pinnules. No madreporic plates exist, but certain holes lead from the water-vascular system into a ramifying system of vessels, whence others open to the exterior.*

The skeleton of the Crinoids is composed of a number of ossicles with a very definite arrangement. The topmost segment of the jointed stalk is termed the centro-dorsal plate; in the Comatulidae, which lose their stalk when adult, this persists as the central aboral plate, and bears several whorls of cirrhi which have a root-like appearance. The stalked forms, such as *Pentacrinus caput Medusae*, also have numerous cirrhi, arranged in whorls on their stalks. From the centro-dorsal piece five radial plates radiate; these are continued by second and third radial ossicles, and the last of these bears two brachials (Fig. 141). These brachials form the first of a series which form the axis of each of the ten arms.

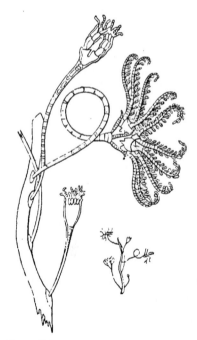

FIG. 138.—Pentacrinoid larval forms of *Comatula*. Natural size and magnified.

The growing point of the arm forks at short intervals, and one branch of the fork alternately, right and left, remains small and constitutes a pinnule, a method of branching which occurs in plants, and is termed by botanists scorpioid dichotomy.

The arms and their pinnules have a grooved ciliated ventral surface, at the disk the grooves of the two arms· of a pair unite, and the five grooves thus formed run to the mouth. The arms are flexible, and the free Crinoids swim through the sea by the graceful undulations of these processes.

In a transverse section of the arm the following parts may be distinguished: dorsally a large brachial ossicle which is traversed by an axial nerve, the contiguous ossicles being united and moved by a pair of muscles (Fig. 140). Ventral to the ossicle is the body-cavity broken up into four spaces which communicate at intervals. One of these is dorsal, one ventral, and two lateral, the ventral portion is traversed by the sterile generative rhachis. Below these coelomic spaces lies the radial water-vascular vessel which gives off alternating branches to the nonsuctorial tube-feet. At the side of the ambulacral groove some spherical bodies of unknown function are situated, these are termed *sacculi*, and consist of a membrane enclosing a large group of spherules.

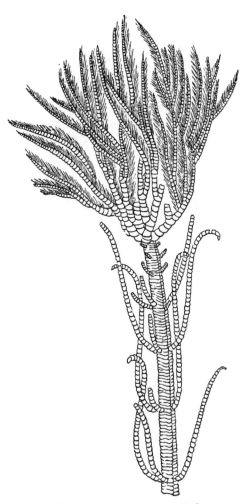

FIG. 139.—*Pentacrinus caput Medusae.*
After Guttard.

The pinnules resemble the arms, with the exception that the generative rhachis has become functional, producing either ova or spermatozoa.   The rhachis, both in the arms and in the

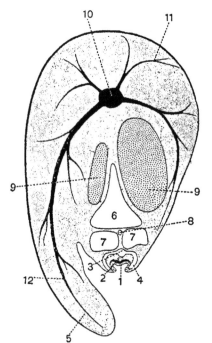

Fig. 140.—Transverse section of a Crinoid arm (partly diagrammatic).   After Milnes Marshall.

1. Ambulacral groove.
2. Ambulacral nerve.
3. Ambulacral water-vessel.
4. Tube-feet.
5. Pinnule.
6. Coeliac (dorsal) canal.
7. Subtentacular (lateral) canal.
8. Ventral canal : contains genital rhachis.
9. Muscles connecting the joints of arm.
10. Axial cord.
11. Its branches.
12. Branch to pinnule.

pinnules, is surrounded by a blood-plexus, and the whole is enclosed by the ventral division of the body-cavity, which is relatively much larger in the pinnules, corresponding with the enlargement of the rhachis.   The generative cells escape through a series of special pores.   At the tips of both arms and pinnules all the sections of the body-cavity communicate with each other.

The mouth is central, and the anus is interradial in position and on the oral surface of the disk; the alimentary canal is coiled, and lined by a ciliated epithelium.   The coelom in the disk is much broken up by strands of connective tissue which support the viscera.   The mouth is surrounded by vascular, water-vascular, and nervous rings, which each give off extensions into the arms (Fig. 141).   The water-vascular ring gives off numerous ciliated canals which open into a series of vessels which communicate with the exterior by a series of ciliated pores which traverse the integument. This system represents the stone canal of Asterids. In one species of

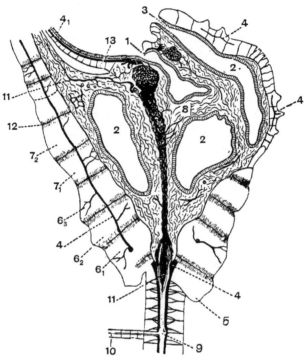

FIG. 141.—A longitudinal section through the plane of the mouth and anus of *Pentacrinus decorus*, Wyv. Th. After Carpenter.

1. Mouth.
2. Alimentary canal.
3. Anus.
4. On the left, the axial cord of the ray; on the right, extensions of nerves from the axial cord into the plated perisome of the ventral side. 4_1, ambulacral nerve. The central nervous mass is shown at 4, near the basal plate.
5. Basal plate.

6_1, 6_2, 6_3. First, second, and third radial plates.
7_1, 7_2. First and second distichal (brachial) plates.
8. The more or less calcified connective tissue in the body-cavity.
9. Central vascular axis of stem.
10. A cirrhus.
11. Genital rhachis.
12. Ligament between the ray joints.
13. Radial water-vessel.

The black plexus of blood-vessels in the centre of the figure is the plexiform gland, containing the genital stolon.

Rhizocrinus there is one canal and one pore in each interradius, but the number is much increased in other Crinoids.

The five genital strands are continuous with a central genital stolon, which here, as in Asterids, has been mistaken for a heart by German authors. Around this stolon are numerous vessels, which in the central capsule of the dorsal nervous system dilate into chambers which give off vessels to the cirrhi. Above they communicate with a plexus of vessels around the oesophagus, this plexus communicates with the distal portion of some of the stone canals.

The chief nervous system is situated dorsally; it consists of a mass of nervous matter lying within the circle of basal ossicles, and giving off a large nerve to the stalk, which supplies branches to all the cirrhi, and five radial nerves, each of which divides into two, and the resulting nerves supply each arm and govern their movements. This system is continued into the pinnules; it is probably connected here and there with the ambulacral system of nerves, whose function seems to be mainly sensory. This dorsal or anti-ambulacral system may be derived from concentrations of a subepidermal nervous system, such as exists in Asterids, which have sunk into the body.

Crinoids are attacked by an order of highly-modified Chaetopods, termed *Myzostomidae*. These occur only on the Crinoidea, and live parasitically either on the disk or arms; their presence often causes local abnormalities of growth, producing swellings sometimes termed galls. The order includes two genera, *Myzostoma* and *Stelechopus*. Extinct Crinoids seem to have suffered from the same parasite.

CLASS IV. ECHINOIDEA (Sea Urchins).

CHARACTERISTICS.—*Spheroidal or heart-shaped Echinodermata, sometimes flattened dorso-ventrally. The calcareous ossicles take the form of definitely-arranged plates usually immovably united by their edges, and of moveable spines. The number of radii always five in recent forms. Mouth and anus present.*

A ciliated ectoderm covers the body of the Echinoids, beneath this is a nerve plexus. The calcareous plates which constitute the shell of the animal are developed in the connective tissue of the integument. The apical series of plates consists of a dorso-central piece surrounded by ten plates; five of them, the radials or ocular plates, bear sense organs, the alternating five, interradial in position, are pierced by the genital pores. The ambulacral plates abut against the radials, and it is between the most dorsal ambulacral and the radial plates that new ambulacral pieces are intercalated.   One of the ambulacra

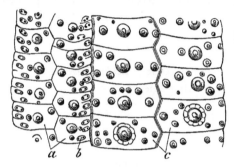

FIG. 142.—A portion of the shell of *Echinus gracilis*.   After Agassiz.

a. Ambulacral plates.

b. Poriferous zone.

c. Interambulacral plates.

is regarded as anterior, and an interradius is posterior ; in those forms in which the anus is not central, it lies in this posterior interradius.   Adopting this orientation, the madreporic pores are usually found on the right anterior genital plate.

Both the ambulacral or radial and the interambulacral or interradial areas are composed of a double row of pentagonal plates, firmly united with all the contiguous plates.   Each of the ambulacral plates is formed by the fusion of several small plates, the pore-plates ; these latter are pierced by two holes,

FIG. 143.—Spine of an *Echinid*.   After Leuckart.

1. Spine.

2. Basal knob.

3. Circular muscle of spine.

4. Ligament.

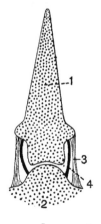

through which two processes from the water-vascular system pass and fuse to form one tube-foot.   Both the radial and interradial plates bear calcareous knobs, upon which long spines are articulated ; these are moved by certain muscles attached to their base, and form important locomotor organs. Pedicellariae, with usually three jaws, are also present.   Some

16

of these are provided with glands which open to the exterior near the tip of the jaws ; the glands are said to secrete a sticky fluid by means of which the Echinoid attaches to itself pieces of seaweed, etc., which screen it from observation.   The smaller

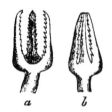

FIG. 144.—Pedicellariae of *Echinus saxatilis*.
After Gegenbaur.

a. Open.   b. Closed.

pedicellariae serve chiefly to clean the surface of the body, and some of them serve as locomotor organs, and to catch passing worms, etc.   They are well supplied with nerves, and some of them have in addition a special tactile organ.

The peristomial area immediately surrounding the mouth

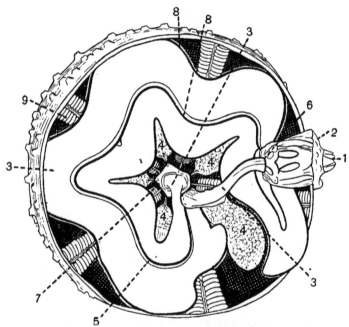

FIG. 145.—View of Sea Urchin, with part of the shell removed to show the course of the alimentary canal (from Leuckart).   After Cuvier.

1. Mouth, surrounded by five teeth (displaced).
2. Lantern of Aristotle.
3. Oesophagus, coiled intestine, and rectum.
4. Ovaries with oviducts.
5. The siphon.
6. Oral vascular ring.
7. Aboral sinus.
8. Blood-vessel accompanying intestine.
9. Ampullae at base of tube-feet.

is soft and membranous, with scattered ossicles. The mouth opens into an oesophagus, surrounded in the Regulares and Clypeastroids by a complicated masticatory apparatus known as Aristotle's Lantern. The oesophagus extends into the inter-radius of the madreporic plate, and opens into an intestine which takes a spiral course, finally opening by the anus, which may be nearly central in position or quite eccentric. The intestine is accompanied by a second tube, the siphon, which may have been pinched off from the intestine, into which it opens at each end. The whole is held in position by numerous strands of connective tissue.

The body-cavity is spacious and is filled with a fluid in which amoeboid corpuscles float, similar to those found in the water-vascular system.

The circumoral water-vascular ring lies at the dorsal end of Aristotle's Lantern. The ring gives off in each inter-radius a diverticulum or Polian vesicle, and in each radius a radial vessel which runs along the inner surface of the ambu-lacral plates; it bears a number of ampullae, which open, as a rule, by two ducts into the tube-feet, these vary much in structure; when suctorial the sucker contains calcifications. In the interradius of the madreporic plate a stone canal, which may be membranous or calcified, passes to an ampulla which opens by the madreporic plate.

The blood system of Echinoids is still involved in obscurity. There is a circumoral ring adjacent to the water-vascular ring, giving off two vessels which run one on each side of the intestine, and there are probably radial vessels, and one or more vessels accompanying the stone canal. Glandular tissue representing the so-called heart of other forms is developed in the wall of this structure.

In the Regulares ten buccal gills are usually found pro-jecting from the peristomial area around the mouth; these are hollow arborescent diverticula of the coelom, resembling in essential structure the dermal branchiae of the Asteroids.

There is a circumoral nervous ring situated in the angle between the base of Aristotle's Lantern and the peristome; this gives off five radial nerves, each of which ends in a sensory prominence of the epidermis, which traverses the ocular plate.

The radial nerves send branches to the tube-feet, from the base of which a nerve passes to the sub-epidermal plexus of nerve

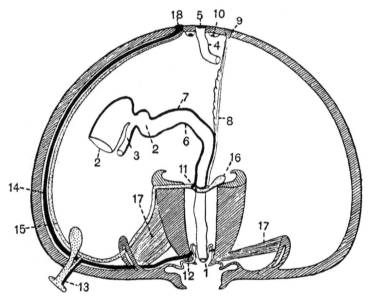

FIG. 146.—Diagrammatic vertical section of a Sea Urchin (from Leuckart).
After Hamann.

1. Mouth.
2. Intestine cut short.
3. Siphon.
4. Rectum.
5. Anus.
6. Ventral vessel on intestine.
7. Dorsal vessel on intestine.
8. Stone canal.
9. Madreporic plate.

10. Aboral sinus containing so-called blood-vessel.
11. Circumoral water-vascular ring.
12. Oral nerve ring.
13. Tube-foot with ampulla.
14. Radial nerve.
15. Radial water-vascular vessel.
16. Polian vesicle.
17. Muscles.
18. Ocular plate.

fibrils, which ramify all over the body just outside the calcifications, and govern the movement of the pedicellariae and spines.

The generative organs typically consist of five arborescent glands, though the number is often reduced, lying interradially, and opening on the genital plates. In the young they are all connected by a circular genital rhachis; they become very conspicuous in the breeding season.

The pore plates of the paired ambulacral areas are in the female *Hemiaster philippii* extended and depressed so as to form four deep oval cups. In these the eggs are deposited and the young develope, being kept in position by some of the spines bending over them.

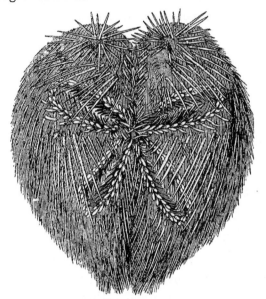

FIG. 147.—*Spatangus purpureus.*

The Echinoidea are divided into three subdivisions :

(i.) **Regulares.**—*Sphaeroidal or flattened circular bodies. Ambulacral and interambulacral areas equal in length. Central mouth and subcentral anus. Complex masticatory apparatus—Aristotle's Lantern—present.* Echinus, Toxopneustes.

(ii.) **Clypeastroidea.**—*Shield-shaped, often flattened bodies. Central mouth, with Aristotle's Lantern. Very broad ambulacra, with their dorsal ends forming a petaloid rosette round the apical plate; small tube-feet. Anus excentric.* Clypeaster, Rotula.

(iii.) **Spatangoidea.**—*Irregular heart-shaped bodies. Mouth and anus excentric. No Lantern of Aristotle. Ambulacra petaloid, and the anterior one unlike the others.* Spatangus, Brissus.

CLASS V. THE HOLOTHUROIDEA (Sea Cucumbers).

CHARACTERISTICS.—*Echinodermata with elongated bodies, usually pentagonal in cross section. The integument is leathery, and contains small scattered calcareous ossicles. The mouth is surrounded by a circlet of retractile tentacles, into which the water-vascular ring sends extensions. The madreporic plate usually opens into the body-cavity. The anus is usually terminal.*

The body of the Holothurians is elongated along an oral-apical axis. The ambulacra are five in number; they may be equally developed, or three of them, the trivium, may be flattened and form a creeping sole upon which the animal rests; the bivium is then convex. This occurs in *Psolus* and in all the Elasipoda. When this specialisation of radii takes place, the tube-feet are modified on the trivium. In other cases the tube-feet are scattered all over the body, and in others—the *Synaptae*—they are entirely wanting. The skin is covered by an ectoderm with an external cuticle; within this is a layer of connective tissue, in which cells laden with pigment and calcareous ossicles are scattered. This layer also includes a nervous plexus. The connective tissue sheath surrounds a muscular layer whose fibres run in a circular direction, and more internally are five radial bands of longi-

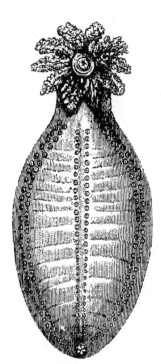

Fig. 148.—*Holothuria papillosa.*

tudinal muscles, one running along each ambulacrum, and lying beneath the water-vascular vessel and nerve; anteriorly these bands are attached to the pharyngeal ossicles, which are radial and interradial in position. The ossicles in the integument are always small in size; they may be simple spicules, or may assume a number of very elegant forms in the different genera. In the Elasipoda they exist in the mesenteries and in the walls of the alimentary canal, as well as in the integument.

The coelom is large, and is lined with ciliated cells; a special section of it surrounds the pharynx, and in the outer walls of this the pharyngeal ossicles are formed; these are notched for the passage of the radial nerve and water-vascular vessel.

In some of the *Synaptae* the alimentary canal runs nearly

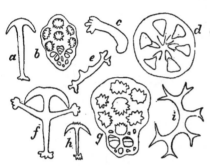

FIG. 149.—Spicules of Holothuroidea. After Semper.

a and *b*. Anchor and anchor plate of *Synapta indivisa*. Semper.
c. Spicule of *Chirodota rigida*. Semper.
d. Wheel spicule of *Chirodota vitiensis*. Gräffe.
e. Spicule of *Thyone chilensis*. Semper.
f, *g*, *h*. Anchor and anchor plate of *Synapta godefroyii*. Semper.
i. Spicule of *Rhopalodina lageniformis*. Gray.

straight from the mouth to the anus, but as a rule it forms a coil with three limbs. The mouth is situated in the centre of

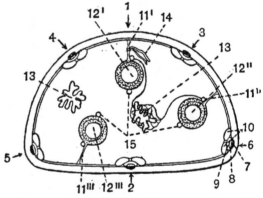

FIG. 150.—Diagram of a transverse section through the body of a Holothurian. From Leuckart.

1. Dorsal surface, dorsal interradius.
2. Ventral surface.
3. Left dorsal radius.
4. Right dorsal radius.
5. Right ventral radius.
6. Left ventral radius.
7. Radial nerve.
8. So-called radial blood-vessel.
9. Water-vascular vessel.
10. Radial muscles.

11', 11", 11'". The mesenteries of the three limbs of the intestine.
12', 12", 12'". The three limbs of the intestine.
13. Respiratory trees, the left surrounded by a rete mirabile of blood-vessels.
14. Two tubules of the generative organs lie to the right of the genital duct.
15. Ventral blood-vessels.

a peristomial area, and is in the Elasipoda directed ventrally. The mesentery of the first descending limb of the alimentary canal is situated in the interradius, in the middle dorsal line,

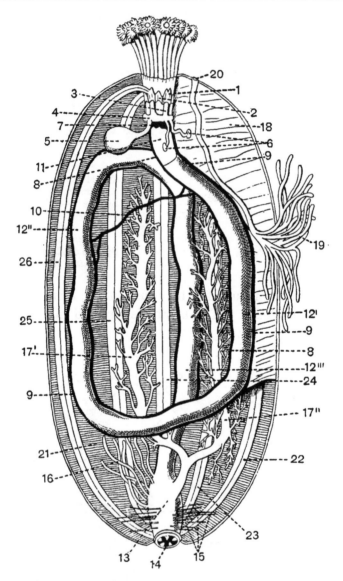

FIG. 151.—View of the internal organs of a Holothurian which has been cut open along the middle dorsal line.   From Leuckart.

1. Radial ossicle of the calcareous ring, into which the longitudinal muscle is inserted.
2. Interradial ossicle of the calcareous ring.
3. Radial water-vascular vessels.
4. Circumoral ambulacral ring.
5. Polian vesicle.

6. Two stone canals ending in Madreporic plates; the upper one is attacned to the dorsal mesentery, the lower one hangs freely.
7. Circumoral blood-vessel.
8. Ventral blood-vessel.
9. Dorsal blood-vessel.
10. Anastomosing branch between different parts of the ventral blood-vessel.

that of the second or ascending limb in the left dorsal inter-radius, and that of the third or second descending limb in the right ventral interradius. The pharynx is followed by a stomach with muscular walls; the intestine forms the longest portion of the alimentary canal; the posterior end of the rectum, or the cloaca, is rhythmically contractile, and takes in and sends out sea water; special muscles run between it and the body-wall.

Certain appendages known as respiratory trees open into the cloaca. These are sometimes regarded as homologous with the two interradial caeca which open into the rectum of Asterids. They are branched structures, usually two in number; each terminal ramification opens by a fine tube into the coelom, and they doubtless serve to introduce sea water into that space. Their function is probably in part respiratory, and they re-semble in structure the similar organs in the armed Gephyrea.

In some species some of the basal ramifications of the respiratory trees are modified into the so-called Cuvierian organs. In these the peritoneal covering becomes glandular, and when the animals are irritated they are discharged into the water through holes torn at their base in the cloacal wall, and swell up into long tenacious elastic threads, which serve the purpose of entangling their enemies. At times, when much disturbed, the Holothuroidea will throw out their whole intes-tine through the anus; but it is probably regenerated.

The water-vascular system consists of an oral ring, which gives off five radial vessels, these run at first upwards and out-wards, and give off branches to the oral tentacles; the tentacles may increase in number with age. The tentacles assist in the pro-cess of feeding, either shovelling in mud or sand—in which case

11. Anterior part of alimentary canal.
12', 12", 12'''. The three limbs of aliment-ary canal.
13. Cloaca.
14. Cloacal opening with five teeth.
15. Radiating muscles of cloaca.
16. Organs of Cuvier.
17', 17". Respiratory trees.
18. Dorsal mesentery with free posterior margin.
19. Generative organs.
20. Opening of generative duct.
21. Circular muscles in body-wall.
22. Right dorsal muscle.
23. Right ventral muscle.
24. Median ventral muscle.
25. Left ventral muscle.
26. Left dorsal muscle.

The tentacular ampullae are omitted; the mouth is in the centre of the divided tentacles.

the intestine, like that of the Gephyrea, is full of sand—or entangling food particles; and in the latter case the tentacles are then thrust into the mouth, which removes any nutritive particles. The radial vessels pass through notches in the radial ossicles of the pharyngeal calcareous ring, and run along the ambulacra giving off tube-feet outside the bundles of muscle fibres. They are absent in one group, and devoid of tube-feet in others. The ampullae of the tube-feet are embedded in the circular muscle layer in the Elasipoda, and in many of this group the stone canal opens on the dorsal surface, and in others it lies in the tissue of the integument; in other subdivisions it is supported by the mesentery, and the madreporic plate opens freely into the coelom. It may or may not have calcareous walls; the fluid in this system contains numerous corpuscles.

The vascular system consists of spaces in the connective tissue not lined by an epithelium. There is a circular space round the pharynx, just behind the water-vascular ring. This

FIG. 152.—Sea Cucumber *Cucumaria crocea* (Falkland Islands) bearing its young. After Sir Wyville Thomson and Murray, "Challenger" Narrative.

communicates with a dorsal and a ventral intestinal vessel, and these two are connected by numerous anastomoses round the walls of the alimentary canal. The dorsal vessel is in connection with a plexus which surrounds the left respiratory tree. There are no radial vessels.

The circumoral nerve ring gives off five radial nerves, and nerves to the tentacles. There is a nerve plexus in the skin and the tube-feet. In Synapta ten auditory vesicles containing numerous otoliths have been described at the base of the radial nerves.

The generative organs are either a single gland situated at the left of the dorsal mesentery, or a double gland, one on each side of it; the glands are continuous with a single duct, which opens in the dorsal middle line close to the base of the tentacles. With one or two exceptions, the Holothuroidea are dioecious.

The young of one species, *Cucumaria crocea*, found near the Falkland Isles, are attached in rows on each side of the dorsal ambulacrum. The early stages of developement apparently take place rapidly, and the embryos are arranged in position by the tube-feet of the ambulacrum.

In the East Indies some species form an article of commerce under the name of Bêche-de-mer. They are dried and sold to the Chinese, who use them in the preparation of soup.

The Holothuroidea are classified as follows :

I. ACTINOPODA.—*Radial canals present in the water-vascular system.*

    a. **Aspidochirotae.**—*The tentacles are peltate in form, respiratory trees are present.* Holothuria, Mülleria.

    b. **Elasipoda.**—*Tentacles as above. The dorsal tube-feet produced often into very long stiff processes. Respiratory trees rudimentary or absent. Stone canal sometimes opens to the exterior.* Deima, Elpidia.

    c. **Dendrochirotae.**—*The tentacles have a dendriform shape.* Cucumaria, Thyone.

    d. **Molpadiidae.**—*The tentacles are simple or pinnate. The radial canals bear tentacles, but no other tube-feet.* Molpadia, Caudina.

II. PARACTINOPODA.—*The tentacles are pinnate. No radial canals, no tube-feet, no respiratory trees.* Synapta, Chirodota.

ARTHROPODA

CRUSTACEA

ENTOMOSTRACA
- Phyllopoda . { Branchiopoda—*Apus, Estheria, Limnadia.* / Cladocera—*Daphnia, Sida, Leptodora.*
- Ostracoda . *Cypris, Cypridina.*
- Copepoda . { Branchiura—*Argulus, Gyropeltis.* / Eucopepoda { Gnathostomata—*Cyclops, Calanus.* / Parasitica—*Chondracanthus.*
- Cirrhipedia . { Thoracica—*Lepas, Balanus.* / Ascothoracica—*Laura.* / Abdominalia—*Alcippe.* / Apoda—*Proteolepas.* / Rhizocephala—*Sacculina.*

MALACOSTRACA
- Leptostraca . *Nebalia.*
- Thoracostraca { Cumacea—*Diastylis, Leucon.* / Stomatopoda—*Squilla.* / Schizopoda—*Mysis.* / Decapoda { Macrura—*Pagurus, Astacus.* / Brachyura—*Carcinus.*
- Arthrostraca { Amphipoda—*Gammarus, Phronima.* / Isopoda { Anisopoda—*Tanais.* / Euisopoda—*Asellus, Oniscus.*

TRACHEATA
- Prototracheata } Antennata.
- Myriapoda
- Hexapoda
- Arachnida.

CHAPTER XVI

ARTHROPODA

CHARACTERISTICS. — *Bilaterally symmetrical Coelomata, with a chitinous exoskeleton. Their body is segmented hetero-nomously. The segments usually bear a pair of jointed appendages, those in the neighbourhood of the mouth are modified in connection with the prehension and mastication of the food. The nervous system consists of a brain or supra-oesophageal ganglion, a ring round the oesophagus, and a ventral, usually segmented, nerve cord. A heart is typically present dorsal to the alimentary canal, blood enters it through a series of lateral ostia ; the coelom is reduced, and to some extent replaced by a haemocoel. The sexes are typically distinct, and the paired genital glands usually open by paired ducts. Cilia are universally absent from the group, with the single exception of Peripatus.*

The Arthropoda may be divided into two large groups, according to the nature of the breathing organs: (i.) the BRANCHIATA, which breathe by gills and are typically aquatic; and (ii.) the TRACHEATA, which breathe by tracheae or lung books, and are typically terrestrial.

The Branchiata include but one class, the **Crustacea.** The Tracheata include four: (i.) the **Prototracheata,** (ii.) the **Myriapoda,** (iii.) the **Insecta** or **Hexapoda,** and (iv.) the **Arachnida.** The first three classes may be grouped together as the Antennata, and opposed to the last class, the Arachnida.

I. BRANCHIATA.

Class CRUSTACEA.

CHARACTERISTICS.—*Aquatic Arthropods, which breathe either through the general surface of their skin or through specialised extensions of the same, the branchiae or gills. Two pairs of antennae are found, and the appendages are as a rule biramous. A limb-bearing thorax is either free or united with the head. The usually segmented abdomen may or may not bear appendages. Some of the limbs are modified to form jaws. The gills are usually extensions of the basal joint of some of the appendages. The whole group, both in its internal and external features, is, with few exceptions, rigidly bilaterally symmetrical.*

The Crustacea are divisible into two series: (A) the Entomostraca and (B) the Malacostraca.

A. ENTOMOSTRACA.

A. *The Entomostraca include many comparatively small and simply-organised Crustacea, the number of whose segments varies within wide limits. A large carapace, which may enclose the whole body, is often present. The demarcation between thorax and abdomen is often shown by the opening of the generative organs. Paired compound eyes and an unpaired simple eye often coexist. There is no masticating stomach. The developement almost always includes a Nauplius stage.*

The ENTOMOSTRACA consists of four orders:

1. PHYLLOPODA.
2. OSTRACODA.
3. COPEPODA.
4. CIRRHIPEDIA.

ORDER 1. PHYLLOPODA.

CHARACTERISTICS.—*Crustacea, with usually elongated and well-segmented body, partially covered by a shield-like carapace, which may be laterally prolonged to form a bivalved shell. The number of segments and appendages varies greatly, but there are never less than four leaf-like lobed swimming-feet.*

The Phyllopods are divided into two sub-orders : (*a*) the Cladocera and (*b*) the Branchiopoda.

a. Cladocera.—*The Cladocera or water-fleas are all small. Their body is laterally compressed, and their carapace takes the form of a bivalved shell, within which the larger part of the body lies concealed. A pair of large biramous antennae are used as swimming organs.*

The Cladocera include many species common in freshwater streams and inland lakes. *Daphnia pulex* and *Daphnia longispina* occur frequently in ditches and ponds in England, and although they are of minute size, 4 or 5 mm. in length, they form a convenient type of this sub-order. The sexes differ both in size and structure, and it will be convenient to describe the female first, and afterwards to mention those points of difference which the male presents.

As is usual in Crustacea, the body of *Daphnia* is divisible into three regions : the head, the thorax, and the abdomen. The large bivalve shell which encloses the body like the valves of a Lamellibranch, is an extension of the dorsal surface of the cephalic segments. The head is provided with a pair of antennules, uniramous, very small, and bearing olfactory hairs ; a pair of antennae which are biramous and very long and are used for swimming ; a pair of mandibles ; and one pair of maxillae. The larva has two pairs of maxillae, but the second pair disappear before the adult stage is reached. The thorax consists of five segments, which are free from the shell. Each segment bears a pair of lamelliform swimming-feet. The abdomen is three-jointed, and carries no appendages ; it is curved forwards ventrally, and terminates in an unsegmented post-abdomen or telson. The abdomen bears dorsally several processes which assist in enclosing the brood pouch, a space left between the dorsal side of the thorax and abdomen and the shell. The post-abdomen on which the anus opens bears two long dorsal tactile setae, and ends in two hooks or styles.

Appendages of *Daphnia*.

1. Antennules, uniramous and small.
2. Antennae, biramous.
3. Mandibles.   4. 1st maxillae.
5. 2nd maxillae, disappear during larval life.
6. 1st thoracic swimming-feet.
7. 2nd   ,,   ,,
8. 3rd   ,,   ,,
9. 4th   ,,
10. 5th   ,,

The head bears a large median compound eye, which is formed by the fusion of two originally lateral eyes, and behind this lies a single simple eye.

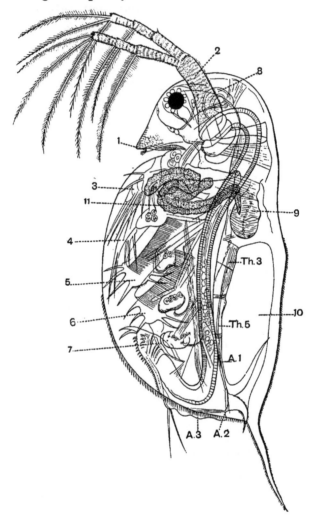

FIG. 153.—Side view of female of *Daphnia similis*. After Claus.

1. Antennules.
2. Antennae.
3. 1st pair of legs.
4. 2nd pair of legs.
5. 3rd pair of legs.
6. 4th pair of legs.
7. 5th pair of legs.
8. Hepatic diverticulum.
9. Heart.
10. Brood pouch.
11. Shell gland.
Th. 3 and Th. 5. 3rd and 5th thoracic segments.
A.1, A.2, A.3. 1st to 3rd abdominal segments.

An upper lip guards the entrance to the alimentary canal, and the mouth lies between the two-toothed mandibles; the digestive tube passes straight through the body, with hardly any change of diameter, to terminate in the anus in the post-abdomen. The only structures which open into it are a pair of small curved caecal processes which are given off near the anterior end, and which are usually regarded as liver diverticula.

The heart, which is much shorter than is usual with Crustacea, is correlated with the small size of the animal. It consists of an oval sac, the muscular nature of whose walls is very evident. The sac is suspended in a pericardium which contains blood; this blood enters the heart through a single ostium on each side, and is forced out by the rhythmic contractions of the organ through an anterior opening. Although there are no blood-vessels with distinct walls, the blood follows a definite course, flowing in channels through the various parts of the body and shell. The blood contains amoeboid corpuscles.

A coiled gland, which ends blindly at its inner end, opens to the exterior in the region of the second maxillae (Fig. 153). This is termed the *shell-* or *maxillary gland*, and it is the characteristic nitrogenous excretory organ of the Entomostraca, as opposed to the *antennary gland* of the Malacostraca. In *Estheria* the gland terminates in a vesicle, the walls of which are lined by flat epithelial cells, and it has been suggested that this may represent a portion of the primitive coelom, just as does the vesicle at the inner end of the nephridium in *Peripatus.* The larvae of some Phyllopods possess an antennary gland as well as a shell gland, but this disappears before the adult stage is reached.

Many species of Daphnidae, e.g. *Sida,* have also a neck gland; these animals swim on their backs, and the neck glands secrete a sticky substance which enables them to attach themselves to foreign bodies.

The brain in *Daphnia* gives off two stout nerves, which pass forward and almost immediately fuse to form a large optic ganglion, which supplies the compound eye; it also gives off nerves to the simple eye, to a curious sense organ composed of an aggregation of ganglion cells in the neck, and to the first or olfactory antennae. A pair of circum-oesophageal commissures surround the oesophagus, and the large swimming antennae

17

are supplied by the first sub-oesophageal ganglia. The right and left strands of the ventral cord remain distinct and separate from one another throughout their course, being united by commissures at the various ganglia. Of these one pair supplies the muscles which move the mandibles and maxillae, and there is a pair for each pair of legs. The last pair of ganglia give off posteriorly a fine nerve, which supplies the sensory spines of the abdomen.

The ovaries lie one on each side of the alimentary canal; they are tubular, and reach as far forwards as the heart, and their walls are continuous with those of the oviducts which open into the brood-pouch between the abdomen and the shell. The ovary is broken up into a series of segments, each of which contains four cells. One of these becomes an ovum and increases in size by absorbing the other three. In the case of the large winter eggs, the contents of two or more segments are absorbed by the cell destined to become the ovum.

The male is but little more than half the length of the female; the first pair of antennae, which are minute in the

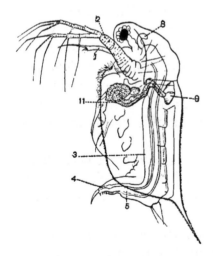

Fig. 154.—Side view of male *Daphnia similis*, magnified to the same extent as Fig. 153. After Claus.

1. Antennules.
2. Antennae.
3. Testis.
4. Ductus ejaculatorius.
5. Rectum.
8. Hepatic diverticulum.
9. Heart.
11. Shell gland.

female, are here of considerable size, though not so large as the second. They are provided with numerous olfactory hairs. The first pair of swimming legs are provided with a claw and a spine which project between the two valves of the shell. The space between the abdomen and the shell is a narrow split, and is not hollowed out to form a brood-pouch, as is the case in the female.

The testes occupy relatively the same position in the body as the ovaries, lying one upon each side of the alimentary canal. They are continuous with the vas deferens, which opens by means of a muscular ductus ejaculatorius upon the post-abdomen.

The small males are much rarer than the females; they are usually to be found in the autumn, but sometimes occur at other times of the year, when the conditions of life become un-favourable. During the summer the female produces a number of summer eggs, which hatch out in the brood-pouch. The enormous fertility of the water-fleas is shown by the fact that in nineteen days a female *D. pulex* produced five broods, the total number of young being 209; and it has been calculated that the descendants of a single individual, which becomes mature in ten days, and then produces broods of fifteen young every three days, would amount to over twelve million in less than two months. The rapidity of developement is rendered possible by the nutriment stored up in the summer eggs, and this is in some species augmented by the secretion of additional food material into the brood-pouch.

Like many other river animals, such as freshwater Polyzoa and Sponges, the *Daphnia* have developed a means of ensuring the existence of the species through the frosts, etc., of winter. This is effected by means of the winter eggs. These eggs are larger than the summer ones, and contain more yolk, a correspondingly large amount of the contents of the ovary being absorbed during their maturation. They are in-capable of developing without fertilisation. The winter eggs pass into the brood-pouch, and a part of the carapace of the mother becomes in this region modified to form a capsule for the eggs, this is called the *ephippium*. At the next ecdysis the ephippium, which usually contains two eggs, is thrown off and floats away. It is a bivalved structure and has a very striking resemblance to the statoblasts of some Polyzoa. The ephippium may be redeveloped by the female even when fertilisation does not occur, but in this case no eggs are laid; it may also be replaced by the ordinary brood-pouch.

The characteristic larval stage of the Entomostraca, the *Nauplius*, is not present in the Cladocera, except in the genus *Leptodora*, and there it is confined to the winter-egg generation.

The Cladocera are as a rule freshwater inhabitants, living in enormous numbers in lakes, ponds, and springs; only a few species flourish in brackish water or in the sea.

b. Branchiopoda.—*The Branchiopoda are Phyllopods of considerable size, with clearly segmented bodies often partially covered by a shield-shaped or laterally compressed shell. They possess from ten to sixty pairs of foliaceous swimming appendages, which bear well-developed gills.*

The Branchiopoda differ from the Cladocera in the fact that they are larger, consist of more segments, and are generally more complicated in their structure. They present considerable differences in the various species. *Estheria* and *Limnadia* bear large bivalved shells which completely enclose the body, *Apus* has a shield-shaped dorsal shell which covers the head and body, but leaves the tail free, whilst *Branchipus* is devoid of any shell.

The flattened leaf-like appendages of *Apus* have been looked upon as a primitive type from which the appendages of other Crustacea may be derived. This type of appendage is almost completely retained in the foliaceous maxillae of *Astacus*. The abdominal appendages, that is, those which are situated behind the genital openings, are the least specialised, and these present an unjointed axis which bears on its inner edge six processes termed *endites*, which bear numerous setae. The axis ends in a sub-apical lobe, and carries on its outer sides two *exites*; of these the distal is large, and has been termed the *flabellum*, the proximal is devoid of setae, and forms the *branchia*.

In the female the appendage of the 11th thoracic segment, which bears the opening of the oviduct, is modified; the 6th endite and the sub-apical lobe are enlarged to form a hollow cup, over which the flabellum closes like a lid, this forms a receptacle for the ova, and the appendage is known as the *oostegopod*.

Appendages of *Apus cancriformis.*

1. 1st pair of antennae (antennules).
2. 2nd   ,,   ,,
3. Mandible.
4. 1st maxilla.

5. 2nd maxilla.
6-16. The 11 pairs of thoracic limbs.
17-68. The 52 pairs of abdominal limbs.

The annulations of the abdomen are much fewer than the number of appendages; the nerve ganglia, however, correspond with the limbs. The last two or three segments carry no appendages. The two nerve cords are separated by a considerable interval, being connected by transverse commissures at the ganglia. The heart is elongated, and extends through the eleven thoracic segments, with a pair of ostia in each. In *Branchipus* the heart extends throughout the whole body.

The Branchiopoda are frequently parthenogenetic, and the males are much rarer than the females; one species of *Apus* has recently been shown to be hermaphrodite, the posterior end of the generative gland producing spermatozoa. They inhabit freshwater lakes and pools, and one genus, *Artemia*, lives in brine pools, in which the salt may be so concentrated as to be fatal to all other forms of life. Their eggs are capable of surviving long periods of drought, embedded in the dried-up mud : a property they share with those of the Cladocera, the Cyclopidae, and the Rotifera, etc.

Order 2. OSTRACODA.

CHARACTERISTICS.—*Small, usually laterally compressed Entomostraca, with an unsegmented body bearing seven pairs of appendages; the abdomen is rudimentary. The whole body is enclosed in a bivalved shell.*

The group Ostracoda includes a great number of genera and species, but it is nevertheless a very homogeneous assembly, the various species differing but little from one another. The bivalved shell which encloses the animal has a striking resemblance to that of some Lamellibranchs; the valves are divaricated by means of an elastic ligament which occupies about the middle of the dorsal surface, and are occluded by an adductor muscle.

The body is not segmented, but head, thorax, and a rudimentary abdomen can be distinguished. The appendages are as follows :

Appendages of *Cypris*.

1. 1st pair of antennae (antennules).   5. 2nd pair of maxillae.
2. 2nd   ,,   ,,   6. 1st   ,,   thoracic limbs.
3. Mandibles.   7. 2nd   ,,   ,,   ,,
4. 1st pair of maxillae.

The antennae are usually adapted for walking or swimming, and in some of the marine forms the shell is notched, so

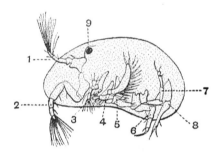

FIG. 155.—Lateral view of *Cypris candida*. After Zenker.

1. Antennules.
2. Antennae.
3. Mandibles.
4. 1st maxillae.
5. 2nd maxillae.
6. 1st pair of legs.
7. 2nd pair of legs.
8. Tail.
9. Eye.

that the antennae can be protruded even when the shell is closed. In *Cypridina* the anterior pair bear olfactory hairs, and in *Cypris* the second pair end in hooked bristles, by means of which the animal can anchor itself.

The mandibles are strong and toothed; they bear a palp, which is usually elongated and leg-like. The second maxilla functions sometimes as a maxilla, sometimes as a leg; there are usually two, rarely three, thoracic legs; the abdomen is devoid of appendages, and is rudimentary; it may end in a caudal fork, as in *Cypris*, or in a plate beset with setae.

The oesophagus expands into a crop, which lies in front of the true stomach. The last-named region of the alimentary canal gives off a hepatic diverticulum on each side, which is prolonged into the cavity of the shell. The anus opens at the base of the abdomen. These animals are entirely carnivorous. A heart is often absent; when it exists, as in *Cypridina*, it lies on the dorsal surface, in the region where the body and shell are in continuity. There are as a rule no special respiratory organs, respiration probably taking place through the general surface of the body. A shell gland is present, and opens in the region of the second maxilla.

There is a cerebral ganglion and a short ventral chain of ganglia, which, however, often fuse together to form a complex ventral nervous mass. The single eye of the Nauplius larva is usually retained, but some forms have in addition a compound eye.

The Ostracoda are dioecious, and the males are easily distinguishable from the females ; the former usually possess some organs for holding the latter ; in *Cypridina* this is found on the second antennae, in *Cypris* on the second maxillae, and they

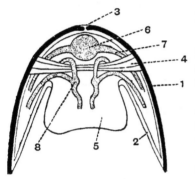

FIG. 156.—Transverse section through the body and shell of *Cypris candida*. After Zenker.

1. Shell.
2. Inner fold of mantle lining shell.
3. Ligament between the two halves of the shell.
4. Adductor muscle.
5. Body.
6. Intestine.
7. Liver tubes extending into the shell.
8. Testes sending prolongations into the shell.

are also provided with more highly developed sense organs. The ovaries and the testes may, like the liver diverticula, extend into the lining of the shell (Fig. 156). The accessory male organs are very complicated, and the spermatozoa attain a great size, being in some cases longer than the body. *Cypris* attaches its ova to water-plants, but *Cypridina* carries them about within the shell of the female until they hatch out.

ORDER 3. COPEPODA.

CHARACTERISTICS.—*Entomostraca with an elongated segmented body, without a dorsal shell. The thorax bears four or five biramous swimming feet ; the abdomen, which consists of four segments, is devoid of limbs. Some species are parasitic, and are then more or less degenerate.*

An enormous number of very variously modified Crustacea are included in the group Copepoda. The free-swimming forms are distinguished by the constant number, and by the persistence, of their paired appendages. The parasitic forms, which are very numerous, undergo every stage of degeneration, and in the extreme cases lose all trace of their Entomostracan affinities in the adult condition. The systematic position of these forms is, however, clearly shown by the history of their developement.

The Copepoda are divided into two sub-orders: (i.) the Branchiura, which includes the two genera *Argulus* and *Gyropeltis*; and (ii.) the Eucopepoda, which embraces the mass of the Copepods, both free-swimming and parasitic.

(i.) The **Branchiura** are known as Carp-lice, *Argulus* being parasitic upon Carp and Sticklebacks. These animals differ considerably from the members of the Eucopepoda. In

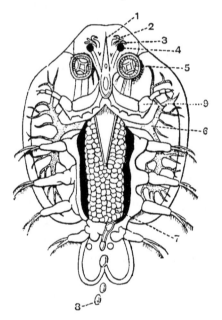

FIG. 157.—*Argulus foliaceus*, female, seen from the ventral side. After Jurine.

1. Pointed rostrum.
2. Antennules.
3. Antennae.
4. Eye.
5. Sucker on anterior ramus of 2nd maxilla.
6. Liver.
7. Ovary.
8. Eggs leaving the body.
9. Posterior ramus of 2nd maxilla.

Argulus the mouth is prolonged into a suctorial tube, within which lie concealed the serrated mandibles and styliform first maxillae: a modification of the mouth parts which is common amongst certain orders of Insects. In front of the mouth is a styliform weapon, which also lies in a sheath. The second pair of maxillae, the so-called maxillipedes, are modified to form organs of adhesion, the anterior ramus taking the form of a sucking disk, the posterior being hooked. Behind these are found four pairs of swimming legs, all of them, except the last, concealed by the body. The abdomen terminates in two caudal plates; the whole body is flattened dorso-ventrally. The male is more active than the female, and possesses copulatory appendages on the last thoracic limbs; the eggs are not carried about in the usual Copepod fashion, but are deposited upon surrounding objects.

(ii.) The **Eucopepoda** are further subdivided into: (a) *the* Gnathostomata *or free-swimming forms with masticating mouth appendages*; and (b) *the* Parasitica, *in which the mouth parts are modified for sucking or piercing, and whose body is incompletely segmented.*

(*a*) GNATHOSTOMATA.—The genus *Cyclops*, which includes many species living in fresh or brackish water in Great Britain, is fairly typical of this subdivision. In shape its body has been compared to that of a split pear with its flat side ventral, tapering posteriorly. The head bears five pairs of appendages, viz. (i.) antennules, (ii.) antennae, (iii.) mandibles, (iv.) 1st maxillae, (v.) 2nd maxillae. Dorsally its terga form a continuous carapace, and are fused with the first thoracic tergum. The thorax consists of six segments, the five posterior being free dorsally as well as ventrally, except in the female, where the last thoracic segment is fused with the first abdominal. The abdomen consists of four cylindrical segments devoid of appendages, the last bears the furcal caudal processes so characteristic of the Entomostraca.

Appendages of *Cyclops*.

1. Antennules.	5. 2nd maxillae (maxillipedes).
2. Antennae.	6-9. Four pairs of thoracic limbs.
3. Mandibles.	10. 5th pair ,, ,,
4. 1st maxillae.	(rudimentary).

The antennule is a well-developed appendage, which acts as a strong oar, in the male it also acts as a clasper, and is correspondingly modified; the antennae are short and four-jointed; the mandibles guard the mouth, one on each side; they and the 1st maxillae have rudimentary palps. The 2nd maxillae are biramous, the split extending to their base; the shell gland or excretory organ opens upon the first joint of the outer ramus. The first four thoracic limbs are flattened from before backward; they are biramous, and their bases are united by a median plate, which extends across the middle ventral line. The fifth thoracic segment bears a pair of rudimentary feet at the sides of the ventral surface. The next segment has a rounded cross section like those of the abdomen, with which it is generally grouped; but as it bears the openings of the genital ducts, it is better to regard it as the sixth and last thoracic. It bears a pair of appendages which are reduced to a small valve-like structure overlying the sexual aperture. In the female this last thoracic segment is fused with the first abdominal. The four abdominal seg-

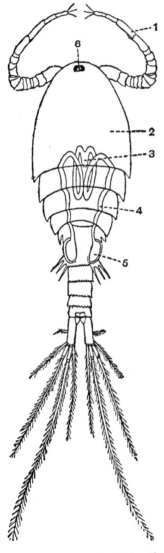

FIG. 158.—Dorsal view of male *Cyclops*. The reproductive organs are diagrammatic. After Hartog.

1. Antennae.

2. Carapace.

3. Testis.

4. Vas deferens.

5. Vesicula seminalis.

6. Eye.

ments are cylindrical in shape, and are entirely devoid of appendages; the last, however, bears the caudal fork.

The mouth leads into a narrow gullet which ends in the anterior end of the stomach, an oval sac extending back to about the second thoracic segment. The chitinous cuticle which covers the body of the animal is invaginated at the mouth, and lines the gullet and the anterior third of the stomach. The rest of this organ is lined by large columnar

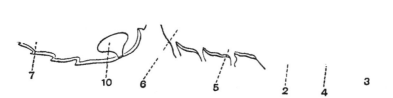

FIG. 159.—Longitudinal vertical median section through *Cyclops*, partly diagrammatic. After Hartog.

1. Cerebral ganglion.	7. Proctodaeum.
2. Ventral nerve cord, ganglion omitted.	8. Ovary.
3. Mouth.	9. Uterus.
4. Stomodaeum.	10. Spermatheca.
5. Cells lining posterior part of stomach.	11. Eye.
6. Intestine.	

cells, some of which contain fat globules, and others granules which pass into the intestine, and so out of the body. These granules have been regarded by some authorities as urinary. The intestine stretches from the stomach to the hind end of the second abdominal segment, and then passes into a rectum, which is lined with a chitinous cuticle continuous at the anus with the general chitinous covering of the body. No liver diverticula open into the alimentary canal of *Cyclops*, although these structures are found in some other Gnatho-stomata, but there are a pair of salivary glands which open by a common duct into the oral face of the labrum, a process which overhangs the mouth anteriorly.

No heart exists in *Cyclops*, but the space in which the alimentary canal lies contains a colourless fluid, in which colourless corpuscles float. The space which contains this fluid is much broken up by connective tissue trabeculae and strands of muscle, which support the alimentary canal and generative organs. The various muscles are striated, and their contraction causes the stomach and intestine to move rhythmically backwards and forwards in such a way as to move the corpusculated fluid forward in the dorsal part of the body and backward in the ventral. At the same time certain dilator muscles attached to the rectum cause this tube to expand and take in through the anus a certain amount of water, which is subsequently ejected again. This rhythmical taking in and ejecting of water has probably a respiratory significance, and the process has been termed anal respiration. It has been observed in numerous Crustacea amongst various classes, *e.g.* in *Argulus* and *Caligus* amongst the Copepoda; in *Daphnia, Moina,* and the larvae of *Apus* amongst the Phyllopoda; in *Gammarus* and *Asellus* amongst the Arthrostraca; and in the Zoaea-larvae of Macrura and Brachyura. The interchange of gases between the circulating medium and the surrounding water may, in the case of *Cyclops,* where the integument is thin, take place all over the body, but in those animals whose cuticle is thick it may reasonably be supposed to take place more readily through the thinly-lined rectum.

The kidney consists of the typical Entomostracan shell- or maxillary gland; it is a simple tube ending blindly at one end, and opening to the exterior on the second ramus of the second maxilla at the other. The functions of the kidney are believed to be partly taken over by the walls of the intestine, which excrete granules supposed to be urinary. The larva of *Cyclops* is said to possess an antennary gland.

The nervous system consists of a supra-oesophageal mass which gives off nerves to the eyes and to the antennules; of two circum-oesophageal cords from which the nerves to the antennae arise; and of a single ventral cord which extends to the fifth thoracic segment, where it bifurcates, and continues through the abdomen as two cords, which ultimately end in the furcal processes. The ventral cord presents no marked distinction into ganglionic and interganglionic regions.

Besides the numerous sensory hairs, each supplied with nerve fibres, the median frontal eye, divided into three parts, is the sole sense organ.

The Copepoda are dioecious; in *Cyclops* the ovary is a median gland situated beneath the first thoracic tergum. The oviduct, which is continuous with the ovary, gives off a series of uterine processes in which the ova accumulate, and finally opens on the posterior thoracic segment, which in the female is fused with the first abdominal. Just at the point of opening the oviducts receive the ducts of the spermatheca; this is a sac lying within the fused segments, and having a special pore through which the spermatozoa make their entrance (Fig. 159).

The testis resembles the ovary in form and position; the vasa deferentia are rather coiled, and may be divided into different sections, they open into two vesiculae seminales, one on each side of the sixth thoracic segment, and these open to the exterior upon the same segment (Fig. 158).

One fertilisation suffices for many broods of ova; the eggs as they leave the oviduct are surrounded by a cement-substance which forms an investment for each egg and keeps them together in an oval mass, which the female carries one on each side of her abdomen. One of the ovisacs may contain from seventy to ninety ova.

The Calanidae differ from the Cyclopidae in the possession of a well-developed heart, which in *Calanella* is produced into a cephalic artery; and in the fact that their antennae are biramous. The Notodelphidae which inhabit the branchial chamber of Ascidians have their antennae modified for attachment, and the posterior thoracic segments of the female are fused together and form a brood-pouch in which the eggs develope.

(*b*) PARASITICA.—The parasitic or semi-parasitic Copepods have their mouth parts adapted for piercing and sucking; their body is usually incompletely segmented, and the abdomen is reduced. The males are often smaller than the females to whom they adhere.

Most of the Parasitica live on fish, hence their common name, fish-lice; they may inhabit the gill-chamber, or the pharynx, or they may live on the skin, in which the female buries half her body, and in some instances, e.g. *Penella*, they even bore into the body and live partly embedded in the tissues.

Chondracanthus gibbosus, which inhabits in considerable numbers the gill-chamber of the Lophius, or fishing-frog, exhibits many of the peculiarities of these remarkable epizoa. The female may be half an inch long; the body is flattened, and has lost most signs of segmentation, it is produced laterally into two folds, which have a crimped or frilled appearance; pos-

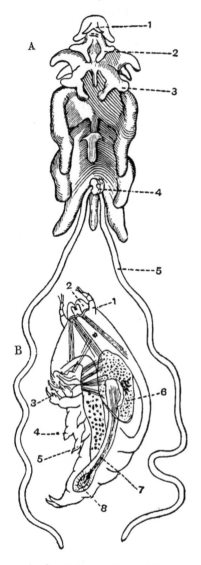

FIG. 160.—*Chondracanthus gibbosus.*

A. Female from under surface.

    1. Antennae, hook-like.

    2. 1st pair of legs.

    3. 2nd pair of legs.

    4. Males adherent to female.

    5. Egg bags.

B. Male, more highly magnified.

    1. Antennules.

    2. Antennae.

    3. Mouth parts.

    4. 1st pair of legs.

    5. 2nd pair of legs.

    6. Testis.

    7. Vas deferens.

    8. Spermatophore.

teriorly it is produced into two long white filaments, the ovisacs. The head forms a kind of hood which bears antennules and antennae, the latter modified into strong hooks, by means of which the animal attaches itself to its host. The labrum which overhangs the mouth does not form a tube enclosing the mandibles, as is often the case in this group. The mandibles and both pairs of maxillae form claws. Behind these are two pairs of bifid lobes representing the first two pairs of feet. The abdomen is represented by a small papilla; just in front of this the oviducts open. The male is much smaller than the female, and retains a much more definite segmentation; it is found clinging on to the female by means of its antennary hooks in the neighbourhood of the opening of the oviduct. It possesses an eye, and its caudal extremity ends in two processes. Both male and female, like so many other parasites, have very largely developed generative organs.

Many of the parasitic Copepods have departed still farther from the Crustacean type than *Chondracanthus*. The Lernaeidae are vermiform, with their mouth parts forming a piercing and sucking tube. L. *branchialis* is found on the Cod. The Lernaeopodidae have their maxillipedes enlarged, and in the females united to form an organ of attachment to their host. In this family the swimming-feet have entirely disappeared. Some species are parasitic on Sharks and on the Salmon.

Order 4. CIRRHIPEDIA.

CHARACTERISTICS.—*Fixed Crustacea whose body is enclosed in a fold of the skin, which is generally strengthened by calcareous valves. The body is indistinctly segmented. There are usually six pairs of thoracic feet, and the abdomen is rudimentary. Hermaphrodite, with mobile spermatozoa. Exclusively marine.* The class Cirrhipedia may be divided into five orders:

1. **Thoracica.**
2. **Ascothoracica.**
3. **Abdominalia.**
4. **Apoda.**
5. **Rhizocephala.**

The last four of these orders are composed of parasitic species, whose habits have involved considerable modifications in their structure.

The Thoracica include the common forms *Lepas* and *Balanus* and many other genera, which have retained more of their

Crustacean characters than their parasitic congeners, but owing to their fixed habit of life they have undergone numerous modifications. Both the parasitic and sessile forms pass through a freely mobile larval stage, and it is not until they reach the last of numerous larval stages that they either attach themselves to some host or settle down upon some foreign body. Whilst this is taking place, the larva, which is now in what is termed the *Cypris* stage, takes no food, but lives upon the nutrient material stored up in a well-developed fat-body, a condition of quiescence which recalls the pupa stage of many insects.

Like so many other sessile animals, the Thoracica fix themselves by their head ; the anterior end of this in the Lepadidae

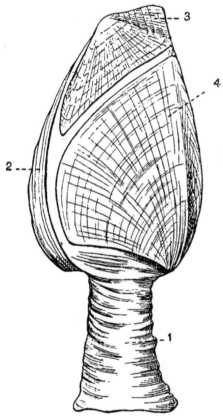

Fig. 161.—Side view of *Lepas anatifera*. After Leuckart.

1. Stalk.

2. Carina.

3. Tergum.

4. Scutum.

grows out into a long stalk which projects beyond the shell, and serves to lodge some of the internal organs of the body. The body is enclosed in a reduplication of the skin, which is

continuous with the body only in the region of the head.   This
mantle is strengthened by five calcareous plates, the *carina*
median and dorsal, two *scuta* near the stalk, and two *terga* at
the free end.   Owing to the disposition of these plates, the
mantle can only be opened along the ventral surface; when
opened the biramous thoracic legs can be protruded, and their
movement sets up currents in the water which bring food to
the mouth.   In *Pollicipes* and *Scalpellum*, genera in which the
stalk is smaller than in Lepas, there are a number of triangular
secondary calcareous plates which arise round the top of the
stalk and form a kind of collar round the body of the animal.
Some of these secondary plates are still further developed in

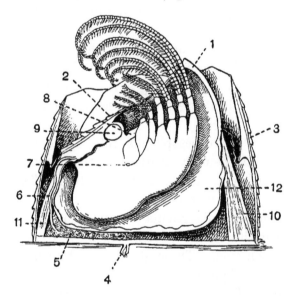

Fig. 162.—View of *Bal-
anus   tintinnabulum*
after removal of the
right shell.   From
Leuckart and Nitsche.
After Darwin.

1. Tergum.
2. Scutum.
3. Outer shell in section.
4. Antennae.
5. Ovary.
6. Oviduct.
7. Opening of oviduct.
8. Labrum.
9. Adductor muscle.
10. Depressor muscle of
    tergum.
11. Depressor muscle of
    scutum.
12. Mantle cavity.

Balanus, which is without a stalk, and here they form a
calcareous tube consisting of six pieces, the median dorsal one
being regarded as equivalent to the carina (Fig. 162).   Within
this tube the body of the animal lies protected by four plates,
the two scuta and two terga.

The head is not marked off from the thorax, except by the
position of the appendages, and the abdomen is reduced to a
short stump, at the end of which the anus is situated.

Appendages of *Lepas*.

1. Antennules.	4. 1st maxillae.
2. Antennae (lost in larval life).	5. 2nd ,,
3. Mandible.	6-11. 6 pairs of thoracic limbs.

The head bears one pair of antennae, the first. Its second joint bears a disk on which the duct of the cement gland opens. In *Lepas* and *Balanus* this appendage is minute, and situated at the base of attachment. The cement gland lies in the stalk of the stalked forms, its secretion hardens into a cement which serves to attach the animal to some foreign body.

The second pair of antennae, although present in the larva, are thrown off at the last moult, and are therefore not found in the adult. Above the mouth is an upper lip or labrum, and on each side of it is a mandible; between the mandible and the labrum is a structure called the palp, which, however, is not the homologue of the ordinary Crustacean palp. There are two pairs of maxillae, the posterior pair being fused together to form a kind of labium.

The thorax bears six pairs of appendages, the biramous nature of which recalls the limbs of the Copepods (Fig. 163). The rudimentary abdomen has no appendages, it terminates in a long penis, which is usually bent forward between the thoracic legs.

The mouth leads by a short oesophagus into a globular stomach provided with certain hepatic diverticula; the intestine passes off from the stomach and ends in a short rectum. The anus is situated dorsally at the base of the penis.

In addition to the hepatic diverticula certain glands have been described lying near the stomach, with which they communicate by a duct on each side. These glands secrete a fluid which has probably some digestive action, they have been termed pancreatic glands, formerly they were described as ovaries.

The Cirrhipedia seem to be devoid of any special circulatory apparatus. The space between the internal organs of the body is largely filled up with connective tissue, but certain cavities occur which seem to be lined by an endothelium, and which have been regarded as truly coelomic. Into these spaces a pair of ducts open by means of funnel-shaped orifices. These ducts have been traced in one or two genera, and have been found to open on to the exterior at the base of the second maxilla, the appendage which bears the aperture of the Entomostracan excretory organ, the shell gland. These tubes are believed to have an excretory function, and if their inner end opens into a genuine coelomic cavity, they would satisfy the conditions of a nephridium and form an interesting transition between these widely distributed excretory organs and the shell gland of other Entomostraca.

The existence of special respiratory organs is a matter of some uncertainty, certain processes at the base of the thoracic legs in L*epas,* and on the inner surface of the mantle in *Balanus,* have been regarded as branchiae by some authorities, but their precise function seems doubtful.

The nervous system of L*epas* is composed of a supra-oeso-phageal ganglion, connected by commissures of considerable length with a ventral chain. This usually comprises five ganglia, the fifth is larger than the others and gives off two pairs of nerves, hence it probably represents the fusion of two primitively distinct ganglia. In the Balanidae the whole ventral nerve cord has fused into a common nerve mass. The supra-oesophageal ganglion gives off nerves to the stalk, to the mantle, and to the eyes; the mouth appendages and the first pair of thoracic limbs are supplied from the infra-oesophageal ganglion. The only sense organs which are definitely known are a pair of rudimentary eyes, which have fused together in L*epas,* but are distinct in *Balanus.*

The Cirrhipedia are peculiar amongst Crustacea, in that with few exceptions they are, as is frequently the case with sessile animals, hermaphrodite (Fig. 163). The testes form a branched gland upon each side of the alimentary canal; the numerous caeca of this gland unite into a vas deferens which dilates into a vesicula seminalis on each side of the body, the two vasa deferentia unite into a single ductus ejaculatorius which traverses the penis. The spermatozoa are motile, a condition not otherwise met with in the Crustacea except amongst the **Ostracoda,** and then only when they have entered the female ducts.

The ovaries are very racemose glands, situated in the *Lepadidae* in the upper end of the stalk, in the *Balanidae* between the membranous or calcareous base of attachment and the mantle (Fig. 162). The oviducts, one on each side of the body, pass up towards the first thoracic limbs and open into a sac; this in turn opens by a duct lined with chitin at the base of the first thoracic appendages. The walls of this sac appear to secrete the cement by means of which the ova

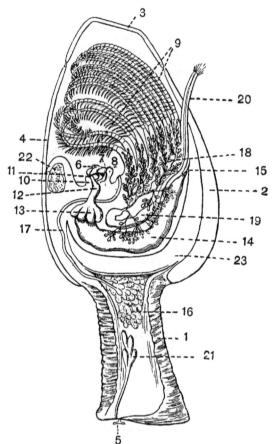

Fig. 163.—A view of *Lepas anatifera*, cut open longitudinally to show the disposition of the organs. From Leuckart and Nitsche, partly after Clans.

1. Stalk.
2. Carina.
3. Tergum.
4. Scutum.
5. Antennae.
6. Mandible with "palp" in front.
7. 1st maxilla.
8. 2nd maxilla.
9. The six pairs of biramous thoracic limbs.
10. Labrum.
11. Mouth.
12. Oesophagus.
13. Liver.
14. Intestine.
15. Anus.
16. Ovary.
17. Oviduct.
18. Testes.
19. Vas deferens.
0. Penis.
1. Cement gland and duct.
22. Adductor scutorum muscle, which closes the shell.
23. Mantle cavity.

are bound together. It will be noticed that the opening of the oviduct is situated much farther forward than in other Crustacea. It lies in the segment which succeeds that upon which the excretory organs open, and it has been suggested that the sac and its duct may possibly represent an original segmental organ.

The ova when laid are cemented together, and form in the *Lepadidae* flattened masses which are attached to a fold of the mantle, and form very obvious structures easily seen when the shell is opened.

In spite of the fact that, with few exceptions, the Cirrhipedia are hermaphrodite, there are certain genera, e.g. *Ibla* and *Scalpellum*, in which males have been described. These are dwarfed forms, often very degenerate, which live usually two or three at a time within the mantle of the normal hermaphrodite or female, and are known as complemental males. Their degree of degeneracy varies within wide limits, and it seems as a rule to increase with the depth of the water in which they live. Three species of *Scalpellum* which inhabit shallow water, *S. peronii*, *S. rostratum*, and *S. villosum*, are provided with males which have retained the division of the body into a capitulum and a peduncle ; in eight other species, which extend to a depth of 700 fathoms, the males have lost this division of the body, but still retain rudiments of shell valves ; in thirteen other species, which were almost all taken from depths of upwards of 1000 fathoms, even these have disappeared. The male of *Scalpellum regium*, one of the most degenerate, is a minute animal about 2 mm. long, attached to the inner surface of the scutum of the female by means of minute antennae which are provided with cement glands. No other appendages are visible. The body is oval, rounded at each end, and covered with a chitinous cuticle, which is produced into short spines. The alimentary canal is rudimentary and functionless, the mouth does not open, and the mouth appendages are gone. There are a supra-oesophageal ganglion, connecting cords, and one infra-oesophageal ganglion ; the eyes have disappeared. The generative apparatus alone is well developed, but the female organs have completely disappeared, the testis and vesicula seminalis are single.

The remarkable genus *Scalpellum* presents yet a further complication in its sexual arrangements. Its species may be arranged in three groups :

A. Those which are truly hermaphrodite, and in which no complemental males exist. Ex. *Scalpellum balanoides*.

B. Those which are hermaphrodite and yet possess complemental males. Ex. *Scalpellum villosum, peronii, vulgare, rostratum*, etc.

C. Those which are unisexual, the females large and normal, the males minute and living parasitically in the mantle of the female. Ex. *Scalpellum ornatum, regium, vitreum*, etc.

The genus *Ibla* presents only the latter two of these degrees of sexual differentiation.

The remaining groups of Cirrhipedes are parasitic, and have undergone considerable degeneration.

(ii.) The Ascothoracica consist of three species: *Laura gerardiac, Synagoga mira*, and *Petrarca bathyactidis*. These are parasitic or semiparasitic in the Actinozoa. They possess a large lateral carapace, into which the digestive and reproductive organs extend, a condition of things characteristic of Ostracods; in other essential respects they resemble the Cirrhipedes.

(iii.) The Abdominalia include the two genera *Alcippe* and *Cryptophialus*. They are unisexual, and in the former the male is dwarfed. They live parasitically, boring into the calcareous shells of Molluscs and Cirrhipedes.

(iv.) The Apoda are composed of only one genus, *Proteolepas*. It has a maggot-like body consisting of eleven segments, and has no thoracic or abdominal limbs. The mouth is a sucking one, and its appendages are present; the alimentary canal is rudimentary. It lives within the mantle of other Cirrhipedes, and seems to be truly parasitic, living on the juices of its host. It is hermaphrodite.

(v.) The Rhizocephala comprise a few genera which are parasitic chiefly on Crustacea. They have reached an extreme stage of degeneration, their body being rounded, and without any trace either of segmentation or of appendages. No alimentary canal exists in *Sacculina*, which is frequently to be found on the abdomen of *Carcinus moenas* or other crabs; the alimentary canal is not present even in the *Nauplius* larva. This genus, like so many other parasitic Cirrhipedes, attaches itself to its host in the *Cypris* stage, fixing itself on the carapace or legs just at the origin of a hair where the skin is soft. It usually chooses some young individual in which the integument has not yet completely hardened. The *Cypris* fixes itself at first by means of its first pair of antennae, it then moults and throws off its skin, and the cellular body of the *Sacculina* now migrates through the cavity of its antenna into the body of the crab. It then makes its way to the intestine and comes to rest in that region of the body where it will ultimately appear at the outside. In addition to the globular body, which is eventually pushed outside the skin of the crab's abdomen, the parasite has a short peduncle which passes through into the body of the host and gives off an immense network of roots which ramify through all the tissues of the crab, and extend even into its limbs. It is by means of these processes that the parasite absorbs its nutriment. This drain upon the resources of the crab does not seem to affect its health, but its growth is arrested, and as a consequence it does not cast its skin, an operation which would naturally be fatal to the *Sacculina*. Fortunately the latter appears to live only three years, and when it dies the crab resumes its growth.

B. MALACOSTRACA.

CHARACTERISTICS.—*The Malacostraca include the more conspicuous and more highly differentiated Crustacea. The number of segments and pairs of appendages is constant. With one exception, there are nineteen segments, each bearing a pair of appendages. The head consists of five, the thorax of eight segments. The abdomen has six segments, and ends in an unsegmented telson. The excretory organ usually opens on the second antenna, and is called the antennary gland, as opposed to the Entomostracan maxillary or shell gland. The Nauplius larva, so characteristic of the Entomostraca, is rare in the Malacostraca. Some of them hatch out from the egg in the adult condition ; the majority, however, pass through a complicated metamorphosis, the larva leaving the egg in the form of a Zoaea, which is characterised by the presence of the seven or eight anterior pairs of appendages, a swimming tail, and two lateral compound stalked eyes, as well as a median Nauplius eye. The proctodaeum and stomodaeum form a larger part of the alimentary canal of the adult than is the case in the Entomostraca.*

The MALACOSTRACA include three orders :

1. LEPTOSTRACA.
2. THORACOSTRACA.
3. ARTHROSTRACA.

Order 1. LEPTOSTRACA.

CHARACTERISTICS.—*This order includes but three recent genera: Nebalia, Paranebalia, and Nebaliopsis; they have bivalved shells, eight free thoracic segments with limbs of the Phyllopod type, and eight abdominal segments, with two caudal processes.*

Nebalia forms an interesting transitional form between the primitive Phyllopoda on the one hand, and the Malacostraca on the other. It is of small size, with a laterally compressed body, partially enclosed in a bivalved cephalothorax which springs from the head. This cephalothorax extends over all the thoracic and four of the abdominal segments; it is closed by a special transverse muscle like that of the Ostracoda.

The appendages are:

1. Antennules.
2. Antennae.
3. Mandibles.
4. 1st pair of maxillae.
5. 2nd pair of maxillae.
6-13. 8 pairs of thoracic limbs.
14-19. 6 pairs of abdominal limbs.

The second antennae in the male, which is a smaller animal than the female, are produced into processes as long as the body. The mandibles bear a three-jointed palp, a structure not found in Phyllopods. The eight thoracic segments are very short from behind forwards; they each bear a pair of foliaceous appendages, which closely resemble the typical Phyllopod limb. Amongst these the female carries her eggs. The eight abdominal segments are considerably longer than the thoracic; the first four bear large swimming-legs, consisting of a broad protopodite and an exopodite and endopodite; the next two segments bear similar but much smaller appendages, whilst the last two are limbless, so that the number of appendages is the typical Malacostracan nineteen, though there are twenty-one segments. The last segment bears two anal furcae, again an Entomostracan feature.

Another primitive arrangement presented by *Nebalia* is the fact that the ventral nerve cord is to some extent continuous with the hypodermis. In the abdomen the two cords run apart from one another, and are connected by transverse commissures. The eyes are stalked.

The characteristic Malacostracan excretory gland is found

opening on the second antennae. The Entomostracan shell gland, however, is found in the larva, and rudiments of it persist in the adult.

The Leptostraca are exclusively marine; they appear to be the survivors of what was once, judging from the numerous allied fossil forms, a large order. They possess great tenacity of life, and are able to live in water which is so full of the products of decomposition as to be fatal to most other animals. The order includes three genera,—*Nebalia, Paranebalia,* and *Nebaliopsis,*—and it is probably cosmopolitan.

ORDER 2. THORACOSTRACA.

CHARACTERISTICS.—*Malacostraca in which all the thoracic or the anterior thoracic segments are fused with the head to form a cephalothorax, which is enclosed by a dorsal shield or carapace. The eyes are compound, and, with few exceptions, are borne on moveable stalks.*

This order contains four sub-orders, the members of which have attained very different degrees of developement. Whilst some of them in their adult condition have a marked resemblance to the larvae of the more highly developed groups, others, as for instance the Decapods, are the largest and in some respects the most highly differentiated of the Crustacea.

The sub-orders which compose the order Thoracostraca are:

1. CUMACEA.
2. STOMATOPODA.
3. SCHIZOPODA.
4. DECAPODA.

Sub-order 1. CUMACEA.

CHARACTERISTICS.—*The carapace is small, and four or five of the thoracic segments are free. There are two pairs of maxillipedes, and six pairs of other thoracic legs; of the latter the two anterior at least are biramous or Schizopod-like. The abdomen in the female is devoid of appendages, except the sixth segment, which bears a pair of caudal styles; the male has a varying number, 2, 3, or 5 pairs of abdominal swimming-feet in addition.*

This is not a large group, and its relation to the other Thoracostraca is the subject of much difference of opinion. The best-known genera are *Diastylis* (= *Cuma*) and L*eucon*. Among the points of interest presented by these animals are the following :—the second antennae, rudimentary in the female, are in the male as long as the body, a sexual distinction which exists also in *Nebalia*. The mandible is without a palp, as in the Phyllopoda. There are only one pair of gills, which are of large size, and are borne by the second pair of maxillipedes. The two eyes lie close together, or have fused into a single mass, which, contrary to the usual rule in the Thoracostraca, is sessile. The brood-pouch, in which the eggs undergo their developement, is formed from processes, probably epipodites, of the fourth, fifth, and sixth thoracic legs, an arrangement which is also met with in the Schizopoda and the Arthrostraca.

Little is known about the habits of these animals; they live together, usually on a sandy bottom, at considerable depths, they rest during the day and come to the surface and move about at night.

Sub-order 2. STOMATOPODA.

CHARACTERISTICS.—*The carapace is short, and leaves a variable number of thoracic segments free. Portions of the head bearing the eyes and antennae are also free and moveable. There are five pairs of maxillipedes, and three pairs of biramous thoracic feet. The abdomen is large, and the abdominal appendages carry the gills.*

The Stomatopoda are animals of considerable size, the larger species of *Squilla* attaining a length of 7 or 8 inches or more. The carapace is short, and does not cover the three thoracic segments which bear legs; the five segments in front of these, which carry the five pairs of maxillipedes, are much shortened. The portion of the head bearing the eyes, and that carrying the antennae, are free and moveable, and the ventral portions of the segments covered by the carapace are also capable of a certain amount of movement upon one another. The six abdominal segments are large, and end in a broad telson. The first antenna consists of a basal shaft bearing three long multiarticulate flagella; the second antenna has a large squama or scale.

The mandibles bear a palp, and the two pairs of maxillae are small and weak. The maxillipedes are the most charac-

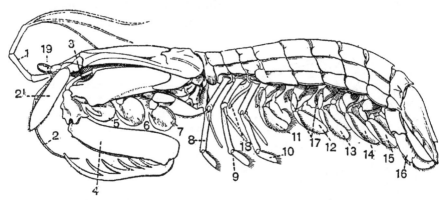

Fig. 164.—*Squilla mantis*, seen from the left side. After Leuckart and Nitsche.

1. 1st antenna.	8-10. 1st to 3rd biramous swimming-legs.
2. 2nd antenna.	
2¹. Scale, exopodite, of 2nd antenna.	11-16. 1st to 6th abdominal legs, the anterior five pairs bearing branchiae.
3. 1st maxillipede.	
4. 2nd maxillipede.	
5. 3rd maxillipede.	17. Branchiae.
6. 4th maxillipede.	18. Penis.
7. 5th maxillipede.	19. Eyes.

teristic appendages of this order; there are five pairs of them, turned forward towards the mouth (Fig. 164). The anterior pair are thin and feeble, but terminate in a small pair of sub-chelae which help to hold the prey; the second pair are the largest appendages of the body. Their terminal joints are strong and toothed, and shut down upon the penultimate like the blade of a knife into its handle; this arrangement has been termed subchelate, and it exists in all the maxillipedes of this group. The three succeeding maxillipedes are smaller, and terminate in small rounded subchelate joints. The three free thoracic segments carry biramous swimming-feet. The six abdominal segments also bear each a pair of swimming-feet, which are remarkable for carrying the gills on the external ramus. The first pair are modified in the male in connection with reproduction.

The elongated condition of the heart, which stretches from the thorax through the abdomen, is doubtless correlated with this position of the branchiae; in each segment it gives off a pair of lateral arteries, and in front it ends in ophthalmic and antennary arteries (Fig. 165).

The liver extends into the abdomen, and is divided into ten pairs of caeca. A pair of glandular diverticula open into the rectum near the anus; they are regarded as excretory.

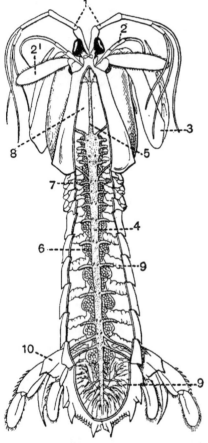

FIG. 165.—A *Squilla mantis,* with the shell removed from a portion of the dorsal surface to show the internal organisation. After Leuckart and Nitsche.

1. 1st pair of antennae.

2. 2nd pair of antennae.

2^1. Scale, exopodite, of 2nd antenna.

3. 2nd maxillipede.

4. Heart, with thirteen pairs of ostia, and numerous paired vessels supplying the limbs.

5. Median unpaired vessel supplying the eyes and antennae.

6. Ovary.

7. Oviduct opening at base of (8) in Fig. 164.

8. Stomach.

9. Liver diverticula from the intestine, these lie under the ovary.

10. 6th abdominal appendage.

The generative organs also lie in the abdomen, the oviducts, however, reaching into the three free thoracic segments, being situated between the alimentary canal and the heart. The vas deferens opens on the last pair of thoracic appendages, the oviduct on the last but two. An antennary gland does not appear to be present.

Unlike most Crustacea, the Stomatopoda do not carry their eggs about with them, but deposit them in the burrows in which they live. The eggs require for their developement a current of water, which is produced by the action of the abdominal swimming-feet of the parent. The group is exclusively marine; the members of it inhabit shallow water, and either live in crevices in coral rock, or burrow in the sand. They often rest in the mouths of their burrows with nothing but their eyes exposed above the sand; in this way they lie in wait for their prey, which they seize with astonishing rapidity by means of their powerful subchelate maxillipedes. They are extremely active in their movements, and difficult to catch, retiring to the inner end of their burrows at the slightest alarm.

Sub-order 3. SCHIZOPODA.

CHARACTERISTICS.—*Small Thoracostraca; the carapace is large and soft, the eight thoracic limbs are biramous. . Those modified to form maxillipedes do not differ markedly from the others. The eyes are stalked.*

The Schizopoda retain throughout their life a condition which resembles that of the last of the series of larval forms which take part in the metamorphosis of some macrurous Decapods. This larva is known as the *Mysis* stage, and is called after the opossum-shrimp, one of the best-known Schizopods.

The members of this sub-order differ markedly from the Stomatopoda, in which group the maxillipedes attain the greatest importance, for in the Schizopods each of the eight pairs of thoracic appendages are biramous legs, none of them being modified to form maxillipedes, although the two anterior pairs may tend in a slight degree to resemble the mouth appendages. The first antennae are biramous, and in the male are provided with a curious comb-like structure covered with olfactory hairs. The second antennae, a pair of mandibles, and two pairs of maxillae complete the appendages of the head.

The carapace is large, and attached to the body only by a narrow area of fusion in the dorsal middle line; in some forms, as *Siriella*, it leaves most of the thoracic segments uncovered.

The abdomen consists of six segments and a telson; the first five abdominal appendages in the female are usually small, but are better developed in the male; the sixth is flattened and lamelliform, and assists the telson in forming the swimming caudal fin. In *Mysis* an auditory sac is situated in the base of the endopodite of this last abdominal pair of limbs. In the family Euphausiidae there is a remarkable series of luminous organs, situated, as a rule, a pair on the peduncles of the eye, another on the basal joint of the second and seventh pairs of thoracic legs, and one in the middle ventral line on the first four abdominal segments. These organs emit at times a beautiful phosphorescent light, and they seem to be quite under the control of the animal. Their use is unknown. *Euphausia* is further remarkable amongst the Malacostraca for having a *Nauplius* stage in its ontogeny.

Sub-order 4. DECAPODA.

CHARACTERISTICS.—*Malacostraca with usually stalked eyes. The thoracic segments are fused with the cephalic, and with rare exceptions they are entirely covered by the carapace. The posterior five pairs of thoracic limbs are uniramous and seven-jointed, and some of them are chelate.*

The Decapoda include a great number of species which superficially exhibit considerable differences of form and habit, but in essential features the group is very homogeneous. It may be split up into two divisions.

Group 1. MACRURA.

CHARACTERISTICS.—*Abdomen long, four or five pairs of abdominal limbs, and a well-developed caudal fin.*

This includes the Carididae (shrimps and prawns), Astacidae (cray-fish and lobsters), Paguridae (hermit-crabs), etc.

Group 2. BRACHYURA.

CHARACTERISTICS.—*Abdomen short and reduced, and bent up and applied to the ventral surface of the thorax. In the male the abdomen is narrow, and rarely bears more than two pairs of feet; in the female it is broader, and has four pairs of appendages.*

This group includes the crabs, which are classified in a great number of families.

In Decapods the region of the head is usually marked off from the thoracic portion by a fold in the carapace, the cervical groove; and in the larger forms the cephalothoracic shield is frequently divided up into areas corresponding to the position of various internal organs. The last thoracic segment occasionally remains moveable on the others.

The sides of the carapace enclose the branchial chamber, and are known as *branchiostegites*; the number of gills

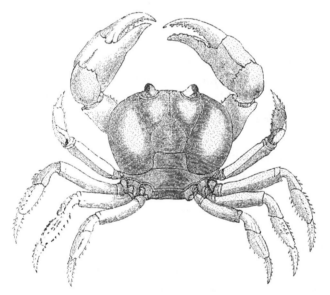

FIG. 166.—*Gecarcinus ruricola* (land-crab of Monserral, West Indies).

varies. The water-supply enters through the slit-like opening between the edge of the branchiostegite and the body in the Macrura, but in the Brachyura there is a special narrow opening situated in front of the first pair of walking-legs. The current is maintained by a specialised portion of the second maxilla, the *scaphognathite*, which flaps to and fro. This process either represents the epipodite, or the epipodite and exopodite fused. The two pairs of maxillae to some extent retain the foliaceous character of the primitive Phyllopod appendage, but all the other appendages depart widely from this type.

A few Decapods, as *Birgus latro*, allied to the hermit-

crabs, and *Gecarcinus* amongst the Brachyura, have forsaken
their natural element, and have come to live on land.   They
retain their gills, and the gill cavity contains both air and
water.   In *Birgus*—a curious animal which is said to climb
palm-trees at night—there is, however, a distinct modification
of structure.   The lower part of the gill space contains the
numerous but small gills; this is shut off from the upper part
by the infolded edge of the branchiostegite, and the cavity
thus formed is a true lung, since it is full of air, and its
walls are produced into numerous vascular folds, which receive
impure blood from the body and return it oxygenated to the
heart.   Such a change, from water-breathing animals to
terrestrial air-breathing animals, is paralleled in the Pulmonata, amongst which an intermediate form exists in the
tropical water-snail *Ampullaria*, which has both a well-
developed branchial cavity with gills and a well-developed
pulmonary chamber, and uses them alternately to breathe
water or air.

The obscure question as to the nature of the body-cavity
and the homology of the antennary or green gland in
Crustacea has recently had some light thrown upon it by
Weldon's researches on *Palaemon serratus* and other Decapods.
If the carapace of a *Palaemon* be removed a delicate sac will
be found occupying the dorsal part of the cephalothorax, and
extending from the anterior end of the head to the generative
gland, to which it is closely attached.   This cavity, termed
the nephro-peritoneal sac, is lined by epithelial cells, and
exhibits many of the relations of a coelomic body-cavity.
At its anterior end the sac gives off on each side a duct,
which passes down and opens into the urinary bladder, thus
putting the nephro-peritoneal sac in communication with the
exterior through the excretory organ.   The duct is lined by
glandular cells, and gives off numerous caecal processes, which
branch and ramify in the neighbouring tissues; in addition to
these there is also a structure known as the "end-sac," the
cavity of which also opens into the urinary bladder.   The
walls of this end-sac are produced inwards into its lumen,
dividing it up into a number of chambers lined with glandular
cells; these walls are well supplied with blood-vessels, and

the septa contain numerous capillaries, so that the whole organ has been compared with the glomerulus of the Vertebrate kidney, and doubtless subserves the same functions. It is by no means certain that the above-mentioned sac is coelomic in nature, and in his last paper Weldon is inclined rather to regard it as an enlarged portion of a nephridial system, such as exists in *Mysis*, and not as a remnant of a primitive coelom.

The history of the developement of *Palaemonetes varians*, recently described by Allen, throws some light upon the nature of the body-cavity in this group. The body-cavity of this animal consists of four regions : (i.) a dorsal sac in which the cephalic aorta lies, (ii.) a central cavity containing the liver, intestine, and nerve cord, (iii.) two lateral cavities containing the proximal ends of the shell glands, and (iv.) the cavities of the limbs containing the distal ends of the same glands. Of these, the first, or dorsal sac, is truly coelomic, its cavity being homologous with that of the dorsal portions of the mesoblastic somites of *Peripatus* (*vide* p. 308). The central and lateral cavities, and those of the legs, represent a pseudocoel. The nephro-peritoneal sac described by Weldon arises in this genus from an enlargement of the tube of the green gland.

In the long-tailed Decapods it is usual to find six pairs of abdominal ganglia, but in *Pagurus*, the hermit-crab, there is only one large one. In these animals the abdomen is soft and distorted into the shape of the interior of the Mollusc's shell which affords it shelter; the abdominal limbs are usually rudimentary, but the chelae or first pair of thoracic limbs are large, and can be closed down over the mouth of the shell, acting as a kind of operculum.

In the Brachyura a great concentration of the nervous system has taken place, and the ventral ganglia have all fused together into a common oval mass. The Decapods, in addition to the main chain of ganglia, possess a very extensive and complex system of visceral nerves.

Besides the olfactory hairs on the first pair of antennae, and the compound stalked eyes, it is characteristic of the Decapods to possess auditory organs, situated at the base of the antennules. These take the form of hollow sacs open to the exterior; their walls support auditory hairs connected with the auditory nerve. In the watery contents of the sac certain particles of sand are suspended, and act as otoliths. In connection with this it is interesting to note that certain genera can produce stridulating noises either, as in *Alpheus*, by clapping together the two claws on the larger leg, or, as in *Palinurus*, the rock-lobster, by rubbing the second joint of the antennae against a portion of the carapace.

ORDER 3. ARTHROSTRACA.

CHARACTERISTICS.—*Malacostraca with sessile eyes. No cephalothoracic shield; seven, rarely fewer, free thoracic segments, and as many pairs of legs. Heart elongated.*

The appendages of the Arthrostraca are arranged as follows : two pairs of antennae, one pair of mandibles, two pairs of maxillae, and one pair of maxillipedes, seven pairs of free thoracic legs, and six pairs of abdominal legs. The free thoracic segments are six in number in *Tanais*, five in *Anceus*, the anterior having in these genera fused with the head. The abdomen may be reduced to a mere process, but usually it is fully developed, and ends in an unsegmented telson.

The antennary gland is usually present, and opens on the base of the antenna. The compound facetted eyes are always sessile. The ova are usually carried about in brood-pouches formed by processes borne on the thoracic legs (*oostegites*). The young do not as a rule pass through a larval stage.

The Arthrostraca are divided into two sub-orders, characterised as follows :

Sub-order 1. AMPHIPODA.

CHARACTERISTICS. — *Arthrostraca with laterally compressed bodies ; the thoracic appendages carry the gills ; abdomen elongated ; it bears three anterior pairs of swimming-feet, and three posterior pairs of backwardly directed feet adapted for jumping.*

As a rule, the Amphipods are small, but a few, some of them living in Arctic seas or at great depths in the ocean, attain several inches in length. They inhabit both salt and fresh water, and progress by swimming or jumping. The males may usually be distinguished by the developement of their olfactory hairs on the first antennae, by the absence of

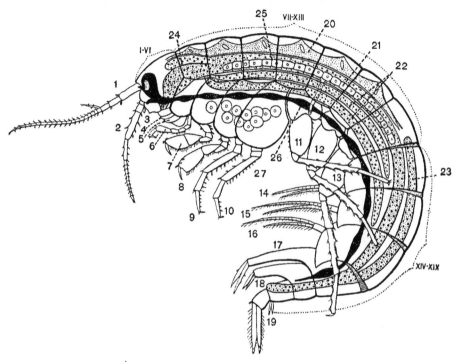

FIG. 167.—*Gammarus neglectus.* Female bearing eggs seen in profile. From Leuckart and Nitsche, after G. O. Sars.

I.-VI. Cephalothorax.
VII.-XIII. Free thoracic segments.
XIV.-XIX. The six abdominal segments.
1. Anterior antennae.
2. Posterior antennae.
3. Mandibles.
4. 1st maxillae.
5. 2nd maxillae.
6. Maxillipede.
7-13. Thoracic limbs.
14-16. Three anterior abdominal limbs for swimming.

17-19. Three posterior abdominal limbs for jumping.
20. Heart with six pairs of ostia.
21. Ovary.
22. Hepatic diverticula.
23. Posterior diverticula of the alimentary canal.
24. Median dorsal diverticulum.
25. Alimentary canal.
26. Nervous system.
27. Ova in egg pouch, formed from lamellae on the coxae of the three anterior thoracic limbs.

oostegites, and by the presence of strong prehensile hooks on the anterior thoracic feet.

Appendages of Amphipods.

1. Antennules.
2. Antennae.
3. Mandibles.
4. 1st maxillae.
5. 2nd ,,

6. Maxillipedes (fused).
7-13. Thoracic legs.
14-16. Abdominal legs (turned forward).
17-19. ,, ,, (turned backward).

The mandible carries a three-jointed palp, and the maxillipedes fuse in the middle line to form an under lip. The bases of the thoracic limbs of the females bear on their inner surface a process, probably the epipodite, which projects towards the middle line, and with its neighbours forms the brood-pouch.

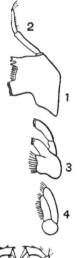

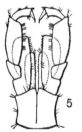

FIG. 168.—The mouth appendages of *Gammarus neglectus*. From Leuckart and Nitsche, after G. O. Sars.

1. The mandible.

2. Its palp.

3. 1st maxilla.

4. 2nd maxilla.

5. Maxillipede of each side together forming an under lip.

A little distal to this, but on the same surface, is a vascular process, the gill. The heart stretches through the thorax, and has usually a pair of valves in the second, third, and fourth thoracic segments. A well-developed "fat-body" is often present, lying amidst the viscera.

Certain Amphipods, as *Gammarus, Orchestia, Talitrus, Caprella*, etc., have two caeca which open into the alimentary canal at the junction of the mid and hind gut. These caeca, which stretch either backwards or forwards, contain numerous concretions in their walls. They have been compared to the Malpighian tubules of the Tracheata, but there is no certain evidence to show that they excrete nitrogenous waste matter from the system, and although they open close to its end, they belong to the mid gut and not to the proctodaeum.

The abdomen is reduced to a limbless process in the LAEMODIPODA, a group which includes *Cyamus ceti* parasitic on

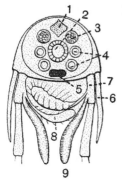

FIG. 169.—*Gammarus neglectus.* Section through the third thoracic segment. From Leuckart and Nitsche, after G. O. Sars.

1. The heart.
2. The alimentary canal.
3. The ovary.
4. The hepatic diverticula.
5. The nervous system.
6. The epimeron.
7. The thoracic appendage bearing—
8. The lamina forming the brood-pouch (oostegite), and
9. The branchiae.

the skin of whales, and in the family CAPRELLIDAE which comprises curious elongated thin animals of a very grotesque appearance, mostly found living amongst Polyzoa and Hydroids. The family PHRONIMIDAE includes some forms which have an enlarged head; the female *Phronima* is a good deal larger than the male, and is usually found swimming about in a barrel-shaped investment formed by the transparent colonies of the compound Ascidian *Pyrosoma*; other allied forms live in jelly-fish.

Sub-order 2. ISOPODA.

CHARACTERISTICS.—*Arthrostraca with usually a dorso-ventrally depressed body; thorax of six or seven free segments; abdomen often reduced, it bears lamelliform appendages, the endopodites of some of which function as gills.*

A very typical example of an Isopod is presented by the little brownish-gray animal *Asellus aquaticus*, which inhabits lakes, large ponds, and marshes. Its body is flattened, and composed of three regions—the head, the thorax, and the abdomen; of these the thorax is far the largest, and consists of seven free segments and one fused with the head. The abdominal segments have all fused together with the exception of the first, which is very small, and bears rudimentary appendages.

Appendages of *Asellus*.

1. Antennules.
2. Antennae.
3. Mandibles.
4. 1st maxillae.
5. 2nd „

6. Maxillipedes ⎱ thoracic.
7-13. Walking legs ⎰
14-19. Abdominal legs
(only five in the female).

The appendages are nearly the same as in Amphipods; there are two pairs of antennae, the second being much the larger; both are uniramous. The mandibles carry a palp. There are two pairs of maxillae and a pair of maxillipedes, and seven pairs of walking thoracic limbs, of which the anterior are directed forward, the posterior backward. The most anterior of all are prehensile organs, and lie forward against the maxillipedes; they are enlarged in the male, and take no part in locomotion. There are six pairs of abdominal appendages in the male, and five in the

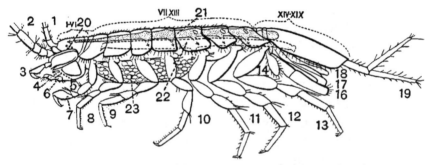

FIG. 170.—*Asellus aquaticus*. Side view of female. From Leuckart and Nitsche, after G. O. Sars.

I.-VI. The six anterior segments forming the cephalothorax.
VII.-XIII. The seven free thoracic segments.
XIV.-XIX. The six abdominal segments, partly fused.
1. The anterior antennae.
2. The posterior antennae.
3. The mandible (palp).
4. The 1st maxilla.
5. The 2nd maxilla.
6. The maxillipede.

7-13. The thoracic limbs.
14. 1st abdominal appendages.
16. 3rd abdominal appendages.
17. 4th abdominal appendages.
18. 5th abdominal appendages.
19. Last abdominal appendage.
20. Eyes.
21. Heart, with three pairs of stigmata.
22. Ovary and oviduct opening on base of 5th thoracic limbs.
23. Brood-pouch.

female; in the former the second pair are modified in connection with the opening of the genital duct, this pair are absent in the female. The three following pairs act as branchiae; the exopodite and endopodite are both squarish, lamelliform structures, the former lying over and protecting the latter, which has very

thin walls, and acts as the branchia. The last pair of appendages project backward like caudal processes.

The mouth is situated far forward, and the short oesophagus
lined with chitin has a nearly horizontal course. The stomach
is small, and lies almost entirely in the cephalothorax; it is
also lined with chitin, which is thickened at parts, and it contains a number of hairs and teeth, which may act as strainers.
The intestine is a wide tube running quite straight to the
anus, which lies on the ventral surface of the posterior end of
the body. The liver consists of two pairs of lateral caeca,
which extend back as far as the intestine. They open by a
common duct on each side, at the junction of the stomach with
the intestine.

There is a cerebral ganglion connected by commissures
with two fused pairs of sub-oesophageal ganglia, then come seven
pairs of ganglia corresponding with the seven free thoracic segments, and finally a complex ganglion formed of three or four
ganglia fused together, which supplies the abdomen. The commissures which connect the various ganglia are separate and
distinct.

The elements which form the eyes are arranged on each
side of the head in four patches (Fig. 172); there do not
seem to be any auditory organs. Olfactory hairs exist on the
small anterior antennae.

The heart is an elongated vessel situated in the dorsal
middle line; it appears to be closed behind, but anteriorly it

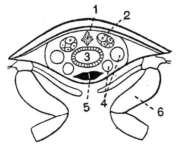

FIG. 171.—*Asellus aquaticus.* Section through
third thoracic segment. From Leuckart and
Nitsche, after G. O. Sars.

1. Anterior end of heart (aorta).
2. Ovary.
3. Alimentary canal.
4. Hepatic diverticula.
5. Nervous system.
6. Base of thoracic limbs with inward projections
(oostegites) which form the floor of the brood-
pouch.

is prolonged into an aorta. Three pairs of ostia situated in
the last three thoracic segments open into the heart. The
widest part of the heart is situated in the abdomen; the aorta

gives off a vessel which supplies the head and its appendages; both the aorta and the heart also give off lateral branches, which supply each segment. No antennary gland is present.

Asellus, like the majority of Isopods, is dioecious. The ovaries consist of two sacs, entirely independent of one another, which occupy the thorax, and extend into the abdomen. A little behind their middle point, in the fifth free thoracic segment, a

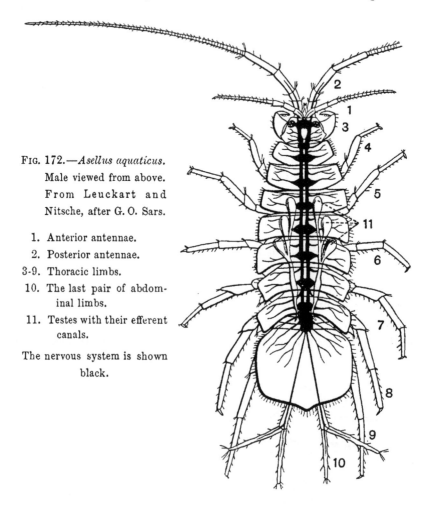

Fig. 172.—*Asellus aquaticus.*
    Male viewed from above.
    From Leuckart and
    Nitsche, after G. O. Sars.

    1. Anterior antennae.
    2. Posterior antennae.
    3-9. Thoracic limbs.
    10. The last pair of abdom-
       inal limbs.
    11. Testes with their efferent
       canals.

The nervous system is shown
       black.

short oviduct descends from the ovary, and opens on the sternum of that segment; this oviduct seems also to function as a receptaculum ovorum; at their exit the ova pass into the brood-pouch formed by oval lamelliform processes (oostegites) borne by each of the four pairs of anterior walking limbs.

The testes lie in the five posterior thoracic segments. They are independent of one another, and consist of three sacs, which open one after another into a narrow vas deferens, the latter ends in an elongated opening on the last thoracic segment (Fig. 172).

The Isopoda are carnivorous; many of them are parasitic, but seldom completely endoparasitic, living chiefly on the skin or in the mouth of fishes, or in the branchial chamber of Decapods. The group Anisopoda, which includes the genus *Tanais*, approaches most nearly the Amphipods. *Tanais* has a small carapace, which includes the first of the thoracic segments which bear walking legs; its body is not much depressed, and generally it has an Amphipod appearance. The abdominal legs are adapted for swimming, and do not function as gills. The heart also lies in the thorax.

Two different forms of male are described in *Tanais dubius*: one provided with a great developement of olfactory hairs on the first antennae, the other with large clasping appendages.

The other group of the Isopoda, the Euisopoda, have a relatively small abdomen, which carries branchial feet. The group contains several families, some of which are parasitic, and some of which are terrestrial in their habits.

The CYMOTHOIDAE are partly parasitic on fish, partly free-living; their mouth parts are adapted either for biting or sucking. They are remarkable amongst Malacostraca for being hermaphrodite and protandrous, the young animals producing spermatozoa, the older animals ova.

The SPHAEROMIDAE, which live in salt and brackish water, are interesting from their habit of rolling themselves up in a ball like the Oniscidae or wood-lice. The IDOTEIDAE have their last abdominal appendages modified to form an operculum, which protects the preceding branchial limbs; the same pair of appendages in the ASELLIDAE are styliform, and project backward; many of the last-mentioned family are freshwater inhabitants; one species bores holes in wood submerged in the sea.

The BOPYRIDAE are a very remarkable family of Isopods, which live parasitically in the branchial chamber of prawns.

They are dioecious and markedly dimorphic, the dwarf male living attached to the body of the female. The female is somewhat disk-shaped, asymmetrical, and with but slight traces of segmentation, and no eyes. The male is elongated, segmented, and provided with eyes. The mouth parts are rudimentary ; in the female the seven thoracic legs bear oostegites, which form a brood-pouch. In those species with a piercing mouth the upper and lower lips form a suctorial tube, within which the mouth parts are enclosed in the form of piercing stylets. A very remarkable fact connected with the parasitism of the Bopyridae is that the presence of certain species of this group has the effect of materially altering the reproductive organs of its male host. For instance, a male *Pagurus* infested by *Phryxus paguri* has its abdominal append-ages of the female type, and its testis is degenerate and the spermatozoa imperfect. The same is true for many other species.

In the family ENTONISCIDAE the more complete parasitism has involved a greater departure from the ordinary Isopod type. Their body is a limbless sack, which lives in an in-vagination pushed into the bodies of Cirrhipedes, Paguridae, and Crabs. The invagination retains its opening to the exterior. The sexual forms of *Portunion* are very remarkable ; the large deformed parasite is a protandrous hermaphrodite, functioning in its adult stage as a female; besides this form there are small males which remain in a larval stage, and also degraded complemental males. It seems that out of the numerous larvae, those which obtain the most advantageous position in the host will become the hermaphrodites and will ultimately produce eggs ; those which occupy the next best position become the males, which retain their larval features, whilst the most disadvantageously situated larvae degenerate and form the complemental males. The genus *Cryptoniscus* is commonly parasitic on the Rhizocephala, which are themselves parasitic on other Crustacea ; it is stated that if the Rhizo-cephala die the Isopods still continue to receive nutriment through the root-like processes of their dead hosts, as if they were parts of its own body.

The family ONISCIDAE, the wood-lice, have become terrestrial in their habits, living on the land, usually in damp places; in several respects their structure approaches that of the Insects; the mandible has no palp, and the exopodites of two or more of the abdominal limbs are sometimes provided with tubular air passages, which are respiratory; they show, however, no detailed resemblance to the tracheae of the higher Arthropods. The first antennae are in many species quite rudimentary. Many of them, like the SPHAE-ROMIDAE, possess the power of rolling themselves up in a ball when disturbed.

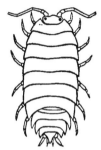

FIG. 173.—*Oniscus asellus*, the common Wood-louse.

CHAPTER XVII

TRACHEATA

Class I. PROTOTRACHEATA

Onychophora (Peripatidea), *Peripatus*

CHARACTERISTICS.—*Animals with soft caterpillar-like bodies, with a pair of antennae, a pair of jaws, a pair of oral papillae, and numerous pairs of walking legs, each bearing claws. Respiration by means of tracheae, the stigmata of which are numerous and scattered. Nephridia present.*

The Prototracheata include a single genus, *Peripatus*, which is morphologically one of the most interesting animals known; its developement, made known to us chiefly by the researches of Sedgwick, is also of the highest morphological import, and our recently-acquired knowledge of it has solved many problems of Arthropod structure. Its distribution is also interesting; it has been found in New Zealand, Australia, the Cape of Good Hope, South America, the West Indies, and possibly in Sumatra.

Peripatus capensis is a very beautiful-looking animal, of a velvety green colour dorsally, shading off into an orange-brown on the under surface. The female is larger than the male, and when adult may measure 65 mm., the male being about three-quarters as long. The skin is thrown into numerous ridges, which are beset with a number of papillae, each bearing a spine.

Appendages of *Peripatus capensis.*

1. Antennae.
2. Jaws in buccal cavity, armed with two cutting blades.
3. Oral papillae, with orifice of slime glands.
4.-20. Seventeen pairs of clawed walking-feet.
21. Anal papillae.

A pair of large antennae are borne on the head; these are composed of a series of annuli, each ring consisting of a number of fused papillae, and consequently beset with rows of short

FIG. 174.—Large adult example of *Peripatus capensis* of natural size. From Moseley.

bristles. At the base of each antenna, on the dorso-lateral part of the head, an eye is situated. At each side of the head is placed an " oral papilla," at the free end of which the slime glands open. Within the buccal cavity lie a pair of jaws, each armed with a pair of sickle-shaped toothed cutting blades, which are cuticular in origin.

The remaining appendages are seventeen pairs of walking-legs, which, with the exception of the fourth and fifth pairs in both sexes and the last pair in the male, resemble one another closely. Each leg is of a truncated conical shape, and bears rings of papillae; it terminates in a foot provided with two cuticular claws. On the inner surface of each leg, near its base, is the slit-like opening of a nephridium. The anus is terminal, and guarded on each side by the last pair of appendages, the anal papillae. The genital opening is sub-terminal and ventral.

The skin consists of a layer of hypodermal cells which secrete the cuticle; within this is a double stratum of circular muscle fibres crossing one another slightly obliquely, and surrounding a layer of longitudinal fibres. The latter are arranged in bundles: two dorsal, two lateral, and three ventral. A sheet of dorso-ventral muscle fibres runs from the space between the dorsal and lateral bands to the outer side of the median ventral band, dividing the body-cavity into three longitudinal spaces: a median which contains the alimentary canal the slime glands and the generative organs, and two lateral which lodge the nephridia, the salivary glands, and the nervous system. The lateral space is continuous with the cavities of the feet. There are special muscles which move the limbs, and others in the walls of the alimentary canal; throughout the body the muscle fibre is unstriated, except in the case of the muscles attached to the jaws, where the fibres are distinctly striated.

The buccal cavity, which contains the two jaws, opens behind into a muscular pharynx, but it is continued ventrally somewhat behind this opening, and the recess so formed receives the united duct of the two salivary glands. These are glands of considerable size, and they extend a long way down the body, lying in the lateral portion of the body-cavity; they have a

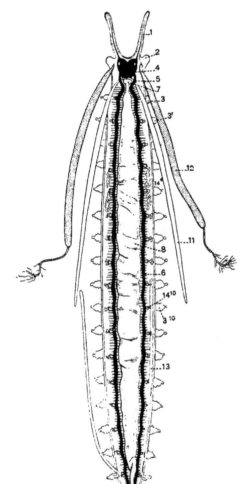

Fig. 175.—*Peripatus capensis,* male, dissected to show the internal organs. × 2. After Balfour.

1. Antennae showing antennary nerve.
2. Oral papilla.
3, 3^1, 3^{10}. 1st, 2nd, and 10th leg of right side.
4. Brain and eyes.
5. Circum-oesophageal cord.
6. Ventral nerve cord of right side, showing the transverse commissures.
7. Pharynx.
8. Stomach.
9. Anus.
10. Male generative opening.
11. Salivary glands.
12. Slime glands and reservoir.
13. Enlarged crural gland of the 17th leg.
14^4, 14^{10}. 4th and 10th nephridia of right side.

remarkable origin, since the history of their developement shows that they are modified nephridia, a condition of things which recalls the modified segmental organs which open into the oeso-phagus in some of the Oligochaets. The pharynx is very muscular, and has a triradiate lumen; it is strikingly unlike the same region of the digestive tube in any other Arthropod, but shows considerable resemblance to the pharynx of many of the Chaetopods. It communicates by a short oesophagus with a large stomach, which extends from the level of the second pair of legs to nearly the posterior end of the body. The stomach lies quite freely unsupported by any mesentery; it ends in a short muscular rectum, which passes to the anus. The pharynx, oesophagus, and rectum are all lined with chitin con-tinuous with that covering the body.

The vascular system consists of a tubular heart, which lies between the two dorsal muscle bundles along the whole length of the body. Paired ostia situated two in each segment, on the dorsal surface of the heart, put the cavity of this organ in communication with the surrounding pericardial space. The pericardial membrane which limits this space is attached on each side to the body-wall; it forms a kind of meshwork with numerous holes, by means of which the central division of the body-cavity communicates with the pericardium.

The space spoken of above as the body-cavity in which the chief· organs of the body lie, is not a true coelom. That is, it does not communicate with the exterior by means of nephridia, the lining of its walls does not give rise to the generative cells, and it does not arise as a definite split in the primitive mesoblastic somites. A true coelom is found in *Peri-patus*, but it has a different fate, and the space mentioned above as the body-cavity has nothing to do with it. The latter space originates either as a split between the endoderm and ectoderm, or as a hollowing out of spaces which appear in the thickened somatic walls of the somites. Speaking broadly, the central compartment and the heart arise in the former manner, the lateral spaces, with the cavities in the appendages and the pericardium, arise in the latter. These various cavities all communicate with one another. Thus the heart and the various portions of the body-cavities form a series of spaces which have been termed haemocoelic. They have nothing to do with the coelom, but are "a series of enormously dilated vascular trunks, of which the heart is the narrowest, and alone possesses the property of rhythmically contracting."

The heart is suspended in the pericardium by a number of strands of tissue, and around these a curious collection of cells, which has been compared to the " fat-body " of insects, is aggregated. This tissue, from its distribution, has probably some action on the blood, and it is possible that it represents functionally the lymphatics of Vertebrates and the botryoidal tissue of Leeches.

The respiratory system is tracheal. Each trachea consists of a simple opening to the exterior,—the stigma,—a short tube which dilates at its inner end, and a number of very minute tracheal tubules which arise from the dilated inner end of the tube, and which are distributed to the tissues of the body. The tracheal tubules are very fine, and for the most part unbranched; they exhibit slight traces of spiral striation. The stigmata are more or less evenly scattered over the whole surface of the animal, opening between the papillae; there are, however, a double dorsal and a double ventral row, and some on the anterior and posterior face of each leg; there is also a large median ventral stigma just in front of the mouth; this supplies part of the nervous system.

A nephridium occurs in the lateral section of the body-cavity at the base of each foot (Fig. 175). Each nephridium consists of the following parts : (i.) the external opening which lies near the base of each leg; this leads by a short duct into (ii.) an expanded vesicle or collecting portion lined with large flat cells; from this a coiled tube (iii.) passes, which is lined in various regions with four different kinds of epithelial cells. This tube ends in a funnel-shaped internal opening, the lips of which are continuous with the walls of a vesicle. The study of the developement has shown that the lumen of this vesicle, like the nephridium itself, is part of the original coelom, which has become separated off from the rest (Fig. 176). The first two somites, those of the antennae and claws, do not develope nephridia; the nephridia of the third, the oral papillae, are converted into the salivary glands; the next three segments have small, and the fourth and fifth pairs of legs, enlarged nephridia; the remaining twelve are normal, and resemble one another.

The nervous system consists of a pair of supra-oesophageal ganglia, united by circum-oesophageal commissures with two ventral cords. Anteriorly each supra-oesophageal ganglion is prolonged into a nerve, which runs to an antenna (Fig. 175). Several nerves run to the skin, and laterally the optic nerves arise, and supply the eyes. A pair of sympathetic nerves emerge posteriorly from the brain; they run back in the wall of the pharynx, and unite on the oesophagus; both in their origin and distribution they closely resemble the sympathetic nerves of Chaetopods.

Posteriorly the supra-oesophageal ganglia are continuous with the circum-oesophageal commissures, which in their turn pass into the ventral nerve cords. These latter lie in the lateral division of the body-cavity, and are consequently separated by a wide interval, but, like the pedal nerves of the Isopleurous Gasteropods, they are connected by a number of transverse commissures, nine or ten in each segment, the first of these lying immediately behind the mouth. There are seventeen ganglionic enlargements on the ventral cords, corresponding with the seventeen pairs of legs, and the circum-oesophageal commissures bear two ganglionic swellings, which supply nerves to the jaws and oral papillae. The ganglion cells are not confined to the swellings, but are distributed all along the cord. Each ventral ganglion gives off two large nerves, which pass into the legs, and a number of smaller nerves, which pass to the body-wall, etc.

One of the most interesting features of the nervous system of *Peripatus* is that the ventral nerve cords pass dorsally at the posterior end of the body, and fuse together above the anus (Fig. 175), like the united visceral and pedal commissures in Chaetoderma.

The eyes lie at the base of the antennae; beneath them is an optic ganglion; in their minute structure they resemble the eyes of Chaetopods or Gasteropods rather than those of Arthropods.

The female is larger than the male, but the only external structural difference is the existence of a small white papilla

20

on the ventral surface of each of the two seventeenth legs
in the male; the enlarged crural gland of this segment opens
at the apex of this.   The genital organs are all included in
the central division of the haemocoel.   The unpaired external
opening is ventral, and a little in front of the anus; on each
side of it is a genital papilla, which represents the appendage
of the twenty-first somite, and shows its homology with the
legs by sometimes bearing claws.   The male organs consist on
each side of a testis, a prostate gland, and a vas deferens.
The two vasa deferentia unite into a single duct which opens
at the genital pore; at the same spot two glandular tubes,
accessory glands, also open.   The crural glands of the seventeenth pair of legs are very much enlarged in the male, but
their exact function is not known.   The spermatozoa are
filiform, and are united into bundles or spermatophores; these
are deposited by the male on any portion of the body of the
female.   It is unknown how the spermatozoa reach the ova, but
they are always to be found in the cavity of the ovary.

The ovary is unpaired, but is divided by a septum into two
tubular halves; it is provided with two oviducts, which dilate
into uteri.   The ovary lies between the fifteenth and sixteenth
pairs of legs; a receptaculum seminis is present, and cilia have
been detected on its walls.   The occurrence of cilia is remarkable, as they are not found elsewhere in the Arthropoda.
Peripatus capensis is viviparous; the fertilised eggs pass into the
uteri in April, but are not hatched till the May of the following year; the period of gestation is thus thirteen months, and
for the first month the ova of one generation, and the nearly
mature embryos of the previous generation, co-exist in the uterus.

The ducts of the generative glands are formed from the
same portion of the twenty-first mesoblastic somite which
gives rise to the nephridia in other segments; they may therefore be regarded as modified nephridia, or, in other words, the
generative ducts are nephrodinic.   The generative glands
themselves, the ovary and testes, are formed from parts of
the true coelom of the sixteenth to the twentieth somite,
and thus, as in so many other animals, the generative cells
arise from the lining of the coelom, and pass to the exterior
through modified nephridia.

Certain glandular structures, known as crural glands, occur in every pair of legs except the first; they consist of a vesicle placed in the lateral portion of the body-cavity, and lined with columnar cells; this communicates with the exterior by a short tube. The crural gland of the seventeenth pair of legs is in the male enormously enlarged, and reaches forward as a tubular diverticulum as far as the ninth pair of legs; its exact function is unknown.

A pair of slime glands lie in the central division of the body-cavity; each communicates with a reservoir which opens by a duct on one of the oral papillae. They excrete a sticky slime, which is ejected with some force when the animal is irritated.

In *Peripatus Edwardsii,* a South American species, the crural glands are absent in the female, and only occur in the male in two segments; the genital opening is between the last pair of legs, and there are no genital papillae; the ovary is double, and the spermatophores are $1\frac{1}{2}$ inch long. The stigmata are irregularly scattered. In *Peripatus Novae Zealandiae* the tracheae are said to be branched.

The importance of *Peripatus* is twofold: in its structure it combines features of two or three different phyla, and the history of its developement has done much to explain the more peculiar characteristics of Arthropod structure, and to bring these animals in line with what is known of other groups.

Both its distribution and its structure point to its being a very archaic form. The arrangement of its nervous system, the sympathetic nerves, the muscular pharynx, the structure of the eyes, the serially repeated nephridia, the comparatively short stomodaeum and proctodaeum, the thinness of the cuticle, and the hollow nature of the appendages, are all features which are naturally associated with the Chaetopoda; some of these features are also met with in the more primitive Mollusca. On the other hand, the segmented appendages, the modification of some of the appendages as mouth organs, the presence of antennae, the tracheal nature of the respiratory organs, the structure of the heart and of the generative organs, are all features shared in common by the Tracheata and many by the Arthropoda in general. Thus *Peripatus* combines some of the more important characteristics

of two, if not three, of the most important phyla of the animal
kingdom.   Its developement is no less interesting; it would
be beyond the scope of this book to dwell upon the early
stages of the embryology, but some of the later stages
have cleared up so many difficulties in the structure of
adult Arthropods that a short reference to them must be
made.

The morphological nature of the body-cavity, and of the
heart, which opens freely into it in all Arthropods, has always

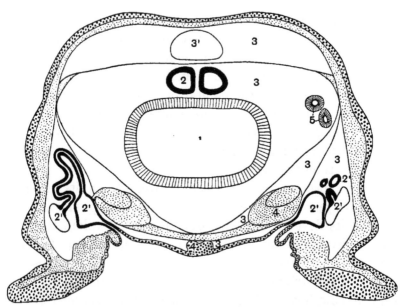

FIG. 176.—Diagram of the arrangement of the coelomic and haemocoelic cavities of
          Peripatus, at the time of its birth.   After Sedgwick.

1. Enteron, alimentary canal.   3[1]. Haemocoelic cavity of heart.
2. Coelomic cavity of generative gland.   4. Nervous system.
2[1]. Coelomic cavity of nephridia.   5. Slime glands.
3. Haemocoelic cavity, general "body-
     cavity."
        The true coelom is everywhere surrounded by a thick black line.

been a difficulty; the developement of *Peripatus* shows us that,
in this genus at all events, it has no connection with the true
coelom, but is a haemocoel, and arises partly by a hollowing
out in the tissue of the mesoblastic somites and partly from
the persistence of a split between the ectoderm and endoderm.
The continuity of the walls of the generative organs with the
genital ducts is explained by the fact that the generative

gland is a persistent portion of the original coelom, and the ducts are in all probability modified nephridia, which open into this portion of the coelom. The ova and spermatozoa are formed from cells lining the coelom, as is so commonly the case in the Coelomata. A small portion of the coelom also persists as a vesicle or swollen termination to the nephridia, which therefore correspond in their relations with the nephridia of other animals, and are tubes opening from the coelom to the exterior. Nephridia have not yet been described in many of the Arthropods, so that, although their presence in *Peripatus* is extremely interesting, the details of their relationship are not so important for the elucidation of Arthropod structure as are the origin of the body-cavity, heart, and generative organs ; and if we may assume that these latter organs arise in other Arthropods in essentially the same way as they do in *Peripatus*, we shall have an explanation of some of the most difficult problems of Arthropod morphology.

Specimens of *Peripatus capensis* are found beneath stones and bark, especially amongst rotten wood, on the slopes of Table Mountain. They are animals which avoid the light, and require moist surroundings. They move slowly, testing the ground with their antennae, which are very sensitive ; the body is borne above the ground by the numerous legs. When irritated, they can expel their slime, which is very sticky, to a distance of almost a foot. They are carnivorous, their food consisting of small insects, etc. The female will produce from thirty to forty young, which are born in the spring, and may be seen at this time of year crawling over the body of their mother.

CHAPTER XVIII

TRACHEATA

Class II. MYRIAPODA

Myriapoda $\begin{cases} \text{Chilopoda—\textit{Lithobius, Scolopendra, Scutigera.}} \\ \text{Diplopoda (Chilognatha)—\textit{Iulus, Polyxenus, Pauropus.}} \end{cases}$

CHARACTERISTICS.—*Tracheate Arthropoda with a distinct head and a number of similar somites; no distinction of thorax and abdomen. A pair of antennae, mandibles, and maxillae, and numerous six-. or seven-jointed clawed legs present.*

Of all the Tracheata, with the exception of *Peripatus*, the Myriapoda exhibit least signs of specialisation in their external structure. The segments posterior to the head, which in the GEOPHILIDAE may amount to some hundreds, are all very similar, and they are not externally grouped together into any regions such as thorax or abdomen.

The Myriapoda are divided into two orders : (i.) the Chilopoda and (ii.) the Diplopoda or the Chilognatha.

Order 1. CHILOPODA.

CHARACTERISTICS. — *Myriapoda with dorso-ventrally compressed body. Antennae long, with many segments. The second pair of postoral limbs form the poison claws. One sternum and one pair of legs to each segment. Genital orifice posterior. Stigmata lateral, tracheae branching and anastomosing.*

Lithobius forficatus is the commonest English centipede ; it is found all over Europe, in the summer living under stones, leaves, etc., and in the winter hiding itself in the earth. It is about one to one and a half inch long, and of a chestnut-brown colour. In external appearance it is very like its larger subtropical congener Scolopendra.

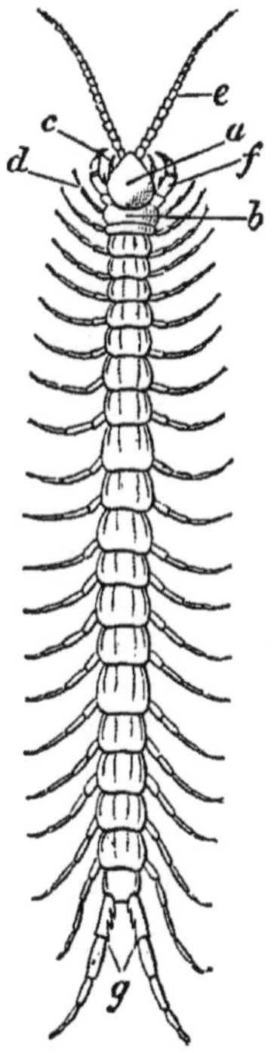

The head bears the multiarticulate antennae, with about forty segments, and just behind their insertion, the eyes, which are composed of twenty-five to forty single eyes grouped together. Immediately behind the head, and articulated with it, is the very narrow segment which bears ventrally the poison claws. Behind this are fifteen similar segments, nine of which are larger than the others, and somewhat irregularly arranged; each segment is covered dorsally by a squarish tergum; the difference in size of the segments is not apparent on the ventral surface, the sterna being all of the same size. The terga and sterna are connected laterally by a soft flexible skin which bears the stigmata.

Appendages of *Lithobius forficatus.*

1. Antennae.
2. Mandibles.
3. Maxillae.
4. Limb-like appendages.
5. Poison claws.
6-20. Fifteen walking-legs.

The appendages are: (i.) the antennae; (ii.) a pair of mandibles with a toothed cutting edge, they are jointed, but, like those of Insects, are devoid of a palp (Fig. 178); (iii.) a pair of maxillae, consisting of a palp and a fused median portion; (iv.) a pair of limb-like appendages turned forward, with their bases in contact. Behind this are (v.) the poison claws, a pair of stout large claws which contain in their last two joints the poison gland; the duct of this opens on the convex side of the apex. In *Scolopendra* a large basilar segment succeeds the head; this consists of four embryonic segments fused, and bears iv. and v., and sometimes a pair of walking legs, but the latter are frequently lost in the adult (Fig. 177). They do not occur in *Lithobius.*

The remaining fifteen segments bear each a pair of seven-jointed legs, arising on the lateral margin of the ventral sur-

FIG. 177. — *Scolopendra morsitans.* After Buffon.

a. Cephalic tergite.
b. Basilar tergite.
c. First postcephalic appendage (=third postoral).
d. Third postcephalic appendage.
e. Antenna.
f. Second postcephalic appendage (=poison claw).
g. Last pair of appendages enlarged and turned backwards.

face, and terminating in a claw. The fifteenth pair are often directed backwards. The anus is terminal, and the genital pore in both sexes is just in front of it, on the ventral surface.

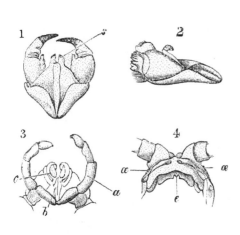

FIG. 178.—Mouth parts of *Scolopendra morsitans*. After Buffon.

1. The poison claws or 4th postoral appendages.
   s. Median cutting processes formed by the anterior edge of the basilar sterna.
2. One of the mandibles with its cutting edge to the left.
3. The maxillae and 3rd postoral appendages.
   a. Limb-like appendages.
   c. Palp-like maxillae.
   b. Small process formed by their fused bases.
4. Ventral view of head with jaws removed.
   œ. Eye.   *e*. Labrum.

The alimentary canal runs straight through the body without any loops. It consists of a short stomodaeum, which receives near the mouth the secretion of a pair of salivary glands, of a wide mesenteron, and of a rectum, which opens by the terminal anus. Two long white tubes—the Malpighian tubules—open into the posterior end of the mesenteron; they are longer than the body, and are consequently somewhat coiled. The Malpighian tubules secrete uric acids and urates, and thus function as organs for the excretion of nitrogenous waste matter.

The **Chilopoda** are carnivorous, *Lithobius* living upon flies, insect larvae, and earthworms, which are quickly killed by the poison from the poison claws; these also assist by holding the prey whilst the jaws tear it in pieces.

The alimentary canal occupies a considerable portion of the body-cavity; this space is to some extent occluded by the " fat-body " which in Myriapods and Insects forms large masses of tissue abundantly supplied with tracheae. The body-cavity is full of blood, which is kept in circulation by the contractile heart. The latter organ stretches all along the dorsal surface, lying immediately beneath the integument. It consists of fifteen chambers, one in each segment : the first of these, which lies just behind the cephalic shield, gives off a median and two lateral vessels; the former supplies the head and mouth parts, the latter form a ring round the oesophagus and unite to form a ventral vessel which lies above the nerve cord. These vessels open into lacunae, which ultimately communicate with the body-cavity, and from this the blood returns to the heart through paired valvular openings in each chamber. The heart is supported by a pericardium, to which are attached certain muscles, the *alae cordis*; these arise from the body-wall. The blood is colourless, and contains numerous white corpuscles.

A pair of stigmata are found on the third, fifth, eighth, tenth, twelfth, and fourteenth leg-bearing segments, situated on the soft tissue between tergum and sternum, immediately behind the base of each leg. Each stigma opens into a swollen sac, which gives off two main tracheae and a number of smaller ones, which supply the neighbouring parts. The arrangement of the larger branches varies in the anterior and posterior ends of the animal, but two main branches are usually present, one running dorsally, the other ventrally; they divide into smaller and smaller branches, and ultimately form a network which ramifies through every organ of the body. The tracheae are kept open by a spiral thickening of their chitinous lining.

The ventral nerve cord bears sixteen ganglia, each of which gives off three nerves on each side to the legs and adjacent muscles. The first of these supplies the poison claws; the cord is connected by circum-oesophageal commissures ending in the bilobed cerebral ganglion which gives nerves to the eyes, antennae, and mouth organs. When a basilar segment is present, the corresponding ganglia are fused together. In some **Chilopoda** the first few ganglia of the ventral nerve cord are more or less completely fused; some writers have held that this indicates a division of the segments of the body into thorax and abdomen. Besides the eyes, sensory hairs are found on the antennae, and in *Scutigera* there is a peculiar sense organ on the ventral part of the head consisting of a pouch lined with sensory hairs, each of which is connected with a nerve fibril.

Lithobius is dioecious, and the generative organs usually attain their full size during the spring. The testis consists of a blind tube, the lining cells of which give rise to the spermatozoa. The tube is narrow anteriorly, then somewhat wider, and finally it narrows posteriorly into a duct, and opens by it

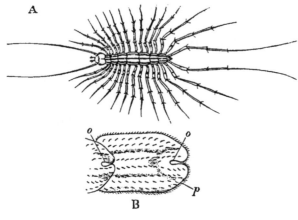

FIG. 179.

A. *Scutigera rubro lineata.* After Buffon.
B. Tergum and part of a second tergum of the same, enlarged to show the position of the stigmata, *o, o.   p,* Hinder margin of tergum.

into a circular canal which surrounds the rectum, and receives the opening of two tubular vesiculae seminales.

Ventrally this canal enlarges into a chamber which receives the ducts of two pairs of accessory glands of considerable size ; it opens to the exterior, on the ventral surface of the last segment, immediately in front of the anus. The spermatozoa are filiform.

The ovary is a single tube, dorsal to the alimentary canal, and reaching as far forward as the head, it is continuous with the oviduct; this divides into two branches, one of which passes on each side of the rectum, and the two unite again in a chamber which, like the corresponding structure in the male, receives the secretion of two pairs of accessory glands.   A pair of hollow vesicles, the receptacula seminis, also open into this median chamber.   The latter opens to the exterior just in front of the anus, the opening being provided with a pair of minute jointed claws.

The exotic species of Scolopendra may attain the length of six inches, and their bite is extremely painful, and even dangerous. *Cryptops* and *Geophilus* have no eyes : the latter possess a great number of segments. Some members of the family GEOPHILIDAE are phosphorescent, and secrete from certain glands on the ventral surface a luminous slime; since this is produced by both male and female, and neither of them has eyes, the secretion is regarded as a means of frightening or warning off enemies. The male *Geophilus* spins a web, and drops a spermatophore in the middle of it, and the female comes and fertilises herself. *Geophilus longicornis* is common in Britain; the female coils herself up and sits on her eggs, and is stated not to leave them until they are hatched. *Cryptops hortensis*, with twenty-one pairs of legs, is also British.

Scutigera has very long antennae and legs, the latter increasing in length at the posterior end (Fig. 179). This genus has facetted compound eyes, and its tracheal system is peculiar; the stigmata are median and dorsal, one for each segment, opening in a notch at the posterior edge of each tergum. Certain species produce a rattling noise by rubbing their legs together.

Order 2. DIPLOPODA (Chilognatha).

CHARACTERISTICS.—*The members of the second sub-order of the Myriapoda are characterised by their bodies being cylindrical or subcylindrical, and their antennae short, of seven segments only. Except at the anterior end, the segments bear two pairs of legs, and the bases of the legs are near together; there are also two pairs of stigmata on every segment. The generative aperture is at the base of the second or third pair of legs.*

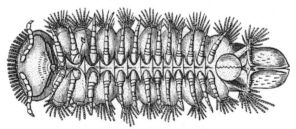

FIG. 180.—Ventral view of *Polyxenus lagurus* (after Bode) much enlarged, actual length a little over $\frac{1}{12}$th of an inch.

a. Position of genital openings.

The Diplopoda are all vegetable feeders, and have no poison claws. The genus *Polyxenus* is somewhat intermediate between the two main subdivisions of the Myriapoda, for although it is a vegetarian, and its generative organs open anteriorly, it resembles the Chilopods in the bases of the legs

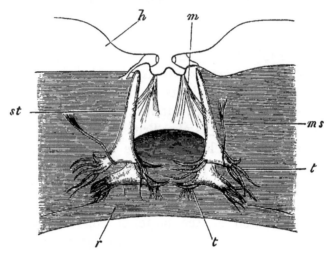

Fig. 181.—Inner view of the sterna of a single segment of *Iulus londonensis*, much enlarged to show the structure and arrangement of the tracheal organs. After Voges. The two pairs of tracheae are seen *in situ*, the posterior pair overlapping the anterior.

h. The posterior margin of the body ring (tergum).
r. Anterior border ; between the two lie the two terga.
st. Tubular chamber of tracheae.
t. Fine tracheae given off from it.
ms. Respiratory muscle attached to tracheal sac.
m. Ventral body muscle.

being somewhat apart, and in the character of its filiform spermatozoa, which are contained in spermatophores, as is usually the case in the last-named order. Its body, which is about $\frac{1}{12}$ inch long, is covered with tufts of hairy scales, probably defensive.

The commonest genus of Diplopoda is *Iulus*; a good many species of this genus occur in Britain, the most frequent being *Iulus terrestris*, which is sometimes an inch or more long, and is often to be found curled up under bark or stones. In this animal the tergum forms an almost complete ring, interrupted in the ventral line only by two sternal plates, one in front of the other. In all the segments after the fourth post-oral each of these plates bears a pair of legs, so that each segment with its single tergum corresponds with two pairs of legs, and has also two pairs of stigmata, one in front of the base of each leg, two nerve ganglia, and two cardiac chambers; these facts, together with others derived from embryology, show that each tergum corresponds with two primitive somites. The legs have five joints and a terminal claw.

Segments and Appendages of *Iulus.*

1. Antennae.
2. Mandibles.
3. Maxillae, fused across the middle line. } Cephalic.
4. Leg-like appendages, modified in male.
5. One pair of legs.
6. This segment bears no legs.
7. One pair of legs.
8. The succeeding segments bear two pairs of legs.

The cephalic appendages comprise a pair of antennae, a pair of mandibles with broad crushing surfaces, and a pair of maxillae which have fused across the middle line and which form a bilobed plate. The appendages are borne on the head; the first post-cephalic segment has a broad tergum, and bears ventrally a pair of leg-like appendages, which are turned forward, and probably assist in the act of feeding; the same appendage is in some species modified in the male, and forms a blunt hook-like process. The second post-cephalic segment bears a normal pair of legs, the third has none, and the fourth one pair, the succeeding segments have two. The generative orifice in both sexes is on the third segment, just behind the bases of the second pair of legs. In the male the seventh post-oral segment bears only one pair of legs and a complex copulatory apparatus not connected with the internal organs. This apparatus is of great systematic value. The female *Iulus,* like that of *Geophilus,* watches over its eggs.

Some genera of the Polyzonidae, e.g. *Siphonophora* (Fig. 182), have the anterior part of the head and the mouthparts prolonged into a sucking or piercing snout. Most of the Diplopoda possess a series of glands which open to the exterior by a row of lateral *foramina repugnatoria,* one on each segment; these emit an offensive fluid; in one species, *Fontaria,* this fluid breaks up into prussic acid and benzaldehyde (oil of bitter almonds).

Except in *Glomeris,* where the tracheae are branched as in the Chilopods, the respiratory organs resemble those of *Peripatus.* The stigma opens into a tracheal sac, from which a number of short unbranched tracheae arise. There are four of these on each segment. The GLOMERIDAE have shortened, broad bodies, and curl themselves up like wood-lice, and in this condition are very often mistaken for the latter.

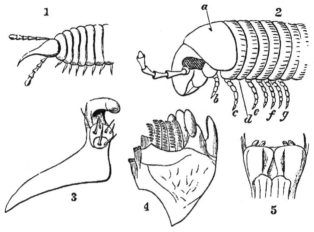

FIG. 182.

1. Head and anterior somites of *Siphonophora portoricensis*. After Koch.
2. Diagram of the arrangement of the anterior somites and appendages of a female *Iulus londonensis*. After Moseley.
    a. Modified tergum of the 1st postoral somite (dorsal plate or collum).
    b. A short single appendage of the same somite, of four joints and a claw only, turned towards the mouth.
    c. Single appendage of the 2nd somite, of five joints and a claw like the remaining appendages.

d. 3rd or generative somite, devoid of appendages and sterna, but bearing the generative apertures.
    e. Single appendage of 4th somite.
    f, g. Dual appendages of succeeding somites.
3. Hook-like postcephalic appendage of male of same attached to its plate of support (=one half of modified sternum ?)
4. Mandible of same.
5. The four-lobed plate formed by the fused single pair of maxillae.

Pauropus and *Scolopendrella* are genera which are sometimes placed in two different sub-orders, the Pauropidae and the

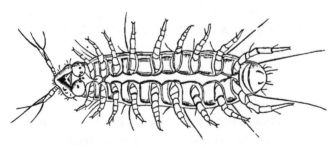

FIG. 183.—Enlarged view of *Pauropus huxleyi*. After Lubbock.

Symphyla, which rank with the Chilopoda and Diplopoda. *Pauropus* is very small; it has ten segments, and *Eupauropus* but five, each of which in the last-named genus bears two pairs of legs. The last segment of the antenna of *Pauropus* bifurcates and bears three flagella (Fig. 183).

Scolopendrella, the genus which constitutes the sub-order Symphyla, has obvious resemblances to certain members of the most primitive sub-order of the Insecta, the Thysanura, and in spite of the fact that no clear division of thorax and

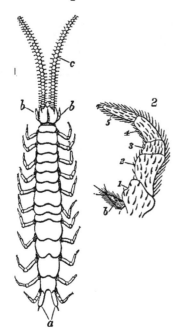

FIG. 184.—*Scolopendrella immaculata,* highly magnified. Slightly altered from Packard.

    a. Caudal stylets.

    b, b. First postcephalic appendages.

    c. Antennae.

2. One of the functional legs further enlarged (from Wood Mason), showing the five joints and terminal pair of claws.

    b. Inner rudimentary leg of same somite.

abdomen exists, and that practically all the segments bear legs, and there are only two pairs of mouth appendages, it has been proposed to class this animal with the Thysanura. *Scolopendrella* is but 5 or 6 mm. long; behind the head there are fifteen terga, but only thirteen sterna exist, and twelve pairs of legs; the latter are five-jointed and terminate in two claws (Fig. 184), as in *Campodea,* the genus of Thysanura it most closely resembles. At their base is a small process which gives the limb almost a biramous character. A pair of caudal styles bear the apertures of certain silk glands.

APTERA . . . { COLLEMBOLA—*Podura, Sminthurus.*
{ THYSANURA—*Lepisma, Machilis.*

ORTHOPTERA . { DERMAPTERA—*Forficula* { Cursoria—*Blatta, Periplaneta.*
{ ORTHOPTERA GENUINA { Gressoria—*Mantis, Phasma.*
{ { Acridiidae—*Acridiun*
{ Saltatoria { Locustidae—*Locusta.*
{ Gryllidae—*Gryllus,*
{ *Gryllotalpa.*

NEUROPTERA . *Termes, Aeschna, Ephemera, Phryganea.*

LEPIDOPTERA { MICROLEPIDOPTERA—*Tinea, Tortrix.*
{ { Geometrina—*Cheimatobia, Fidonia.*
{ { Noctuina—*Plusia, Agrotis.*
{ MACROLEPIDOPTERA { Bombycina—*Bombyx, Cossus.*
{ { Sphingina—*Sphinx, Acherontia.*
{ { Rhopalocera—*Papilio, Vanessa.*

COLEOPTERA . { PENTAMERA—*Dytiscus, Carabus.*
{ HETEROMERA—*Meloe, Lytta.*
{ PSEUDOTETRAMERA—*Curculio, Scolytus, Cerambyx.*
{ PSEUDOTRIMERA—*Coccinella.*

HEMIPTERA . { HETEROPTERA—*Notonecta, Reduvius, Acanthia.*
{ HOMOPTERA—*Cicada, Aphis, Coccus.*
{ PARASITICA—*Pediculus.*

DIPTERA . . { APHANIPTERA—*Pulex.*
{ DIPTERA GENUINA—*Musca, Tipula, Cecidomyia.*

HYMENOPTERA { PHYTOPHAGA—*Sirex, Lophyrus.*
{ ENTOMOPHAGA—*Ichneumon, Cynips, Pteromalus.*
{ { Formicariae—*Formica, Oecodoma.*
{ ACULEATA { Vespiariae—*Vespa, Polistes, Eumenes.*
{ Apiareae—*Apis, Bombus.*

CHAPTER XIX

TRACHEATA

CLASS III. INSECTA

CHARACTERISTICS.— *Tracheata whose body is divided into three distinct regions : head, thorax, and abdomen. The head carries the antennae and three pairs of mouth appendages. The thorax is composed of three segments, each with a pair of legs, and usually the posterior two segments bear each a pair of wings. The abdomen is devoid of limbs, and consists of a varying number of segments ; ten may be made out in some species, but the number is often less.*

The class Insecta includes an enormous number of species, probably far more than the whole of the rest of the animal kingdom put together. The single order of beetles—Coleoptera —contains more than 120,000 described species, and there is reason to believe that the flies—Diptera—are as numerous or even more so. At the present date, the total number of named species of insects must be very nearly a quarter of a million.

The principles on which this enormous amount of material has been classified and brought into order rest upon (i.) the structure and arrangement of the mouth parts, (ii.) the characters of the wings, (iii.) the relation of the first thoracic segment —the prothorax—to the rest of the thorax, and (iv.) the degree of metamorphosis.

The arrangement of the mouth organs is intimately connected with the food of the insect; by the modification or suppression of some of the three pairs of oral appendages or parts of them, a very great diversity of structure is produced,

21

which is of the utmost value in any system of classification. The character of the wings, and the relation of the prothorax to the rest of the thorax, are connected with the powers of flight. In some sub-orders wings are entirely absent, and in others, although they are fully developed, they may be thrown off, as in the ants, and in the workers amongst the white ants or Termites. The degree of metamorphosis which an insect undergoes in passing from the egg to the adult, though possibly a good criterion for phyletic relationship, is of less use for practical purposes of classification, inasmuch as it assumes the life-history of the insect to be known, and this is by no means always the case.

The Insecta are divided into eight orders—

1. APTERA.
2. ORTHOPTERA.
   NEUROPTERA.
3. LEPIDOPTERA.
4.
5. HEMIPTERA.
6. COLEOPTERA.
7. DIPTERA.
8. HYMENOPTERA.

Before considering the subdivisions and characteristics of these orders, it will be advisable to obtain an insight into the structure and anatomy of some fairly typical insect form, and the common cockchafer, *Melolontha vulgaris*, one of the Coleoptera, both on account of its size and its frequency, will form a convenient type.

The cockchafer is about $\frac{3}{4}$ to 1 inch long, and the chief divisions of an insect body into head, thorax, and abdomen are easily recognised. The head bears a pair of antennae, and three pairs of mouth appendages. The antennae differ in the two sexes; they consist of ten segments, the first of which is known as the *scape*. In the male the last seven joints, and in the female the last six joints, are flattened out into a series of plate-like processes, which have given the name Lamellicornia to the subdivision of the Coleoptera to which the cockchafer belongs. They are much longer and larger in the male than in the female, and in both, each lamella bears an enormous number (some thousands) of shallow pits lined by specialised sense cells connected with nerves, which apparently function as olfactory organs.

Immediately behind the base of the antennae lie the compound eyes.

Appendages of *Melolontha.*

1. Antennae.
2. Mandibles, without a palp.
3. 1st maxillae.
4. 2nd ,, = Labium.

5. 1st pair of legs, prothoracic.
6. 2nd ,, mesothoracic.
7. 3rd ,, metathoracic.

The oral appendages comprise the typical insect mouth parts, a pair of mandibles, and two pairs of maxillae. The anterior part of the head forms the *clypeus,* and this is continued into a hinged portion, the *labrum,* which overhangs the mouth. On each side of this orifice lies a mandible, a biting-jaw of pyramidal shape, whose opposed edges bear a number of teeth. The first pair of maxillae are behind the mandibles;

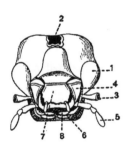

FIG. 185.—View of the posterior surface of the head of *Melolontha vulgaris.* After Strauss-Durckheim.

1. Eyes.
2. Opening into head for oesophagus, nerve cord, etc., to pass through.
3. Base of the cut-off antennae.
4. 1st maxillae.
5. Maxillary palp.
6. Labrum.
7. Labium (=fused second maxillae).
8. Labial palp.

each consists of a basal piece which articulates with the head, the *cardo,* this bears another joint, the *stipes,* and the stipes terminates in an inner piece with one tooth, the *lacinia,* and an outer piece, the *galea,* with a row of teeth. A maxillary palp is inserted into the distal end of the stipes. The second pair of maxillae have fused together and formed a plate-like lower lip, the *labium.* This consists of a *mentum* and *sub-mentum,* and it carries a pair of labial palps. The function of both pairs of palps is sensory.

The thorax is made up of three segments, called respectively the *pro-, meso-,* and *meta-thorax;* each of them carries on its ventral surface a pair of legs. The anterior pair are directed forward and the two posterior backward. Each leg consists of the following segments, a *coxa,* a *trochanter,* a *femur,* a *tibia*, and the five-jointed *tarsus*, the distal joint of which carries two claws.

The dorsal part of the prothorax forms a broad shield, the *pronotum*. The mesothorax bears dorsally the anterior pair of wings, which are horny and hard, and are termed *elytra*; they afford a protective covering to the membranous posterior wings, and to the abdomen as far as the eighth segment. The meta-thoracic wings are membranous, they stretch out at right angles to the body and are used for flight, at other times they are folded under the elytra. The superficies of these wings is divided into a number of small areas or cells by the presence of chitinous tubules, in which tracheae and nerves ramify; these "cells" are of great importance in Insect classification.

The third division of the body, the abdomen, is by far the bulkiest; it comprises eight segments, each composed of a

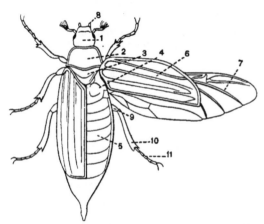

FIG. 186.—A male *Melolontha vulgaris*, seen from above, slightly enlarged. After Vogt and Yung.

1. Head, stretched forward.
2. Prothorax.
3. Mesothorax, scutellum.
4. Metathorax.
5. Abdomen.
6. Anterior wing (*elytron*) of right side, turned forward.
7. Posterior wing of right side, expanded.
8. Maxillary palps.
9. Femur of third right leg.
10. Tibia of third right leg.
11. Tarsus of third right leg.

dorsal plate, the *tergum* or *notum*, and a ventral plate, the *sternum*, the soft integument which connects the sides of these successive plates is pierced by the apertures of the six pairs of abdominal stigmata. The eighth tergum is prolonged into a long bluntly pointed process, which overhangs the openings of the alimentary canal and generative organs (Fig. 186).

The alimentary canal of the cockchafer is about six times as long as the body, and is therefore necessarily thrown into loops more or less coiled (Fig. 187). This is unusual amongst Arthropods (though common in Insects), in which the digestive tract is as a rule a straight tube, running directly between mouth and anus. The mouth is overhung by the labrum, and has the mandibles and first maxillae on either side; behind it is bounded by the labium or fused second maxillae; the mandibles cut and crush the food, which is held in position by the maxillae. The passage of the food into the mouth is assisted by a hairy prominence on the anterior surface of the labium.

The mouth leads into an oesophagus which pierces the nerve mass, and then swells into an inconspicuous crop; this

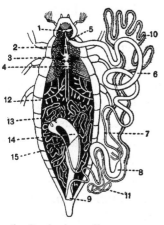

FIG. 187.—View of male *Melolontha vulgaris*, from which the dorsal integument and heart have been removed to show the internal organs. After Vogt and Yung.

1. Cerebral ganglion.
2. 1st thoracic ganglion.
3. 2nd and 3rd thoracic ganglion.
4. Fused abdominal ganglia.
5. Oesophagus.
6. Mid-gut.
7. Small intestine.
8. Colon.
9. Rectum.
10. Malpighian tubules, brown portion, with caeca.
11. Malpighian tubules, distal end.
12. Tracheae with vesicles.
13. Testes, opening into coiled vasa deferentia.
14. Penis.
15. Single vas deferens.

opens into a brown mid-gut, in which the processes of digestion are mainly carried on. The mid-gut passes into a finer tube, the small intestine, which receives the Malpighian tubules, and this in its turn passes into the colon, which has on its inner side six longitudinal muscular ridges; this opens through the rectum to the exterior. With the exception of the mid-gut, the alimentary canal is lined with a thin layer of chitin, continuous at the mouth and anus with the exoskeleton, and this is cast at the ecdysis of the exoskeleton. In *Melolontha* neither salivary glands nor hepatic diverticula are described.

The Malpighian tubules are four in number; they are very

long, and are closely applied to the outside of the alimentary canal (Fig. 187). Each of them consists of a free end, which is usually white, and of a brown portion which bears numerous small side diverticula, giving the tubule a feathered appearance; they open into the small intestine. The cells lining the tubule contain crystals of uric acid, which are excreted through the rectum.

The tracheal system, which carries air to every part of the body, communicates with the exterior by eight pairs of stigmata. The first two of these are situated close behind the base of the first and second legs, between the pro- and meso-, and meso- and meta-thorax respectively. The remaining six are found on the soft integument which unites the terga and sterna of the seven anterior abdominal segments. Each stigma is surrounded by an oval ring of chitin, and the opening can be closed by the action of certain muscles. It leads into a large trachea, which in the first thoracic stigma swells into a considerable vesicle; from this two branches pass off and enter the head. The most dorsal of these unites with its fellow of the opposite side, and the single trunk gives branches both to the eyes and to the brain; the other branch also unites with its fellow, and supplies the antennae and mouth appendages. A third branch arises from the same enlargement and runs to the anterior pair of legs, and several smaller branches supply muscles. Besides these, a stout branch runs backward and opens into the main trachea of the second stigma; this bears in its course many oval tracheal vesicles (Fig. 187), and gives off branches to the elytra; two or three other branches with vesicles also pass backwards, and one of them supplies the second pair of legs. The main trachea from the second stigma gives branches to the membranous wings, the third pair of legs, and longitudinal branches which open into those of the adjacent stigmata.

The abdominal tracheae are very regularly arranged; the six stigmata on each side open into a trachea, which immediately divides into a dorsal and a ventral branch; these arch round, pass backward, and meet together again in the main trachea of the next segment. The dorsal and ventral arches of the last stigma unite to form a ring, and at the point of union give off a trachea to the generative organs. Each of the dorsal loops of the system thus formed breaks up into six or eight smaller tracheae, which pass dorsally, giving off fine branches to the various viscera, and frequently terminating in vesicles; the ventral loops also give off smaller branches, and each of the last six gives rise to a long trachea, which passes backward and opens into a large swelling in the fifth abdominal segment; similar branches come from both sides of the body, thus the right and left tracheal system is in communication within the body.

Each trachea consists of a tube of very thin transparent chitin, which is strengthened and kept expanded by a spiral thickening of the chitin; this gives the trachea its characteristic spiral appearance. The chitin is secreted by a layer of polygonal cells, with conspicuous nuclei, which surround the tracheae. The vesicles are simply oval swellings on the tracheae. The finest branches ramify between the cells which compose the various tissues, and thus in a tracheate animal the cells are directly supplied with air, and are not dependent upon the blood for their supply of oxygen. Owing to the complete intercommunication of the various tracheae the whole system could be filled with air from any one stigma, so that if anything rendered the supply from some of them inefficient, it could be made right by the others. The vesicles when charged with air doubtless serve to render the body of the beetle lighter during flight.

The heart of the cockchafer lies in the median line immediately beneath the dorsal integument; it consists of a tube closed behind but open in front, with contractile muscular walls. In the abdominal section of this tube there are eight pairs of ostia, with valvular lips; through these the blood enters the heart, and by its contractions is propelled forwards. The ostia mark the limits of the eight chambers into which the heart is sometimes said to be divided; anteriorly it is continued into a vessel, the so-called aorta, which passes as far forward as the head and then suddenly ends with an open mouth. The heart is lodged in a space, the pericardium, whose dorsal wall is formed by the terga of the various segments, and the ventral by a pericardium or pericardial membrane. The alary muscles, which correspond in number with the cardiac chambers, and have a wedge-shaped outline, are attached at one end to the integument, and at their broader extremity to the pericardial membrane; when they contract the latter is depressed. This membrane consists of connective tissue, pierced by numerous oval apertures; when it is depressed the blood in the body-cavity passes through it, and at the same time the diastole of the heart taking place, the blood enters through the eight pairs of ostia, and at the systole is forced forward and so out of the open mouth into the body-cavity again. In this way the blood, which is a colourless fluid with amoeboid corpuscles, is kept in circulation.

The body-cavity of Insects is to a great extent occluded by the various viscera, but in addition to the alimentary canal, generative organs, etc., there is a considerable amount of a tissue, known as the fat-body, which is formed primitively from mesoblast cells lining the integument. This fat-body is especially abundant in the larvae, where to some extent it acts as a storehouse for reserve material, particularly in those Insects which pass through a protracted pupa stage; it is also found in mature Insects, and is usually present to a greater or less extent on the pericardial membrane. In the Tracheata, where the air is directly conveyed to the cells of all the tissues, the blood has to a great extent lost its respiratory function; it is still, however, of the utmost importance. It bathes all the internal organs, and these, as is usually the case when organs are surrounded by nutrient media, do not form solid compact masses, but are branched and subdivided as much as possible. The food which has been digested in the alimentary canal is thus distributed by the blood into which it passes, the fats are stored up by the fat-body, and the nitrogenous excreta, the urates or uric acids, are either conveyed straight to the Malpighian tubules, or are stored up in the cells of the fat-body. From time to time these cells break down, and then the stored-up urates are taken by the blood to the Malpighian tubules, and from them pass out of the body. The body-cavity in Insects, as is probably the case in all Arthropods, is a haemocoel, and the true coelom is probably confined to the lumen of the generative organs. The developement of a tracheal system of breathing, and the consequent absence of a respiratory function of the blood, has taken place concurrently with the formation of a new method of ridding the body of its waste nitrogenous matter. In most other Coelomata this is effected either by means of tubules ultimately ending in flame cells, or by nephridia, and these two methods are connected by intermediate forms; but in the true Tracheata the nitrogenous matter leaves the body by caecal diverticula of the alimentary canal, formed usually from the proctodaeum, but sometimes from the mesenteron. These Malpighian tubules receive the matter they are to excrete either at first hand from the blood, or the urates, etc., have been stored for a time in the cells of a mesoblastic tissue, the fat-bodies, and, when these disintegrate, the nitrogenous matter is carried by the blood to the tubules, and thence passes through the rectum to the exterior.

The nervous system of *Melolontha* is very concentrated, instead of the double nerve cord enlarging into a ganglionic

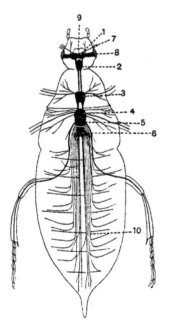

Fig. 188.—View of nervous system of *Melolontha vulgaris*. After Vogt and Yung.

1. Cerebral ganglion.
2. Sub-oesophageal ganglion.
3. 1st thoracic ganglion.
4. 2nd thoracic ganglion.
5. 3rd thoracic ganglion.
6. Fused abdominal ganglia.
7. Nerves to antennae.
8. Optic nerves.
9. Origin of sympathetic nerves.
10. Abdominal nerves, a pair to each segment, which split into an anterior and posterior branch.

mass in each segment, the ganglia are to a great extent fused into a central mass, from which the nerves radiate.

The supra-oesophageal ganglion occupies a considerable part of the head; it consists of two well-marked lobes separated by a slight groove.   Each lobe is continued laterally into a stout optic nerve, which supplies the compound eyes; besides this it gives off a nerve to the antenna of its own side, which branches abundantly in the lamelliform segments, and a third nerve to the labrum or upper lip.   Two short commissures encircle the oesophagus, and unite in the sub-oesophageal ganglion, which also lies in the head; this ganglion supplies nerves to each of the mouth appendages, the mandibles, and first and second maxillae.   From the sub-oesophageal ganglion two commissures pass backward, and enter the first or prothoracic ganglion, which gives off nerves to the thoracic muscles and first pair of legs.   Close behind this is a large ganglionic mass, formed by the fusion of the meso- and meta-thoracic ganglia, its double origin being shown by a transverse groove (Fig. 188).   This compound ganglion supplies nerves to both pairs of wings, and to the posterior two pairs of legs.

The abdominal ganglia are all fused into one mass, which is withdrawn into the thorax, and lies in the metathorax immediately behind the ganglion of that segment; indeed, it seems as if part of the abdominal nervous system is absorbed into the last-named ganglion, which is said to supply nerves to the first abdominal segment; the remaining seven pairs of nerves arise from the abdominal mass, and pass backwards to their respective segments, where they split into an anterior and a posterior branch.

In addition to the central nervous system, there is a small sympathetic system, which consists of the following parts.   A pair of fine nerves arise from the supra-oesophageal ganglion and fuse in the middle line, and so form a minute triangular ganglion lying in the head (Fig. 188); from this a median unpaired nerve passes back and forks over the crop, and the branches unite into a small ganglion again.   There are also two pairs of minute ganglia which innervate the heart and tracheal system, these are situated just behind the supra-oesophageal ganglion.

Like all other Insects the cockchafer is dioecious, and the female may be distinguished from the male by having six lamelliform segments in its antennae, whereas the male has seven.

The testes lie in the fourth and fifth abdominal segments; they consist on each side of six small flattened bodies, each of which has a short duct. The six ducts unite into a single vas deferens, which is much coiled; just before the vasa deferentia of the two sides unite they are rather swollen, and form vesiculae seminales, they then receive the secretion of two coiled accessory glands. The united vasa deferentia open into an extremely large and complicated ejaculatory apparatus, which can be protruded just below and in front of the rectum.

The ovaries consist of six tubes upon each side; their inner tapering ends are united into a strand of tissue which is attached to the tergum of the first abdominal segment; the ova arise from the endothelial cells which line these tubes. The cells are undifferentiated at the inner end of the tube, but as they approach the oviduct they assume more and more the character of the ripe ova; between each two eggs is a mass of cells whose function is to afford nourishment to the ova, which attain a considerable size. The six tubules on each side unite into an oviduct, and the two oviducts fuse and form the vagina, which opens just in front of the rectum. A small accessory gland and a spermatheca are present, and, in addition to these, a large bursa or sac, into which the penis is introduced during fertilisation.

The adult cockchafers may be seen flying about in the dusk during the months of May and June; they live upon the leaves of deciduous trees, and at times do a good deal of damage by denuding the branches of their foliage. The female deposits her ova, in clumps of about thirty, several inches below the surface of the ground. Each egg gives rise to a larva with a brownish, hard, chitinous head, and a white body of twelve segments, the last two of which are swollen into a "sack." The three segments immediately succeeding the head are each provided with a pair of four-jointed legs. The larva creeps through the earth, and lives on roots, in this way often causing considerable loss to the agriculturist. The larva lives three years, and in this time grows to a considerable size; at the end of the third summer it burrows to a depth of about two feet in the ground, and there forms a spherical cell; in this it turns into a brown chrysalis. The pupa thus formed is a *pupa libera*, that is, its appendages are free, and not hidden under a covering, as is the case with the Lepidoptera. The pupal stage lasts till the following spring, and in the interval the individual has undergone great changes ; its nervous system, with a ganglion in each segment, has become concentrated, its wings have developed, and its appendages have assumed their adult form. The perfect insect or imago emerges from the pupa some little time before it appears on the surface, but during the month of May in its fourth year the mature cockchafer makes its way above ground, and is found hanging underneath the leaves of the trees which serve it as food.

The life-history of *Melolontha* affords a good example of a complete metamorphosis, with its larval, pupal, and imaginal stages. The egg gives rise to a larva which has little or no resemblance to the adult insect, and the change from the vermiform larva to the winged insect is effected during the period of quiescence which constitutes the pupal stage. Many insects undergo a similar metamorphosis, whilst the young of others are but miniatures of their parents. Intermediate conditions between these two extremes are not uncommon, and the variations which the life-history of the various orders of insects present are of use in the classification of this class.

Order 1. APTERA.

CHARACTERISTICS.—*Wingless insects whose body is covered with scales or hairs. The segments of the thorax are not fused together. The mode of progression is either running, or springing with the aid of an apparatus borne on the ventral side of the abdomen. There is no metamorphosis.*

The Aptera form the most primitive order of insects. This order consists of a few genera, which are grouped in two suborders, (i.) the Collembola and (ii.) the Thysanura, differing considerably from one another.

Sub-order 1. **Collembola.**

The members of this group are widely distributed, but very inconspicuous. Specimens of them may be found under stones or dried leaves, on roofs, etc., and some of them live on the surface of the water, upon which they move actively.

The head bears a pair of antennae with few joints (4-8). The prothorax is usually compressed; there is no trace of wings, and no evidence that the group ever possessed them. Each of the three thoracic segments bears a pair of legs with four or five joints.

The abdomen is short; the Sminthurinae have apparently only two or three segments, the Podurinae six. On the ventral side of the first abdominal segment is a structure known as the "ventral tube"; this in *Podura* and *Lipura* is a simple tubercle, but in other genera it takes the form of a tube, which in *Sminthurus* is divided into two halves at its end, from each half a long delicate tube can be protruded at the will of the animal. The ventral tube is essentially a protrusible part of the integument, it may be compared with somewhat similar structures in the Thysanura. Its function is possibly an adhesive one. The springing apparatus consists of a forked process borne on the fifth, or, as is stated in *Podura*, on the fourth segment; this process is directed forwards, and in those species which jump the best, it is retained in position by two chitinous hooks which form the "catch"; these hooks in *Tomocerus* are borne on the third abdominal segment. The spring acts by the process pressing violently against the ground, and the insect is thus propelled into the air; the process is then folded under the abdomen again, and retained in position by the catch. This saltatorial apparatus is absent in some species, as *Lipura, Anura*, etc.

The nervous system consists of the usual chain of ganglia, but the number in the abdomen is reduced. In *Sminthurus* and others there is only one abdominal ganglion. Eyes may be absent or present, in the latter case they consist of two little groups of at most eight simple eyes.

The mouth appendages are, as is usual in insects, a pair of mandibles and two pairs of maxillae; the first pair are provided with palps, and the second are partially fused into an under lip. All the mouth appendages can be withdrawn into a cavity, and in this respect the Collembola resemble the Myriapod *Scolopendrella*. The alimentary canal is a straight tube stretching between the mouth and anus. It is at present unsettled whether any salivary glands or Malpighian tubules exist.

The heart, which lies in the middle dorsal line in *Macrostoma*, is said to have five pairs of valves. The blood is yellowish and corpusculated. It seems doubtful if a tracheal system exists in many of the Collembola, but in the larger Sminthurinae a pair of stigmata open on the under side of the head, a very unusual position; from these bundles of tracheae radiate.

Like all other Insects, the Collembola are dioecious; there is very little external difference between the sexes. The developement is direct.

Sub-order 2. **Thysanura.**

The Thysanura are of larger size than the Collembola, and have to a much greater extent the appearance of insects. One of the most familiar genera is *Lepisma*, often termed the "silver-fish," a quickly-running, silvery-gray insect, which infests old chests of drawers and disused cupboards. It is a nocturnal insect, and hides away during the winter. The genus *Machilis* is found in woods, etc., or on rocky sea-coasts, where it lives between stones or in clefts of the rock, but it loves to run on warm sunny places. *Campodea* is found under fallen leaves or stones in shady places and in loose earth. *Japyx* also shuns the light; it is found widely distributed in Europe, but not in cold places.

The number of abdominal segments is always ten, and the abdomen ends in three long many-jointed processes or cerci. The three thoracic segments each bear a pair of legs; in *Machilis* the coxae of the two posterior pairs of legs bear peculiar processes, which externally resemble certain paired processes found on the ventral surface of the abdomen in different species of Thysanura. These processes, however, are regarded by some authorities as the representatives of abdominal limbs; in *Machilis* they are found on the second to the ninth segments, in *Japyx* on the first to the seventh, in *Lepisma* only on the eighth and ninth—in the last-mentioned insect the coxal processes are wanting. Similar processes occur internal to the base of each of the twenty-two pairs of legs in *Scolopendrella*.

Some very remarkable spherical protuberances of the integument near the middle line are found in *Machilis* and *Campodea*, projecting between the abdominal appendages. They are twenty-two in number in the former genus, a pair projecting behind the sterna of the first, sixth, and seventh segments, and two pairs behind those of the second, third, fourth, and fifth. These protuberances appear to be extended by the forcing into them of some of the blood. They have special muscles which retract them, and as a rule they are found in the retracted condition. They probably serve as respiratory organs. They are absent in *Lepisma* and *Japyx*.

The number of ganglia in the abdomen is eight, except in *Campodea*, where only seven have been described. In the last-named genus the nervous system is in intimate connection with the hypodermis. *Machilis* has a pair of large compound eyes. The Thysanura are all provided with salivary glands and Malpighian tubules, and the heart has nine pairs of ostia. The tracheal system is fairly well developed; in *Machilis* a pair of stigmata are found on the meso- and meta-thorax, and on each of the abdominal segments from the second to the ninth ; the tracheae do not anastomose.

The primitive position of the Aptera is shown (i.) by the absolute absence of wings, (ii.) by the direct developement,—this is, however, shared by several other orders of insects,—(iii.) by the presence of abdominal appendages, (iv.) by the very slight developement of the cuticle, and (v.) by the general resemblance of some of the genera to the larvae of higher forms. This is recognised by the application of the term campodiform to the larvae of most insects with direct developement, and to some of those, *e.g.* certain families of beetles, with indirect.

ORDER 2. ORTHOPTERA.

CHARACTERISTICS.—*Insects with direct developement. Prothorax free. Biting mouth parts. Wings usually unequal, the anterior pair small and hard, the posterior membranous.*

The insects grouped together in the order Orthoptera are all of a fair size, and compared with the beetles and flies are comparatively few in number. They may be classified in two groups: (i.) the Dermaptera, comprising the earwigs, and (ii.) the Orthoptera genuina, which include the cockroaches, grasshoppers, locusts, crickets, etc.

1. Dermaptera.

FORFICULIDAE. — This family consists of the earwigs, the most familiar of which in our country is the genus *Forficula*. The body is elongated, the head flattened, with filiform antennae, round eyes, and no ocelli. The prothorax is free, the anterior wings short and horny; the posterior, which are folded longitudinally and transversely beneath them, are membranous. The abdomen has nine segments, and terminates in a pair of forcep-like processes which have been homologised with the cerci anales of other forms. These insects are nocturnal in their habits, concealing themselves in flowers and fruit during the day. The female watches over her young.

The remaining groups of the Orthoptera—the Cursoria, the Gressoria, and the Saltatoria—are usually grouped together as the Orthoptera genuina.

2. Orthoptera Genuina.

I. CURSORIA.—This sub-order comprises the various species of cockroach which are found all over the world. The body of these insects is flat and oval, the pronotum is large, the antennae long, the legs are adapted for running, and the tarsus is five-jointed; a pair of ringed cerci anales are present.

There are about 800 species of cockroaches: some, as for example *Polyzosteria*, are wingless; others, as *Heterogamia*, have the females wingless; whilst *Blatta* (*Phyllodromia*) and *Periplaneta*, the species found in Europe, bear wings in both sexes, except *P. orientalis*, the female of which is wingless. Cockroaches avoid the light, and are nocturnal in their habits and fond of heat. The eggs are laid in a capsule or ootheca, which is variously shaped in the different species; it is often carried about by the female protruding between the terminal segments of the abdomen for some days before it is deposited.

II. GRESSORIA include two very remarkable families of Insects, the Mantidae and the Phasmidae. Their legs are adapted for walking.

The Mantidae have their anterior pair of legs modified to form predatory organs. The toothed tibia can be folded down against the femur, as the blade of a pocket-knife into the handle. This subchelate appendage is used in capturing other insects or spiders for food. The prothorax, which bears these enlarged appendages, is very much elongated. The abdomen is elongated and oval. The commonest colour of these insects is green. The eggs are laid in regular clumps on sticks or stones. *Mantis religiosa* is found in South Europe; the devotional attitude in which it sits, with the anterior legs raised, has obtained for it the name of the praying insect; many legends and superstitions centre around it.

The Phasmidae are mostly tropical insects of large size which feed on leaves. They are slow in their movements, and escape observation by their very extraordinary resemblance to various natural objects amongst which they live. The genus *Phyllium* of the East Indies mimics various forms of leaves, the veins on the wings resembling the venation of the leaf, and in some cases the legs bear flattened leaf-like expansions which increase the resemblance; others have holes in their wings and a dried appearance which simulates that of a tattered, withered leaf. The genus *Phasma* includes many species of an elongated shape which closely resemble dried twigs; one species attains the length of 12 inches. *Ceroxylus laceratus* is covered with tufts of processes which give it the appearance of a mossy twig. The whole family affords a very striking example of protective resemblance.

III. SALTATORIA.—This division includes all those forms which have the legs modified for jumping, such as the grasshopper, locusts, and crickets. It may be divided into three families:

1. **Acridiidae** or **Grasshoppers.**—The body is compressed,

22

with a large vertical head. The antennae are short, and the hind legs are enlarged for jumping; the tarsus has three joints;

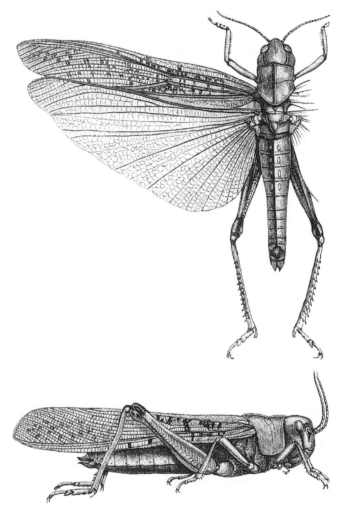

Fig. 189.—*Pachytylus migratorius.* Natural size.

they produce a chirping noise by rubbing their tibia against a vein on the anterior wing, and their auditory organ is situated on the first abdominal segment. The female has no projecting ovipositor. The eggs are laid in cocoons in packets of 50 to 100 at a time.

Some of the grasshoppers are very voracious, and as they move in immense swarms, they occasion very considerable damage. One species, *Acridium migratorium*, which is found all over Europe and Asia, moves as a locust swarm, and devours every green thing which it comes across.

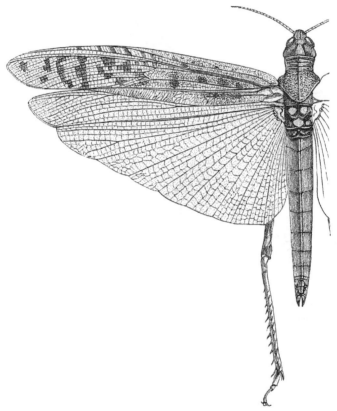

Fig. 190.—*Acridium peregrinum.*   Natural size.

Some species attain a large size, measuring four or six inches in length.

2. **Locustidae.**—The Locusts are usually of a green or brown colour; they have long filiform antennae, and their wings lie vertically along the side of the body. The auditory organ is situated upon the tibiae of the anterior pair of legs; the tarsus is four-jointed. The female bears a pair of long, sabre-like processes which form the ovipositor; this bores into the ground, and the eggs are deposited by it. The stridulating noise of the male is caused by drawing the rough vein of the left anterior wing over a file-like structure situated on a vibrating membrane at the base of the right.

Locusta viridissima is one of the commonest European forms.

3. **Gryllidae.** — The crickets have a shorter and more

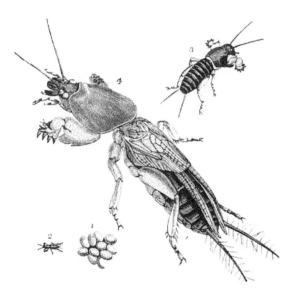

FIG. 191.—Mole cricket (*Gryllotalpa vulgaris*).

1. Eggs.

2. Young just hatched.

3. Larva after first moult.

4. Adult, nat. size.

cylindrical body than those of the two preceding families. Their antennae are long and filiform. The male produces an

FIG. 192.—Rocky Mountain Locust (*Caloptenus spretus*).

a. Females in various positions ovipositing.
b. Egg pod extracted from the ground with the end broken open.
c. A few eggs lying loose on the ground.

d, e. The earth partially removed to illustrate an egg mass already in place and one being placed.
f. Shows where such a mass has been covered up.

irritating noise by rubbing the short anterior wings against the posterior. As in the preceding family, the auditory organ is on the proximal end of the anterior tibia. The tarsus is three-jointed. The females are usually provided with a straight ovipositor, and the males attach a spermatophore containing semen to the genital orifice of the female.

Gryllus campestris is the field cricket, and *G. domesticus* the house cricket. *Gryllotalpa vulgaris*, the mole cricket, leads mainly a subterranean life; it is of a brown colour, and has its two anterior legs short and thick, and adapted for digging. The female lays from 200 to 300 eggs in a nest the size of a hen's egg, situated some inches below the surface of the ground.

The Orthoptera genuina, from their habit of moving in immense swarms and devouring all the green parts of plants

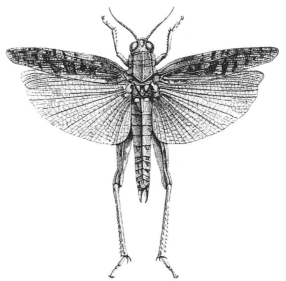

Fig. 193.—*Caloptenus italicus.* Natural size.

which come in their way, are Insects of great economic importance. The females usually lay their eggs in waste and inaccessible places termed in America "mauvaises terres"; the young larvae when hatched make their way to more cultivated districts. As an example of the enormous number of these insects, it may be mentioned that in the spring of 1882 over 1300 tons of locusts' eggs were destroyed in the island of

Cyprus, and over 12,000 tons of locusts. Various species compose the locust swarms in different parts of the world.

Acridium peregrinum does much harm in India and Algeria. Swarms of *Caloptenus spretus*, another of the Acridiidae, have been known to clear off every green thing over 300 square miles in Colorado in less than six weeks. In Minnesota 300 egg capsules, each containing thirty eggs of this species, were found on an average in every square foot. These figures give some idea of the astounding numbers of these locust swarms.

ORDER 3. NEUROPTERA.

CHARACTERISTICS.—*Insects with membranous wings, both pairs alike, with the veins forming a more or less close network. The mouth parts are, as a rule, of the biting type. Metamorphosis complete or incomplete.*

The Neuroptera form a rather heterogeneous collection of Insects, which, however, resemble each other in the character of their wings. Many of the subdivisions of this order have but little in common, and it is difficult to group them into sub-orders; it will therefore be advisable to consider a few of the more important forms under the designation of their families.

Family 1. TERMITIDAE.—The white ants flourish most abundantly within the tropics; certain genera, however, as *Calotermes*, and some species of *Termes*, occur in subtropical and temperate climates.

The antennae are short; the abdomen of nine segments is oval and flattened, unlike the linear abdomen of most of the Neuroptera. The wings when at rest are unfolded, and lie flat upon the back (Fig. 194).

The TERMITIDAE, like the more highly organised ants and bees, live in communities, and the individuals have undergone considerable modifications, correlated with their particular functions in society. The sexual forms are winged; the apterous members of the community are either larvae, which, unlike the larvae of the Hymenoptera, take an active share in the work of the nest, or are neuters, *i.e.* individuals with sexual glands, which, however, do not reach maturity, and which are apparently functionless. The neuters may be workers with small rounded heads, or soldiers with greatly enlarged heads and most formidable mandibles.

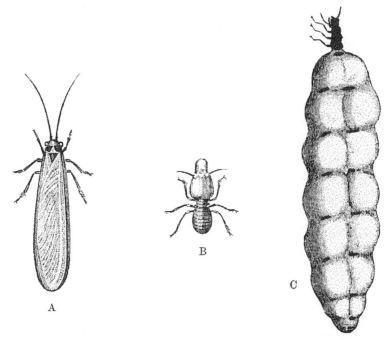

FIG. 194.—White Ant (*Termes bellicosus*).

A. Winged male.　　　　　B. Soldier.　　　　　C. Queen.

The communities live as a rule underground; some species construct nests of such size and strength as to serve as watch-towers for the big game of South Africa. These nests are 12 to 15 feet in height. Others live on tree-trunks, etc., and cover the branches with little tunnels built of pellicles of clay which they bring up from below, and line with excrement. They invariably work in the dark, building the tunnels from the inside; if anything breaks down part of their work, the workers disappear, and the powerful "soldiers" take their place and defend their home.

Those of the larvae which are not destined to become workers or soldiers acquire wings, and a few weeks afterwards they leave the nest, and pair whilst flying in the air. They then fall to the ground and lose their wings; in this condition a number of them perish, but some of them, guided by a few neuters, will succeed in founding a new nest. In this a special chamber is set apart for the queen, whose body swells enormously. It may attain a length of more than 3 inches, and is distended by an enormous ovary. The queen lays eggs at the rate of 80,000 to 90,000 a day, and these are carried away and cared for by the workers.

In *Termes lucifugus*, found in South Africa, the larvae which mature in the spring become kings and queens, those which mature in the summer become complementary kings and queens, and replace the functional ones if occasion arises. The king dies in the autumn, but, although the queen ceases to lay eggs during the winter, she survives, and resumes the egg-laying in the spring.

The nests of *Calotermes* are the most incomplete; there is no special chamber for the queen, and their home consists of passages tunnelled in trees.

Family 2. THRIPSIDAE.—A family of very small, usually black insects with fringed wings. Their body is long and

FIG. 195.—Corn thrips (*Thrips cerealium*), female. Magnified.

narrow, their antennae long and slender, and their mouth parts suctorial. *Thrips cerealium* does a good deal of harm to wheat crops, others injure flowers, etc. They are sometimes regarded as a separate order of Insects, and called the Thysanoptera; other authorities place them with the Hemiptera.

Family 3. EPHEMERIDAE.—The Ephemeridae or May-flies spend but a short part of their life in the imago condition, at most only a few hours. They are delicate insects with a long body and a ten-jointed cylindrical abdomen which ends in two or three very long anal filaments. The imago takes no food, its mouth parts are rudimentary, and the oral cavity is stated not to open into the alimentary canal. The ducts of the reproductive organs do not unite in either sex, but open independently, one on each side of the ninth abdominal segment in the male, and between the seventh and eighth in the female.

The larvae are aquatic, and feed on other insects, etc.; they have well-developed mouth parts. The tracheal system is closed, and the larvae breathe by means of plate-like gills, which are borne to the number of six or seven pairs on the anterior abdominal segments. The tracheae ramify in these gills, which are moved about in the water, and possibly serve to some extent as locomotor organs. These gills have been regarded by some authorities as structures from which the wings of Insects in general may have originated.

The larvae moult very frequently, in one species more than twenty times, and live two or three years. They pass into a pupa stage, and from this a winged insect emerges, this is the *subimago*, it takes a short flight, and then casts its skin, and gives exit to the imago. The male fertilises the female on the surface of the water, and shortly afterwards the latter drops her eggs into the stream or pond, and then dies.

Family 4. LIBELLULIDAE.—The dragon-flies are large

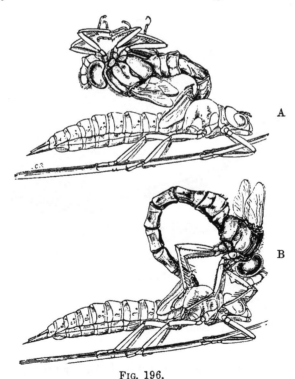

FIG. 196.

A. The anterior portion of the body of *Aeschna cyanea* freed from the puparium.

B. The tail being extricated.

insects with a thin cylindrical abdomen, but an enormous
head and thorax.　　Many of them are strikingly beautiful.
The generative organs of the male open on the ninth abdominal
segment, but there is a kind of vesicula seminalis on the second

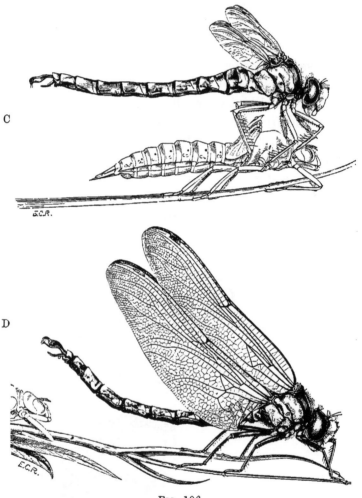

Fig. 196.

C. The whole body extricated.
D. The perfect Insect (*Aeschna cyanea*), the wings having acquired their full
dimensions, resting to dry itself, preparatory to the wings being horizontally extended.

segment of the abdomen, into which the semen is transferred,
and is thence introduced into the females.　　In some species
the female then flies over the water, touching it at intervals

with the end of her abdomen, and depositing an egg at each dip, or in other species she deposits the eggs on submerged water-plants.

The larvae are very voracious, and feed upon other insects, their lower lip or labium is peculiarly modified into a structure known as the "mask," this can be suddenly shot out, and serves to capture food. The larva of *Agrion* has at the end of its abdomen three leaf-like gills, but other species breathe by their rectum, the walls of which are richly supplied with tracheae. The entrance and exit of the water is controlled by three valves, which can be opened or closed at will; in some species the sudden expulsion of water serves to propel the larva through the water. This anal respiration recalls a similar change of function of the posterior part of the alimentary canal in some Crustacea (see p. 268). The pupa stage, which precedes the imago, is in the Libellulidae an active stage, and is sometimes termed the "nymph" (Fig. 196).

Family 5. MYRMELEONTIDAE.—The ant-lion, *Myrmeleo*, in the imago condition has clubbed antennae, a small prothorax, a large mesothorax, and wings of equal size. The larvae live at the bottom of little conical sandy pits, which they excavate; they lie partially embedded in the sand at the bottom of the pit, and

FIG. 197.—Ant-lion, *Myrmeleon formicarius*.

seize with their powerful mandibles any insect, etc., which happens to stray over the edge. The mouth is closed, and the food is sucked in through perforations in the mandibles. It is stated that the proctodaeum of the larva does not open into the alimentary canal, but is modified to form a silk gland, which serves for the spinning of the cocoon in which the pupa envelops itself.

Family 6. In the Family HEMEROBIIDAE the larvae of *Chrysopa* and *Hemerobius*, which are known as Aphis lions, feed on Aphides. They also have the proctodaeum modified to form a silk gland.   The larvae of another genus, *Mantispa*, are parasitic; they pass their life in the ovisacs of spiders, into which they burrow to devour the ova.

Family 6. PHRYGANIDAE.—This family is sometimes placed in a distinct sub-order, the Trichoptera; it comprises the insects popularly known as caddis-flies.   The head bears long filiform antennae, and a pair of hemispherical projecting eyes.   The prothorax is small; the wings have few veins, and are longer than the body; the posterior pair can be folded. The wings are covered, as in the Lepidoptera, with microscopic hairs or scales.   The mouth parts are rudimentary, more especially the mandibles, and the first maxillae and labium are modified.   The resemblance of this family to the Lepidoptera is further marked by the general appearance of the imago, which approximates to that of some members of the Microlepidoptera, and of the cylindrical larva and quiescent pupa; the latter resembles that of a moth, but has free limbs and wings.

The larvae known as caddis-worms live in tubular cases, which they build up of particles of sand, shells, or bits of grass or other plants, the material varying in different species. At times they are wholly retracted within these cases; at other times their head and thorax project, and they walk about carrying the case retained round their abdomen by two recurved hooks.   Like the EPHEMERIDAE and LIBELLULIDAE, some of the larvae have their tracheal system closed, and carry tracheal gills.   They are either carnivorous, and then very voracious, or purely herbivorous.   The case is closed at both ends at the end of larval life, and serves as a cocoon for the pupa, which developes into the imago out of the water. The female deposits her eggs in gelatinous clumps on sticks or stones in the water.

ORDER 4. LEPIDOPTERA.

CHARACTERISTICS.—*Insects with suctorial mouth parts, which take the form of a spirally rolled proboscis.   The four wings are similar, and are covered with minute scales.   The prothorax is fused with the mesothorax.   The metamorphosis is complete.*

The Lepidoptera are a very homogeneous group, containing a large number of species. They are familiar to every one as moths and butterflies.

The head is large, and covered with hairs; it bears compound eyes, and sometimes ocelli are also present. The antennae are straight, but vary a great deal in their details. The mouth parts have undergone very remarkable modifications; the labrum and mandibles are aborted; the first maxillae are each elongated into a very long, grooved, closely-jointed structure, and when this is opposed to its fellow the whole forms a closed tube, which, when at rest, is coiled under the head like a watch-spring; in many species the two halves of this proboscis are held together by a number of minute hooks. The maxillary palps are rudimentary, except in the Tineidae, where they are well developed.

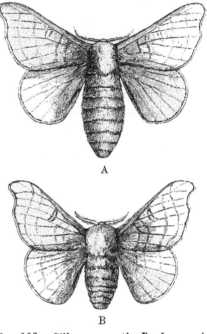

Fig. 198.—Silk-worm moth, *Bombyx mori.*
A. Female.   B. Male.

The second maxillae or labium form the spinnerets in the larva or caterpillar, but they disappear in the imago; their palps, however, persist, and are large and hairy.

The thoracic segments are all fused together, the wings are large, and the ⌐ rior and posterior of each side are occasionally hooked together. In some of the Geometridae the wings are aborted in the females. The scales which give the beautiful colour to the wings are morphologically hairs, which are flattened out, and variously marked. The legs are weak, the tarsus five-jointed. The abdomen has ten segments, some of them concealed, it is covered with hairs.

The internal organs show the following modifications. There are only two thoracic ganglia. The two anterior abdominal

ganglia of the caterpillar abort, and the next four or five persist.   There are eight stigmata, the anterior between the pro-
and meso-thorax, the remaining seven in the abdomen; the
tracheae which arise from the latter bear air vesicles.

The length of the proboscis varies, and is said to correspond
with the position of the nectaries in the various flowers upon
which the Lepidoptera feed.   The nectar is sucked up the
hollow tube formed by the maxillae by the action of a suctorial
stomach which communicates by a short stalk with the oesophagus.   There are as a rule six Malpighian tubules.

The sexes are usually distinct, the male being in many
cases the more beautiful.   A few Lepidoptera are parthenogenetic.   This phenomenon occurs sometimes in the silk-worm
moth *Bombyx mori,* and in some of the Psychidae.   The eggs

FIG. 199.—Larva of *Bombyx mori.*

are usually laid on such plants as the larvae, when hatched, feed
upon.   The larvae are commonly known as caterpillars; these
bear three pairs of five-jointed thoracic feet, and, in addition,

FIG. 200.—Head, pro-leg, and leg of *Bombyx mori,* larva.

two to five pairs of unjointed *pro-legs,* which may occur on the
third to the sixth abdominal segments, and on the last.   These

pro-legs as a rule terminate in a ring or semicircle of chitinous hooks. The larval life may endure from a couple of weeks to three years; it is followed by the quiescent pupa stage.

The pupa may be suspended by the hindmost pro-legs, and this position may be rendered the more secure by a rope of silk round the thorax, as in *Pieris*, or the pupa may be enclosed in a silken cocoon, as in the silkworm, or simply buried in the earth, as in the Sphingidae. The pupa has the limbs of the insect enclosed in a common covering, and is hence known as a *pupa obtecta*, as opposed to the *pupa libera* of the Coleoptera, in which the limbs stand out freely from the body.

Fig. 201.—Cocoon of *Bombyx mori*.

The Lepidoptera may be divided into two sub-orders:

Sub-order 1. **Microlepidoptera.**

CHARACTERISTICS.—*These are usually very small and delicate moths, with, as a rule, long setiform antennae. The caterpillar has eight pairs of legs, terminating in a circlet of hooks —"pedes coronati."*

They are as a rule secluded during daylight. Many of their larvae burrow in the mesophyll of leaves or buds, or form tubular cases by rolling the leaves together.

The following families may be mentioned:

Family 1. PTEROPHORIDAE.—Small moths with a long slender abdomen and long legs. Their wings are hairy, the anterior pair are usually more or less cleft, and the posterior pair are divided almost to their base into three (*Pterophorus*), or into six (*Alucita*), separate lobes. They form no cocoons, but the larva attaches itself by its tail to some leaf or twig, sheds its skin, and becomes a pupa.

Family 2. TINEIDAE.—This is a very numerous family. The Tineids have bristle-like antennae. Both the maxillary and labial palps are well developed. The narrow wings are fringed with hairs. Many of the larvae burrow in leaves, others live together in nests, and they usually spin slight silken cocoons. Many of them are destructive : *Tinea sarcitella* is the clothes moth, *T. tapezella* the fur moth, *T. granella* lays its egg in grains of corn and the caterpillars devour the grain. The genus *Solenobia* is parthenogenetic.

Family 3. TORTRICIDAE.—The leaf-rollers have short palps and oblong anterior wings. They are as a rule larger than the Tineids. The moths fly at night, and lay their eggs on the buds of the trees, which are attacked by their larvae. The caterpillars roll the leaves into cylinders, and in these turn into brown pupae in silken cocoons. *Tortrix viridana* is common on oak-trees. *Retinia buoliana* attacks pine-trees.

Family 4. PYRALIDAE.—The members of this family bear long slender palps. They are as a rule gregarious, and fly in the twilight. The larvae have a glassy appearance, and bear but few hairs. The female of one species, *Aphomia colonella*, creeps into bee-hives and deposits her eggs there ; the larvae, which are found in great numbers, devour the honey, to the great detriment of the hives.

Sub-order 2. **Macrolepidoptera.**

CHARACTERISTICS.—*Lepidoptera of large size, with a complicated system of nervures on the wings. The feet are generally, though by no means always, provided with a semicircle of hooks—"pedes sub-coronati."*

I. GEOMETRINA.

Slender moths, whose large thin wings lie horizontally when at rest. The antennae are bristle-like, and in the male sometimes toothed. The caterpillars have a varying number of pro-legs, usually two pairs, and their manner of moving has given them the name of loopers. When at rest they fix themselves by the hindmost legs and raise the anterior half of the body; in this position they may remain for hours, when frightened they drop, but remain attached to their base by a small thread of silk. They either spin cocoons under leaves, or form brown chrysalids under the earth.

Many are injurious to fruit trees, as *Cheimatobia brumata*, the female of which has rudimentary wings. *Fidonia piniaria* attacks Conifers.

II. NOCTUINA.

The group includes the forms popularly known as owlets; it is the largest group of the Lepidoptera, containing over 2500 species. Most nocturnal moths of fair size belong to it. The antennae are long, sometimes pectinate in the male. The fore-wings are small, and the larger posterior wings are folded under them when at rest. They are usually of a dull colour, and there is almost always a round spot and a kidney-shaped patch in the middle of the anterior wing. There is little variation between the sexes, or between the different species in the moth, but the caterpillars differ considerably. The latter are striped and barred, naked, or more rarely hairy; they usually have five pairs of pro-legs, but some have four. The pupae are usually underground, enclosed in earthen cocoons. The eggs are laid singly, and the larvae are not gregarious.

The Noctuina include numerous families, amongst whom the Plusiidae, the Agrotidae, and the Ophiusidae may be mentioned.

III. BOMBYCINA.

The members of this group are often termed spinners. They are large unwieldy moths, often very beautiful and strange in form. Their body is usually very hairy (Fig. 198), the head is small and sunken, and the mouth parts are reduced and sometimes obsolete. The antennae are setiform, in the male pectinate; the last-named sex are as a rule more brilliantly coloured and more active than the sluggish female. The wings of the female *Orgyia* are reduced, and are absent altogether in *Psyche*. The eggs are laid in groups, and are covered with a woolly substance; the caterpillars have sixteen legs, and are usually hairy. The cocoons are made above ground, the naked larvae forming theirs of silk, the more hairy kinds mixing their hairs with a slighter amount of silk. The sexes are usually very distinct, and the females attract the males from great distances. Parthenogenesis occurs in the family Psychidae.

23

This group contains a number of well-known moths, such as *Lasiocampa quercus,* the oak egger; *Bombyx mori,* the silkworm; *Cnethocampa processionea,* the processional moth; *Cossus ligniperda,* the goat moth, etc.

IV. SPHINGINA.

The hawk-moths or humming-bird moths are large Lepidoptera with short bodies and long powerful wings. Their flight is swift and sustained, and they fly usually at twilight. The antennae are short and taper to a point. The proboscis is very long, and can suck up honey from the depths of a flower without the insect alighting. The sexes are as a rule alike. The caterpillars have sixteen legs, and the last segment bears an anal horn or tubercle. They elevate the anterior portion of their body like a Sphinx, and remain for hours in this position; as a rule they are brightly coloured, and their skin is smooth. The pupae form rough cocoons of earth underground, and the proboscis is usually free. About 400 species are known, many of which are tropical.

Sesia apiformis, the clearwing, has transparent wings and a bee-like appearance; *Acherontia atropos* is the death's-head moth; *Sphinx ligustri* the privet moth.

V. RHOPALOCERA.

The butterflies are mostly brightly coloured, and are diurnal in their habits, loving the sun. The majority are easily distinguished from the moths by their clubbed or knobbed antennae. Their body is small, and the abdomen is, relatively to the rest of the body, considerably smaller than in the moths. The legs are slender and often reduced, rendering walking a matter of some difficulty. The wings are held erect when the insect is at rest, and the anterior is never linked to the posterior by a bristle and socket, as is often the case with moths. The caterpillars have sixteen feet, and are naked or hairy, with varying markings and tubercles. They do not form cocoons, but turn into chrysalids with an angular contour; as a rule these are suspended to a twig or stalk by a silken band which cuts into the thorax; they are sometimes ornamented with bright metallic spots or patches.

A few species, as *Vanessa*, hibernate, but most butterflies pass the winter in the larval or pupal state. The cycle of their developement does not extend over a year, but there may be two generations in a twelvemonth.

The genus *Papilio* contains over 300 species; *P. machaon* is the swallow-tail. The family PIERIDAE contains the numerous "whites." *Vanessa cardui* is the "painted lady," and *V. io* the "peacock," *Apatura iris* the "purple emperor," etc.

ORDER 5. COLEOPTERA.

CHARACTERISTICS.—*Insects with masticating mouth parts. The anterior wings are horny, and in some cases ʃused together. They do not overlap, but meet together in the middle line, ʃforming a straight suture. The prothorax is moveable. The metamorphosis is complete.*

The order Coleoptera has received more attention at the hands of entomologists than any other order of Insects, and the number of species of beetles named and described far outnumbers that of any other group. The beetles form a fairly homogeneous assemblage; and although they vary considerably in size and shape, they do so to a much less degree than the Orthoptera or Hemiptera. It is a comparatively easy matter to recognise a beetle. As a rule they are sombre in hue, but some of them show very beautiful metallic colours, especially after rain. Their outline is usually oval, but it may be linear or almost round.

The head is well developed, and may be free or partially hidden under the projecting prothorax; it bears antennae, which are usually two-jointed and of very various shapes. The antennae are in many cases different in the two sexes. With the exception of a few blind species which inhabit dark caves, beetles usually have a pair of compound eyes; in the GYRINIDAE or whirligigs, which swim half immersed in the water, the eye is divided into two halves, one for seeing in the air and one for the water. Ocelli are as a rule absent. In the weevils and some allied families the head is elongated, and the mouth and oral appendages are at the end of a long snout. The mouth appendages are of the type described in Melolontha; as a rule the maxilliary palp is four-jointed, the labial palp three-jointed.

The prothorax is well developed, the mesothorax small, and the metathorax of fair size. The legs are usually adapted for running, but in some cases they are flattened for swimming or strengthened for digging. The number of joints in the tarsus is usually four or five, but it may be smaller, and this variation forms the basis for grouping the various families into sub-orders.

The anterior wings or elytra, when at rest, meet in a straight line which terminates anteriorly at a small triangular area of the mesothorax termed the scutellum, often invisible except when the wings are opened or the prothorax extended. In some families, as the STAPHYLINIDAE, the wings only extend over the anterior abdominal segments, leaving the larger part of the abdomen exposed. In rare cases they and the hind wings are absent, as in the female *Lampyris*.

The elytra are in some species fused together, and the posterior wings are then feebly if at all developed; flight is therefore impossible. In the more normal forms the beetle, when flying, extends the elytra at right angles to the body, and keeps them in this position motionless (Fig. 204).

The ventral surface of the abdomen is more strongly protected by chitin than the dorsal, which is covered in by the thick elytra. The hinder segments are often invaginated, and form a recess connected with the openings of the generative organs.

One or two genera of the ELATERIDAE, and almost all the LAMPYRIDAE, are provided with phosphorescent organs, usually in both sexes. In the males of the latter order, the light-giving structures shine through the ventral surface of the two posterior abdominal segments; in the former they are placed in the prothorax, and on the suture between the thorax and abdomen; in both cases they consist of numerous fatty cells, with a very rich supply of tracheae and nerves.

Beetles usually lay their eggs in the neighbourhood of the food which will afford support to their larvae. The latter are either free-living, somewhat Myriapod-like grubs, with three pairs of well-developed legs, or are soft, white, almost legless larvae, which live in the earth or burrow in timber, etc.

The pupae have their limbs and wing-cases projecting— *pupa libera.* They may be free, or enclosed in rough cocoons of earth or wood-chips.

The very numerous families of the Coleoptera may be arranged in four groups, corresponding with the number of joints in the tarsus. These divisions have but a slight scientific value, but are useful in dealing with such an enormous number of species as are found in the Coleoptera:

(i.) The PENTAMERA, *with five joints in the tarsus.*

(ii.) The HETEROMERA, *with five tarsal joints on the two anterior pairs of legs, and four on the posterior.*

(iii.) The PSEUDOTETRAMERA, *with one joint of the five-jointed tarsus very small and inconspicuous.*

(iv.) The PSEUDOTRIMERA, *with one joint of the four-jointed tarsus very small and inconspicuous.*

Under each of these subdivisions a few families may be mentioned.

Sub-order 1. **Pentamera.**

Family CICINDELIDAE.—Tiger-beetles; these have very large heads, broader than the thorax, with prominent eyes, long curved mandibles, and slender legs. They are usually of a brown or green colour with a metallic sheen, and are often ornamented with spots or patches. They frequent sunny places, such as the sandy margins of streams, and their larvae are found in tubular passages in the soil. These larvae are provided with two tubercles ending in hooks, which are outgrowths of the ninth segment, and which serve to hold them to their tubular dwellings; the anterior portion of their body projects from the surface of the ground, in order to seize any prey which comes within their reach.

Family CARABIDAE.—A very large family, whose limits are difficult to define. They are predaceous insects, with running legs, and their hind wings are not infrequently absent. They are found amongst grass, or under stones or bark, and as a rule roam about at night. Their larvae are found in the same situations as the beetles; they are rather broad, and their terminal segment is usually provided with two processes.

Family DYTISCIDAE.—Water-beetles, sometimes known as " water-tigers." Some are large oval beetles, others are quite minute, their hind limbs are flattened, covered with hairs, and adapted for swimming. Their antennae are devoid of any sensitive pubescence. The first three joints of the tarsus are in the males of the larger forms modified to form a plate-like organ. The larvae are very voracious; the mouth is closed, but the large pincer-like mandibles are perforated, as in *Myrmeleo*, and the juices of the fish, tadpoles, or other aquatic animals which fall into their clutches are sucked up through these. The genus *Dytiscus* is furnished with nine pairs of dorsal stigmata, and the beetles breathe by coming to the surface of the stagnant water in which they live, expiring the used air through the last large pair of stigmata, and taking in a new supply under their elytra.

Family STAPHYLINIDAE.— This group includes the rove-beetles and devil's coach-horses. They have long linear bodies, with very short elytra, which leave the five or six posterior abdominal segments exposed. They inhabit damp places under stones, manure-heaps, etc., and are often found amongst moss or leaves, or amongst fungi. Many of them live in ants' nests. Some tropical species of this family are viviparous.

Family SCARABEIDAE or LAMELLICORNIA. — This family contains 700 genera and over 10,000 species of beetles. The antennae end in lamelliform plates, such as have been described in *Melolontha vulgaris*. The body is as a rule thick and squarish, the legs often short and fossorial. Many of them attain a gigantic size. The larvae are thick fleshy grubs with a horny head and the posterior segments swollen out, baggy, and incurved.

Family ELATERIDAE.—Skip-jack beetles with serrated antennae, and an elongated body rather flatly arched. There is an articulation between the pro- and meso-thorax, and when the prosternal spine is suddenly brought down into the mesosternal cavity, it causes the beetle, if lying on its back, to be projected into the air, whence it usually falls on its feet. The larvae are known as " wire-worms," and are very destruc-

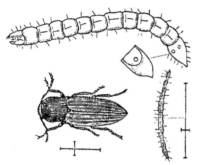

Fig. 202.—*Elater lineatus*, the "skip-jack" beetle, with its larva the "wire-worm." One of the larvae is enlarged to show the markings on the terminal segment.

tive, feeding on the roots of grasses and other plants. They are long cylindrical grubs generally of a reddish-brown hue, and are extremely tough and tenacious of life. *Elater lineatus* is the common skip-jack beetle.

Sub-order 2. **Heteromera**.

Family TENEBRIONIDAE.—An ill-defined family, with many mimetic forms. The elytra are rounded at their ends and cover the abdomen, the hind wings are frequently wanting. The larvae are linear, flattened, and horny, and resemble wire-worms. Many of these beetles shun the light and are sombre in colour, some have an unpleasant smell, and others are covered with a powdery secretion. The larva of *Tenebrio molitor* is known as the meal-worm.

Family MELOIDAE.—The head is bent forward, the legs are long, and the bodies are elongated and soft. The beetles are frequently found on flowers. The larvae are parasitic. The larvae of *Meloe* attach themselves to the bodies of various species of bee ; they are thus conveyed into the hives, where they feed upon the food provided for the larvae of the bees. The larvae pass through a metamorphosis which is more complicated than is usual in insects, this is termed *hypermeta-morphosis*. **Lytta vesicatoria**, the Spanish fly, is used as a vesicant.

Sub-order 3. **Pseudotetramera.**

Family CURCULIONIDAE.—The weevils are easily recognised by the prolongation of their head into a snout; the antennae

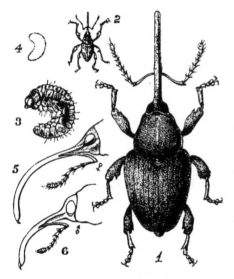

FIG. 203.

1. *Balaninus glandium*, magnified.

2. The same, natural size.

3. The larva, magnified.

4. The same, natural size.

5. Head and snout of the female magnified to show the arrangement of the antennae.

6. The same parts of the male.

are usually bent, and lie partly in a groove at the side of the snout (Fig. 203). The mouth with its appendages is situated at the extremity of this prolongation. Their bodies are often minute and hard; they feign death when disturbed. Their larvae are white, fleshy, footless grubs, with thick jaws; before transformation they spin silken cocoons. The number of species is very great, about 10,000. *Balaninus glandium* lays its eggs in hazel-nuts and acorns; its larva feeds upon the substance of the nut.

Family SCOLYTIDAE.—This family was formerly sometimes called the Bostrychidae; it includes beetles of small and inconspicuous size, whose rounded head is sunk beneath the prothorax, which is large, and forms almost half the body. The larvae resemble those of the preceding family; they have no legs, but their skin is ridged, and bears short hairs. These beetles and their larvae live in societies, boring passages in the wood of trees, on which they feed. In this way incalculable damage is done to forest trees, etc., especially to Conifers. The female lays her eggs in recesses of the passage she has made, and each larva as it hatches out continues the recess into a long tunnel; in this way very peculiar markings are produced, which are characteristic of the various species. *Bostrychus typographicus.*

Family CERAMBYCIDAE.—Often termed Longicorns, they are large showy beetles with oblong cylindrical bodies and long, usually eleven-jointed, recurved antennae. Their eggs are laid in crevices of the bark, and their larvae often bore passages in the wood; they may live one to three years, and then form a cocoon of chips near the mouth of their tunnel. *Cerambyx heros, Saperda carcharias.*

Family CHRYSOMELIDAE.—The leaf-beetles are oval in shape and convex dorsally; they are as a rule small, and of bright colours. The larvae have always three pairs of legs; many of them burrow in the soft mesophyll of leaves; they fix themselves by their hinder end to leaves before pupating. *Chrysomela decemlineata* is the Colorado beetle. *Haltica nemorum,* the turnip-fly, and *H. oleracea,* which attacks cabbages, also belong to this family.

FIG. 204.—*Haltica nemorum* (the turnip-fly).

Sub-order 4. **Pseudotrimera.**

Family COCCINELLIDAE.—The lady-birds are hemispherical in shape, usually of a red or yellow colour, with a varying number of black spots. They lay long yellow eggs, usually in the proximity of plant-lice, which are eagerly devoured by the larvae when they hatch out. The larvae are soft-bodied grubs beset with tubercles; they attach themselves by their pointed tail to leaves, and cast their skin; this they do not throw off, but remain in it during the pupa stage (*coarctate pupa*). The beetles pass the winter under bark, etc.

ORDER 6. HEMIPTERA.

CHARACTERISTICS.—*Insects with mouth parts adapted for piercing and sucking, in the form of a jointed rostrum. Two pairs of wings, which may be alike or may be different. Metamorphosis incomplete.*

This order comprises numerous insects familiarly known as bugs or lice. They present very great variety of form, some of them having short soft bodies with almost every trace of segmentation lost, whilst others are large and hard.

The mouth parts are adapted for taking up fluid. The labium is modified into a jointed sheath which guards the other appendages; at its upper end the hollow structure is closed by the labrum. Within the tube thus formed lie four rigid stylets, which represent the two mandibles with sharpened tips, and the two anterior maxillae of unequal length with serrated edges. The maxillary palps are absent and the labial palps are very small.

The antennae are short and three-jointed, or long and multiarticulate. The eyes are usually small; sometimes two ocelli are present. In the larger species the body is very often flat and angular in outline. There are usually four wings, rarely only two, and sometimes they are entirely wanting. In the former case the anterior wings have their basal half horny and their distal half membranous, whilst the posterior wings are membranous (Heteroptera), or both pairs are membranous (Homoptera).

The legs are usually of the walking type, the tarsus is two- or three-jointed. The lateral margin of the abdomen is greatly developed in some species. The stigmata are usually conspicuous; in the aquatic species there are a pair at the end of the abdomen, often borne at the tip of long processes.

The Hemiptera in most cases emit a fluid with a very disagreeable smell. This is secreted from a pair of pores on the under surface of the thorax, near the coxae of the middle pair of legs. This objectionable fluid is defensive in function. Other members of the group produce considerable quantities of wax, which is secreted by unicellular cutaneous glands.

The young resemble the adults, but are without wings. The males of the COCCIDAE alone form pupae within a cocoon, and thus undergo a complete metamorphosis.

The Hemiptera are divided into three sub-orders:

1. HETEROPTERA.
2. HOMOPTERA.
3. PARASITICA.

Sub-order 1. **Heteroptera.**

CHARACTERISTICS.—*The Heteroptera have the proximal half of their anterior wings horny, the distal half membranous; they lie flat, overlapping one another. Many are apterous. The prothorax is large and free. The proboscis arises from the front part of the head, and when at rest lies against the thorax.*

A few families may be mentioned :

Family NOTONECTIDAE (water-boatmen).—These insects always swim on their back, which is convex, like the bottom of a boat, whilst the ventral surface is flattened. The legs are long, especially the posterior pair, which are flattened for swimming. They fly well, but can scarcely walk; when disturbed they dive beneath the surface, carrying a supply of air for respiration beneath their wings. They remain for some time under water, holding on to aquatic plants, etc.

Family NEPIDAE (water-scorpions).—The members of this family are provided with a pair of long tracheal tubes at the end of their abdomen. Their body is flat and oval (*Nepa*), or elongated and linear (*Ranatra*); their fore limbs are raptorial, their hind limbs adapted for swimming. They are carnivorous, living chiefly on the larvae of aquatic insects and young fish. Their eggs are laid in the water, on stems of plants or under stones.

Family HYDROBATIDAE.—Aquatic insects of oval or elongated form, which run rapidly on the water's surface, and are usually found in colonies. The antennae are four-jointed and unusually long. There are often two adult forms found at the same time—one kind being winged, the other wingless. This family includes the *Halobates*, a marine insect found swimming on the surface of the sea in the tropics. It feeds on dead animals which float on the surface, and is said to attach its eggs to the Sargassum sea-weed.

Family REDUVIIDAE.—A very large and diverse family, including many insects of brilliant colour and some of considerable size. The proboscis is short and three-jointed. They are predaceous, and live for the most part on the blood of other insects, though they occasionally attack other animals. Several species, as *Opsicoetus* (*Reduvius*) *personatus*, attack the bed-bug.

Family TINGITIDAE.—The members of this family are for the most part small; they are found in considerable numbers on leaves and shrubs; their appearance is very characteristic. The anterior wings have the appearance of a network, and processes resembling them project from the sides of the thorax. The proboscis and antennae each have four joints.

Family ACANTHIIDAE. — The family includes *Acanthia lectularia*, the bed-bug, a flat oval insect devoid of wings. The proboscis has three joints, the antennae four. It is very tenacious of life. Many species of the same genus live upon birds.

Family CAPSIDAE.—A very numerous family, whose members are active both in running and flying. They are mostly of a medium size, oval in outline, and convex. Their body is usually soft. Many of them are frequently found in fruit.

Family PENTATOMIDAE.—In this family the scutellum is very large, often equalling in size the area of the abdomen. It is a large family, the members of which are brightly coloured. They are often found on shrubs, and live on caterpillars or leaves.

Sub-order 2. **Homoptera.**

CHARACTERISTICS.—*Both pairs of wings alike; when at rest they lie flat, unfolded, overlapping one another. The wings are often absent. The head is usually continuous with the prothorax, so that there is no neck.*

This group contains a number of species very diverse in their structure, and often with extremely complicated life-histories, in which parthenogenesis and " alternation of generations " play a great part.

Family CICADIDAE (Cicadas).—These insects are well known from the chirping noise they keep up. They are usually of large size, with an extraordinarily broad head fused on to the prothorax, with two large eyes at the angles, and three ocelli. The anterior wings are larger than the posterior. The male is provided with a kind of drum on the under side of the base of the abdomen. These drums are furnished with a curiously ribbed surface, and the characteristic noise of the Cicadas is said to be produced by the vibrations of the ribs set in motion by air forced against them. The females have stout ovipositors; in *Cicada septemdecem* the females lay their eggs in slits which they cut in young twigs, the larva hatches out in six weeks, drops to the earth, and buries itself. It remains underground till the seventeenth year, when it emerges, becomes adult, pairs, and as soon as the eggs are deposited disappears.

Family FULGORIDAE.—The antennae are bristle-like and three-jointed. The insects are very diverse in structure, many of them have the most extraordinary outgrowths of the upper part of the head. These protuberances may equal in size the rest of the insect's body. *Fulgora candelaria* and *F. lanternaria* are stated to be phosphorescent, but this appears doubtful. Some species excrete wax from their abdomen in such quantities that they have a commercial value in China.

Family CERCOPIDAE (Frog-hoppers).—The anterior wings of these insects are opaque. Their three-jointed antennae end in a bristle. The head is triangular, with two ocelli. *Aphrophora spumaria*, the cuckoo-spit, is an insect about ¼ inch in length. It can take very extended leaps. Its larva surrounds itself by a white frothy fluid which it excretes from its intestine. It is common in England on leaves of plants, etc., in fields. The eggs are deposited in punctures in the leaves.

Family APHIDIDAE (Plant-lice).—Small insects with oval or pear-shaped bodies. The wings, when present, are trans-

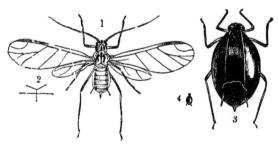

FIG. 205.—*Siphonophora granaria.*

1. Winged aphis.
2. Natural size of same.
3. Wingless form.
4. Natural size of same.

parent, and with few nervures; the anterior and posterior wing of each side are fastened together by a hook. Proboscis two-jointed. Antennae three- to seven-jointed. The sixth abdominal segment often bears a pair of tubes through which a sweet fluid, the honey dew, is excreted. Some excrete also a powdery bloom.

Plant-lice are usually brown or greenish in colour; they live upon cell sap, inserting their proboscis into the tissues of the leaf, stem, or root, and in this way often produce galls. Their life-history is very complex; as a rule males and females coexist in the autumn. The females lay fertilised ova, the winter eggs, from which in the spring incomplete females, with no receptacula seminis, emerge. These give rise to innumerable parthenogenetic generations, but after a certain number of these, males again make their appearance. In many species, however, the male is not yet known.

The genus *Aphis* is very common on plants, and causes great trouble to gardeners. Certain of them are tended and protected by some species of ants, who feed upon the honey dew secreted from their " honey-tubes."

The *Phylloxera vastatrix*, which infests vines, affords a good example of the complicated life-history presented by the Aphididae. The wingless root-dwelling forms—*radicola*—are found with their proboscis firmly fixed in the tissues of the young roots. They do not move about, but lay little clumps of thirty to forty parthenogenetic eggs, which give rise in six to twelve days, according to the temperature, to young larvae. These moult once or twice, creep about a little, and then fix themselves by their proboscis and lay parthenogenetic eggs, like their mother. In this way many agamic generations succeed one another, and the rate of increase is so great that it has been calculated that the descendants of a single insect which laid its eggs in March would number twenty-five millions by October.

FIG. 206.—Root-inhabiting form (radicola) of *Phylloxera vastatrix*, with proboscis inserted into tissue of root of vine. After Girard.

As autumn comes on some of the eggs give rise to larvae

which are provided with the rudiments of wings ; before their last change of skin they creep above ground, and then at the

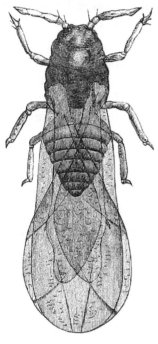

FIG. 207.—*Phylloxera vastatrix*, winged female which lives on leaves and buds of vine, and lays, parthenogenetically, eggs of two kinds, one developing into a wingless female, the other into a male. After Girard.

final moult a winged female emerges and flies away. This form serves to spread the vine disease from one district to

FIG. 208.

a. *Phylloxera vastatrix*, male produced from small egg (c) laid by winged female (Fig. 207).

b. Large egg.

c. Small egg.

d. Wingless female from large egg (b) laid by winged female.

After Girard.

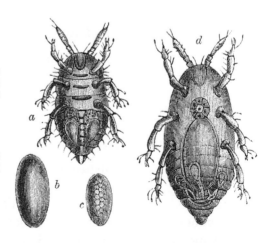

another. It is also parthenogenetic, but lays two kinds of eggs. From the larger of these a complete female hatches out, whilst the smaller produces in eight to ten days a male. This is the first and only appearance of this sex in the life-history of the Phylloxera. The male is devoid of mouth and alimentary canal; it fertilises the female, which soon after lays a single fertilised egg, the so-called "winter egg." This is deposited in some crevice or crack in the bark of the vine; in the spring a "stock-mother" hatches out of this egg and makes her way to the young buds of the vine, and inserts her proboscis into the upper surface of a leaf. The irritation thus set up causes the formation of a hollow gall on the under surface of the leaf, which opens to the exterior on the upper surface. The stock-mother lays eggs, and her offspring—*gallicola*—give rise to new galls, but ultimately some of them descend to the ground, burrow beneath it, and attach themselves to the roots, and thus become radicola.

The complicated life-history of this form may be expressed by the following table :

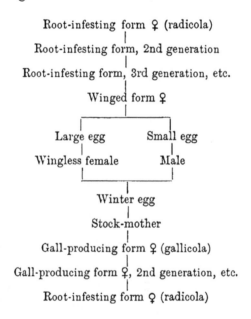

Root-infesting form ♀ (radicola)

Root-infesting form, 2nd generation

Root-infesting form, 3rd generation, etc.

Winged form ♀

Large egg   Small egg

Wingless female   Male

Winter egg

Stock-mother

Gall-producing form ♀ (gallicola)

Gall-producing form ♀, 2nd generation, etc.

Root-infesting form ♀ (radicola)

Family COCCIDAE (Scale Insects, Bark Lice, Mealy Bugs).— The members of this group differ a good deal both in their life-history and in their structure, and very frequently the two sexes of the same species are markedly different in appearance and habits. The antennae are usually long, eight to eleven joints, and filiform; the tarsus is two-jointed. The males alone amongst

Hemiptera have a complete metamorphosis; the adult male has one pair of wings, the anterior, the posterior being replaced by a pair of bristle-like processes, the *halteres*, as in the Diptera. No mouth is present.

The female, after the first ecdysis, becomes almost stationary, she retains her proboscis, which is embedded in the tissues of the plant on which she lives, as a rule she loses her limbs, and almost all trace of segmentation; the successive skins that she casts either remain over her body forming a scale, or she secretes a waxy or woolly covering. The eggs are laid under the body of the mother, which ultimately dies, but remains as a covering for her offspring; under this the young may remain some little time before seeking a convenient place on their plant host to insert their own proboscis.

Some of the Coccidae are of commercial value, *Coccus cacti* furnishes the pigment cochineal, and another species produces shellac. Most of the scale insects, however, are injurious to the plants upon which they live, and cause great loss to horticulturists and fruit-growers. *Aspidiotus conchiformis* is the well-known mussel scale on apple trees, etc., and *A. aurantii* attacks orange groves.

Sub-order 3. **Parasitica.**

CHARACTERISTICS. — *These are wingless Hemiptera, commonly known as lice; they live as ectoparasites on the skin of mammals, sucking their blood.*

The proboscis is fleshy and unjointed, with, as a rule, a circle of recurved hooks round its base. There are two small simple eyes; the antennae have five joints. The legs arise from the edge of the prothorax; they terminate in a hooked claw, which works against a projection of the tibia; this forms an admirable apparatus for clinging on to the hairs of their hosts. The young do not undergo any metamorphosis.

Pediculus capitis is the head louse, and *P. vestimenti* infests the body of all races of man. It is said that peculiar varieties infest the different races; thus those that live on negroes are nearly black, whilst the Chinese have a yellowish, and the natives of South Africa an orange variety. The genus which

24

infests cattle, squirrels, and other animals is *Hacmatopinus*, of this there are several species.

Order 7. DIPTERA (Flies).

CHARACTERISTICS.—*Insects with piercing and sucking mouth parts. The anterior wings are membranous; the posterior are replaced by knobbed processes—the halteres. The prothorax is ʸfused with the rest of the thorax. The metamorphosis is complete.*

The Diptera form one of the largest orders of insects, and when completely worked out it will probably be found to include at least as many species as the Coleoptera. The members of the order are of moderate size, with freely moveable spherical heads. The compound eyes are large, and in the male they sometimes fuse together across the dorsal middle line. There are three ocelli. The antennae are either short and end in a tactile hair,—the *arista*,—as in *Musca*, or are long and filiform as in *Tipula*, where they have thirty-six joints.

The mouth parts consist of a soft sucking proboscis known as the *haustellum*, ending in two swollen lobes. This is partly formed by the labium, which is devoid of palps. In front of this lies the elongated labrum, and between the labrum and labium an unpaired stylet, the epipharynx. The mandibles and first maxillae also form stylets, and the latter are provided with palps. Both the mandibles and maxillae may be rudimentary.

The mesothorax forms the largest part of the thorax. The wings are transparent, with few nervures, and with microscopic hairs and scales. The inner margin of the wing is usually notched, dividing it into two lobes, the *alula* and *squama*. They vibrate with great rapidity, 350 times per second in the fly. In a few cases they are absent.

The posterior wings are replaced by *halteres* (Fig. 209), which are short processes usually ending in a knob. Their function is not definitely determined, but to some extent they appear to act as balancers, and when removed the fly cannot control its flight.

The legs are slender and often long; the tarsus is five-jointed, and terminates in two claws; between these is the *pulvillus*, a fleshy vesicle with tubular hairs, which secrete a sticky fluid.

The abdomen may be short and conical, or long and cylindrical. There is no true ovipositor, but the terminal segments are retracted into the preceding, and can be protruded telescopically.

The nervous system is very concentrated; in some families, as the MUSCIDAE, the HIPPOBOSCIDAE, and the OESTRIDAE, there is, besides the brain, only one large ganglion; this is situated in the thorax, and gives off nerves to the abdomen.

The alimentary canal has a stalked sucking stomach, which opens into the oesophagus.

The sexes do not show much external differentiation; as a rule the eyes in the male are larger, they are sometimes fused together. The eggs are smooth, oval, and slightly curved; they are often provided with a micropyle. In the TIPULIDAE, the daddy-long-legs (Fig. 209), they are mature when the pupal skin is cast, and are laid immediately by the imago.

The larvae are white, fleshy, cylindrical maggots. They may have a distinct head, as in the TIPULIDAE, or a head may be absent, as in the MUSCIDAE. In the first case they have biting mouth parts and are often parasitic, in the second they suck up liquid nutriment. The ova are usually deposited upon the food, either animal or vegetable, in which the larvae burrow and on which they feed. Many of the larvae are aquatic. They change into pupae within the last larval skin, —*pupa coarctata*,—or cast this skin and become moving pupae which swim in water—*pupa obtecta*.

Sub-order 1. **Aphaniptera.**

The PULICIDAE.—This is an aberrant order of Diptera; it comprises the fleas. These insects have no wings, though there are small flat appendages on the meso- and meta-thorax. Their legs are large, and are adapted for leaping. The body is compressed. There is no labrum; the labium is small, with long palps; the mandibles form serrated stylets, and the maxillae have palps. The female lays eight to ten eggs, which are deposited amongst the dust in cracks or crevices; the small larvae have distinct head and jaws.

Pulex irritans is the common flea. *Sarcopsylla penetrans* is the chigoe or jigger of South America; the female burrows into the foot of man or other mammals, and deposits her eggs there. If not removed the larvae hatch out and give rise to ulcers.

Sub-order 2. Diptera genuina.

Family CULICIDAE (gnats and mosquitoes).—The members of this family are provided with very long and slender mouth parts. They are very widely distributed, extending from the arctic circle to the equator; the females lay their eggs on the surface of the water, on which they float in a boat-shaped mass. The larvae live at the bottom of ponds or swamps, eating decaying vegetable matter, and occasionally rising to the surface for air, which they take in through a special respiratory tube. The pupae are curiously curved and somewhat club-shaped; they swim actively about. The males do not leave the neighbourhood of the swamps where they are bred, but the females infest houses, etc. *Culex pipiens* is the common species of gnat in Britain.

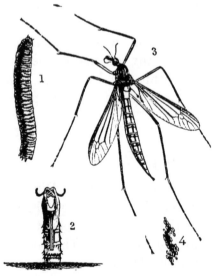

FIG. 209.—*Tipula oleracea* (the daddy-long-legs).

1. Larva.
2. Pupa case.
3. Imago.
4. Eggs, natural size.

Family TIPULIDAE (daddy-long-legs or crane-flies). — The familiar daddy-long-legs has long filiform antennae and very long slender legs. The mesonotum bears a V-shaped mark, and the abdomen is long and cylindrical. Their larvae burrow underground, and do considerable damage to crops by gnawing roots, etc. They are known as leather-jackets, and are of a whitish-brown colour and grub-like appearance. *Tipula oleracea, T. maculosa*, and *T. paludosa* are all common English species.

Family CECIDOMYIIDAE (gall - flies).—This is a family of small Diptera, which give rise to galls in plants. Their wings have few nervures, usually three longitudinal ones; they are rounded at their free ends; the point of attachment is, however, very narrow. The antennae have numerous joints, and are moniliform. The proboscis is short and the legs long. The females have a well-developed ovipositor, by means of which they puncture some plant and deposit their egg therein. This in some cases gives rise to a gall, in which the larvae develope. The pupa stage may also be undergone in the gall, or it may

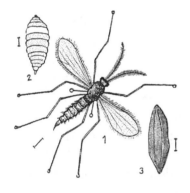

FIG. 210.—*Cecidomyia destructor* (the Hessian-fly).

1. Insect.

2. Larva.

3. Pupa or 'flax seed.'

All magnified.

be free. The larvae of some of the Cecidomyias produce parthenogenetic ova, from which young are born, a phenomenon known as *paedogenesis. Cecidomyia destructor* is the well-known Hessian-fly. *C. salicis* produces galls on willows.

Family TABANIDAE (horse-flies).—*Tabanus* is the horse-fly, a large Dipterous insect. The male does not bite, but lives on the nectar of flowers. The bite of the female, however, is very painful and poisonous. Cases have been recorded of horses on the prairies being worried to death by these insects. The larvae of *Tabanus* are said to live upon snails, those of allied genera live in water.

Family BOMBYLIIDAE.—Very hairy flies, swift in flight,

and often to be found in sunny paths, etc. They lay their eggs in bees' hives, and the fleshy smooth maggots devour the larvae of the bees.

Family SYRPHIDAE.—This is a family of brilliant yellow and black flies somewhat resembling wasps. They have a large head, and the eyes, which are very large, meet across the middle line in the male. They are found in sunny places, hovering motionless in the air, and then making a sudden dart at some flower upon whose nectar they feed. The genus *Syrphus* lays its eggs singly upon flowers, and the larvae feed upon Aphides, which are seized by the mouth parts situated at the end of the first segment ; this segment is then retracted into the second, and the second into the third. The larvae of other species are found in water or in bees' nests ; the rat-tailed larva of *Eristalis* is a well-known object in fresh water. Its body is prolonged posteriorly into a long respiratory tube.

Family CONOPIDAE.—The members of this family also resemble wasps, a resemblance which is probably partly protective, and which furthermore enables many of them to lay their eggs in or on the bodies of the insects they mimic. The *Conopidae* have a pedunculated abdomen, and their proboscis is unusually long. The eggs of *Conops* are probably laid on the soft tissue between the segments of some Hymenopterous insect, the larval and pupal stages are found within the body of the host, and the fly emerges between two of the abdominal segments, often breaking the abdomen of its host in two.

Family OESTRIDAE (bot-flies).—The bot-flies have short, stout, hairy bodies, with minute antennae and rudimentary mouth parts. The eggs are laid on some place whence the parasitic larvae can easily effect a lodgment in its host. The larvae are thick, fleshy, apodal maggots, with eleven segments, which are usually provided with spines or hooks, by means of which they can move about within their host. Those which live in the alimentary canal of mammals have hooks, and attach themselves by these to the walls of the stomach. The stigmata are borne on a horny plate at the posterior end of the animal, the end which is generally turned to the external world. The pupal stage is passed through on the ground.

Gastrophilus equi, the horse-bot, lays its eggs on the hairs of the hips and legs, the horse licks these off, and the larvae hatch out in its stomach; the larvae are found from May to October hanging in clusters on the wall of the rectum. *Hypoderma bovis*, the ox-bot, deposits its egg beneath the skin of cattle; the larvae hatch out, and cause large hollow cysts to form around them; they lie in these cysts with their stigmatic plate directed to the opening. *Oestrus ovis*, the sheep-bot, is said to be viviparous; the female deposits larvae alive in the nostril of the sheep, whence they make their way to the frontal sinuses.

Family MUSCIDAE.—This family, which is very extensive, includes the common house-fly and blue-bottle fly, etc. The antennae are three-jointed, and end in a bristle. The proboscis terminates in a fleshy bilobed process. The wings have four simple longitudinal veins. The abdomen has only five segments visible externally, but others are invaginated. The larvae are white, cylindrical, apodal maggots, with a posterior pair of spiracles, and sometimes a prothoracic pair as well.

FIG. 211.—*Glossinia morsitans* (the Tsetse fly).

Musca domestica and *M. vomitoria* are the house and blue-bottle flies. The tsetse fly, which is so poisonous to cattle in South Africa, is *Glossinia morsitans*. The Tachinidae lay their eggs in caterpillars, within whose body their larvae are parasitic. Several species are stated to be viviparous.

ORDER 8. HYMENOPTERA.

CHARACTERISTICS.—*Insects with biting and licking mouth appendages. The prothorax is fused with the mesothorax. The four wings are membranous, with few nervures, the hind wing on each side bears a row of hooks, by means of which it is hooked to the anterior wing, and the two move together. The metamorphosis is complete.*

The Hymenoptera include the ants, bees, and wasps, insects which, with regard to their social habits and instincts, stand not only at the head of their own phylum, but in front of all other animals, man alone excepted. It is not a very large order, only about 25,000 species are described. All of these are terrestrial, and most of them inhabit the warmer and more temperate parts of the globe.

The head is freely moveable, the eyes large and prominent, and, as is often the case in insects, they are larger in the male than in the female. There are three ocelli. The antennae are often crooked, and then consist of a basal joint or shaft, and ten to twelve shorter joints ; they may be straight.

The mouth is adapted for biting or licking ; there is a small labrum, and the mandibles are large and stout. The maxillae and labium are elongated, palps being present in each case. The nectar upon which bees feed is licked up by the *ligula,* and deposited upon a sheath formed of the labial palps and maxillae, and in the sheath thus formed the food passes up to the mouth. The ligula is the front edge of the labium. In bees it is enormously developed, and divided into three lobes, the two outer of these are termed *paraglossae.*

On the mesothorax are two small scales known as the *tegulae,* covering the base of the wings.

The legs have a five-jointed tarsus, and in bees the tibia and tarsus, especially those of the posterior pair of legs, are covered with short hairs, these help to collect and convey home the pollen grains the bees have gathered during their frequent visits to flowers.

The abdomen is as a rule pedunculated, the two anterior segments forming the peduncle. In the female the abdomen ends in an ovipositor, which in the Aculeata is modified and forms a sting. This complicated apparatus developes from six protuberances on the embryo, four on the last but one, and two on the last segment. These protuberances elongate and form a grooved process with two stylets and a pair of lateral sheaths, besides these there are certain supporting plates. The poison consists of formic acid with some fatty substances ; it is secreted in a special gland, which communicates with a reservoir, this opens at the base of the grooved process.

The nervous system consists of an unusually large and

complex brain, two thoracic ganglia, and five or six abdominal. The alimentary canal is long, with large salivary glands, and a suctorial stomach, or, in the ants, a gizzard. The number of Malpighian tubules is great, 20 to 150.

With few exceptions, the larvae are apodal white grubs, living either parasitically in the bodies of other insects, or in plant galls, or in special cells prepared for their reception. In two families, however, they resemble in appearance and habits the caterpillars of the Lepidoptera. As a rule, the larvae have a retractile head, with short mandibles, maxillae, and labium. The pupae are mostly surrounded by a silken cocoon.

The wax-secreting apparatus consists of numerous glands, mostly unicellular, which open to the exterior by fine chitinous tubes. The wax of bees is secreted in fine transparent plates on the under surface of the abdomen.

The order may be divided into three sub-orders:

1. PHYTOPHAGA.
2. ENTOMOPHAGA (Parasitica).
3. ACULEATA.

Sub-order 1. **Phytophaga.**

CHARACTERISTICS.—*Hymenoptera with a well-developed ovipositor which sometimes functions as a borer. The abdomen is sessile. The larvae feed on plants, and resemble caterpillars.*

Family UROCERIDAE (Wood-wasps).—The insects are large for Hymenoptera, with elongated bodies. The abdomen of the female terminates in a long ovipositor, with which she bores holes into wood to deposit her eggs there. The larvae are white grubs, with three pairs of very small legs and strong mandibles; they eat out passages in the wood, and form silken cocoons mixed with sawdust, in which the pupal stage is passed. The complete insect emerges in these galleries, and eats her way through the bark to the outside world.

Sirex gigas, one of the largest of these insects, attacks in this way the wood of Conifers. *Sirex juvencus* is a smaller species.

Family TENTHREDINIDAE (Saw-flies).—This is a numerous

family of Hymenoptera with short bodies ; the abdomen is sessile. The antennae vary a good deal, but are never elbowed ; their

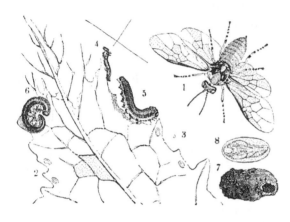

FIG. 212.—*Athalia spin-arum* (Turnip Saw-fly). After Curtis.

1. Saw - fly, magnified, with lines to left show-ing natural size.
2. Eggs in incision in leaf.
3. Egg, natural size.
4, 5, 6. Caterpillars feed-ing on turnip leaf.
7. Pupa case.
8. Pupa.

variations afford a basis of classification. The tibia of the fore-leg is provided with two spines. The ovipositor is short, and retracted when not in use ; it consists of two saw-like processes protected by two valves.

The females deposit their ova in punctures in the epidermis of plants ; one egg is usually laid in each slit. At the same time a drop of fluid is secreted, which is said to cause a flow of sap to the injured part, and this sap is stated to be absorbed by the egg, which increases in size. Only one genus—*Nematus*—forms galls, and these are found in the leaves of the willow tree. *Nematus* resembles the bees and wasps in the fact that its unfertilised parthenogenetic eggs give rise to males. As a rule in the animal kingdom parthenogenetic eggs give rise to females.

The larvae are very like caterpillars. Usually they are found in colonies, and are of a brown or greenish colour. They may be distinguished from the young of Lepidoptera, which never have more than five pairs of pro-legs, by possessing six, seven, or eight pairs. Their heads are also more globular, and the eyes more distinct. When at rest they usually curl round the posterior end of their body (Fig. 212) in the form of a note of interrogation. They are frequently to be found on the leaves of willows, limes, poplars, and conifers. They are said to emit an acid fluid from lateral pores on their body when irritated.

The pupae lie in a cocoon which is found in the neighbour-hood of their food plant, often on the ground. *Athalia spinarum* is the turnip saw-fly; and *Lophyrus pini* causes much damage to young Scotch firs.

Sub-order 2. **Entomophaga**.

CHARACTERISTICS.—*The abdomen is pedunculated. A well de-veloped ovipositor is present in the female. The larvae are usually parasitic in the larvae of other insects, sometimes in plants; they are apodal and aproctous.*

Family CYNIPIDAE (Gall-flies).—Small Hymenoptera with a much-compressed abdomen, the first and second segments of which are large, the others very short. There is a long coiled ovipositor which arises near the base of the abdomen, and is only fully extended when in use. The ova are large, and are said to increase in size during developement. The larvae are short white fleshy apodal grubs, which feed on the galls in which they live. The pupae may be in the gall or under ground. Some are parasitic in Diptera and Aphides.

The oak-apple is formed by a member of this family, *Andricus terminalis*; the common spherical woody galls in the oak, by *Cynips kollari*; and the "spangle galls" in the leaves of the same tree by *Neuroterus lenticularis*. *Rhodites rosae*, another member of this family, gives rise to the beautiful moss or bedeguar galls on rose-trees.

Family ICHNEUMONIDAE (Ichneumons).—Insects with long

FIG. 213.

Aphideus avenae, to the left.

Ephedrus plagiator, to the right.

slender bodies, with an exserted, often very long ovipositor. The abdomen has seven segments. The antennae are slender and many-jointed ; they are seldom elbowed. The ova are deposited on or in the bodies of caterpillars or the larvae of other Hymenoptera. The larvae are white, fleshy, footless grubs which feed on the organs of their hosts, commencing with the fat-bodies and other indifferent tissues. *Rhyssa persuasoria*, which has an ovipositor three inches long, is said to bore through a considerable thickness of wood to deposit its eggs in the body of the larvae of *Sirex*. *Ichneumon laminatorius* lays its eggs in the larvae of *Sphinx pinastri*, *Aphideus avenae* and *Ephedrus plagiator* in the bodies of Aphis.

Family CHALCIDIDAE.—This family includes numerous species of small size and of metallic hue. Their antennae are usually elbowed, and their wings devoid of veins. The differences between the sexes are as a rule very marked. They lay their eggs in the eggs or larvae of almost every kind of insect. Some are parasitic in the parasitic *Ichneumons*, others in the Hessian-fly and in *Musca*, whilst other species, as *Pteromalus*, lay their eggs in the larvae of *Bostrychus* and *Hylesinus*, etc. A few species give rise to galls.

Sub-order 3. **Aculeata.**

CHARACTERISTICS.—*The female is, with few exceptions, provided with a stinging apparatus, or modified ovipositor, at the end of the abdomen, which is only protruded when in use. The abdomen is stalked. The male has, as a rule, thirteen joints in its antennae, the female twelve. Usually cells are constructed in which the apodal and aproctous larvae are reared by the females.*

Group 1. FORMICARIAE (Ants).

Ants have usually a small thorax, and, compared with the thorax or abdomen, a very large head. The first, or first and second abdominal segments form the peduncle. The antennae are elbowed, and have one more joint in the males than in the females. The wings, when present, extend beyond the abdomen, and they have few nervures.

As a rule, the males are smaller than the females. They live in communities, which comprise, besides the winged males and females, a number of wingless workers or neuters. These

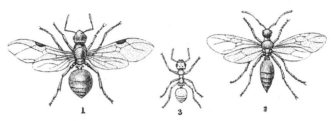

FIG. 214.—*Formica rufa* (Wood-ant).

1. Female.   2. Male.   3. Neuter.

are in reality aborted females, and, like the functional individuals of that sex, are provided with poison glands. The poison consists of formic acid, and is ejected into wounds made either by the sting or the biting mandibles. Some of the workers, as is the case in the Termites, are specialised as " soldiers " with very formidable jaws and large heads. These defend the ant-hills when they are attacked.

The workers or neuters survive the whole year, hibernating during the winter months. Some of the females also may hibernate, but the greater number of both males and females live for a short time only, during the summer. No food is stored up in the ant nest or *formicarium* for winter consumption, as those - individuals which persist through the cold weather become torpid and cease to feed. With the return of spring the females which have persisted lay eggs, and these, or in some species the eggs and larvae of the preceding autumn which have lasted through the winter, develope into a new brood, producing males, females, and workers. The larvae are most carefully tended by the workers, which feed them with semi-digested food from their own stomachs. The sexes pair whilst flying in the air; the males then die, the females cast their wings and either start off to form a new colony or are led back by the workers to the old. The ova are very small. The structures which are commonly called the eggs of the ants are the white oval cocoons.

The nests are usually excavated in the ground or in a

rotten tree, or are built up of clay.   The ants live upon both
animal and vegetable substances, and are very fond of sweet
things. Some of them keep Aphides, and by gently stroking them
with their antennae they induce them to secrete some of their
honey-dew, which the ants greedily lick up.   Certain species
of beetles are also always found living with ants, though the
relationship between them and their hosts is not understood.

Certain species, as *Polyergus rufescens,* and in some countries
Formica sanguinea, make slaves of other and smaller species.
The slaves are carried off either as ova or pupae, and are reared
by the masters, who become so dependent upon their slaves
that if these be removed the colonies of *P. rufescens* perish.

The leaf-cutting ant *Oecodoma* cuts small circular pieces
out of leaves, which are carried home often by another indi-

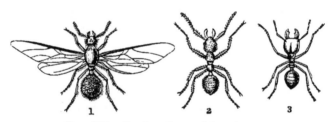

FIG. 215.—Leaf-cutting and foraging ants.
1. *Oecodoma cephalus.*   2. *Eciton drepanophora.*   3. *Eciton erratica.*

vidual, and it is believed that a fungus is grown upon these
leaves, which the ants eat.   The nests of this genus are very
large, and are built of clay.   The leaf-cutting ant does very
considerable damage to vegetation, and it is difficult to keep
them away from any spot.   They have been known to tunnel
under a river to reach a wood.

Group 2. VESPIARIAE (Wasps).

Wasps have slender, almost naked bodies, of a usually
yellow and black colour; the antennae are usually elbowed;
the wings are long and narrow, and longitudinally folded when
the insect is at rest.

There are three families: the VESPIDAE, or social wasps, and
two families of solitary wasps—the EUMENIDAE and MASARIDAE.

Family VESPIDAE.—Social wasps, whose colonies contain

males, females, and workers; the last-named are females which sometimes produce parthenogenetic eggs. These colonies are

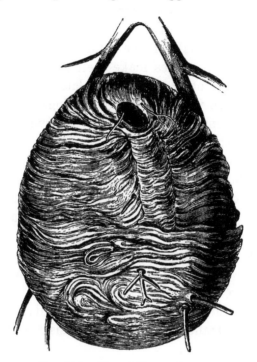

FIG. 216.—Nest of *Vespa sylvestris*.

annual, existing for the summer only. In the autumn all die, with the exception of a few fertilised females, which creep into crevices of trees or under stones, and hibernate through the winter. In the spring the female emerges and commences to build a nest; this is constructed of fragments of wood gnawed up and mixed with saliva, the whole making a papery substance. As soon as two or three cells are finished the female lays an egg in each, and when the

FIG. 217.—*Vespa rufa.*

white apodal grubs hatch out they have to be fed, whilst at the same time the mother is widening and deepening their cells and adding others.

The larval stage lasts about two weeks, and then the grubs

cease to eat and turn to pupae in their cells. The imago emerges in ten days, and sets to work to enlarge the nest. As soon as the perfect insect vacates its cell, it is cleaned out and another egg is deposited in it.

During the first half of the summer, only workers appear, but later males and females make their appearance, the former from the parthenogenetic eggs of the later broods of workers; all these kinds are winged, but the workers are smaller than the males and females. The sexes pair whilst flying, and soon afterwards the males die.

There are seven species of *Vespa* found in Great Britain; of these the hornet, *Vespa crabro*, is the largest. This does not extend north of the Midlands; its nest is larger than the common wasp *V. vulgaris*, but its colonies contain fewer individuals. *V. rufa* has a reddish tinge on its legs; it, like *V. vulgaris* and *germanica*, builds its nest in the ground, whilst *V. sylvestris*, *arborea*, and *norvegica* are tree-wasps, building their nests in trees.

The other European genus of this family, *Polistes*, occurs round the Mediterranean. Its nest consists of a single layer

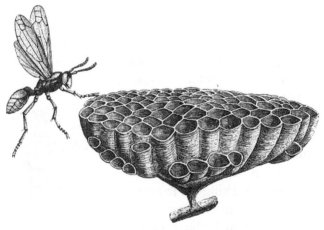

FIG. 218.—*Polistes tepidus* and nest.

of cells, forming a comb which is vertically or obliquely placed, and is not covered in by an outer case, like the nests of the various species of *Vespa*.

Family EUMENIDAE.—*Eumenes coarctata* is the only British

species of this family; it makes small rounded cells of mud, which it attaches to some plant, very generally to heath. In each cell it lays one egg, and stores up some provision of honey to serve as food for the larva; the cell is then closed. Some species of this genus are carnivorous, and when this is the case the mother stings some caterpillar or other insect larva in the ventral nerve cord in such a way as to render it motionless without killing it; the inert

Fig. 219.—*Eumenes smithii.*

larva is then deposited in the cell to serve as food for the grub of the wasp.

Family MASARIDAE.—No members of this family occur in

Fig. 220.—*Masaris vespiformis.*

Britain, and the number of genera is small. They construct nests in banks of earth, with passages leading down to them.

Group 3. FOSSORIA.

These are sometimes termed digging wasps; they have no bend in their antennae, and their legs are elongated. The females construct passages in sand or earth, or sometimes in wood, and lay their eggs at the end of these tubes. They live upon honey and pollen, and either convey fresh food to their larvae every day or store up in the tube a sufficient number of insects or larvae, paralysed by a sting in the nervous

25

system, to suffice for the larva till it pupates. Each species always attacks a particular kind of insect and carries it home as food for its larvae; thus *Cerceris bupresticida* always attacks the larvae of the beetle *Buprestis*, *Sphex flavipennis* attacks *Gryllus*, and *S. albisecta* various species of *Oedipoda*.

Another allied family, the Chrysididae, often lay their eggs in the nests of the Fossoria and other Hymenoptera. They are green or black, and their sting is devoid of a poison-bag.

Group 4. APIARIAE (Bees).

In the bees the body is thick and short, and as a rule hairy. In the workers the tibia and tarsus, especially of the posterior legs, are broadened and covered with hairs like a brush; these serve to gather and carry home the pollen grains. The labium and maxillae are often very long, and can reach to the nectaries at the base of some of the longest flowers. Whilst obtaining food in this way bees very frequently effect the cross fertilisation of the flowers they visit. The anterior wings do not fold. Both the females and workers have stings, which in many cases are provided with recurved spines, so that if inserted into a foreign body they cannot be retracted.

The division of labour which plays such a prominent part in the economy of the higher Hymenoptera reaches the highest pitch amongst the bees. The queen-bee alone in *Apis mellifica* — the honey bee — lays fertilised eggs, sometimes at the

Worker-bee.   Drone.   Queen-bee.

FIG. 221.—*Apis mellifica.*

rate of 3000 a day; she and the workers live through the winter, but the drones all perish in the autumn. The drones are the males of the community; they arise from unfertilised eggs laid by the queen, who can lay at will either fertilised or unfertilised ova.

During the winter the queen-bee and the workers live upon the food stored up in the hive; when spring returns she deposits eggs, first in the cells of the workers and then of the drones. After a time certain large royal cells are constructed, and in each of these she lays a fertilised egg; the larvae which proceed from these eggs receive a richer nourishment and become queens. The drones take twenty-four days to develope, the workers twenty, and the queens sixteen. Before the eldest of the royal pupae gives rise to a queen in the imaginal state, the queen mother with a number of the workers leave the hive and swarm. Thus a new colony arises. The young queens fight until all but one are killed, or the others swarm; the victorious one remains as queen of the hive. Soon after the metamorphosis is complete the queen is fertilised by a drone whilst flying in the air. The drone immediately dies, and the queen, which has only been fertilised once, can continue to lay fertilised ova for several years.

A hive may number as many as twenty to thirty thousand individuals, of which the drones do not form more than one per cent.

Bombus, the humble-bee, makes underground nests which contain from fifty to two hundred individuals. This genus does not make cells, but lays its eggs in a mass of pollen accumulated in the centre of the nest; the larvae eat their way through this, and ultimately turn into pupae. A single fertilised female survives the winter, and inaugurates a new colony in the following year.

CHAPTER XX

TRACHEATA

ARACHNIDA

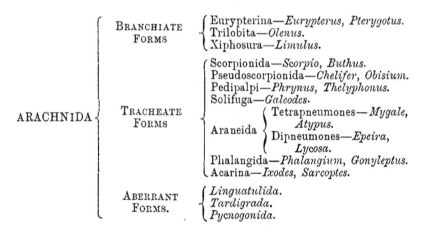

ARACHNIDA

- BRANCHIATE FORMS
  - Eurypterina—*Eurypterus, Pterygotus.*
  - Trilobita—*Olenus.*
  - Xiphosura—*Limulus.*
- TRACHEATE FORMS
  - Scorpionida—*Scorpio, Buthus.*
  - Pseudoscorpionida—*Chelifer, Obisium.*
  - Pedipalpi—*Phrynus, Thelyphonus.*
  - Solifuga—*Galeodes.*
  - Araneida
    - Tetrapneumones—*Mygale, Atypus.*
    - Dipneumones—*Epeira, Lycosa.*
  - Phalangida—*Phalangium, Gonyleptus.*
  - Acarina—*Ixodes, Sarcoptes.*
- ABERRANT FORMS.
  - *Linguatulida.*
  - *Tardigrada.*
  - *Pycnogonida.*

CHARACTERISTICS.—*Arthropoda with a cephalothorax (prosoma) marked off from the rest of the body by the presence of a carapace and by the character of its limbs. The abdomen may be segmented or unsegmented, and is sometimes divided into two regions, the mesosoma and the metasoma, the latter may bear a posterior spine. Four pairs of walking-legs are usually present. The breathing apparatus may take the form of branchiae adapted for aquatic respiration, or of lung-books, or tracheae for breathing air. An internal skeleton in the form of a plate, termed the endosternite, is present in many forms. The genital orifice is usually on one of the anterior segments of the abdomen. They are, with the exception of one order, bisexual.*

The Arachnida fall into three groups—those that breathe water, those that breathe air, and certain aberrant forms,—and the members of groups differ to a very considerable extent.

The Branchiate group includes three classes :

    I. Eurypterina (Merostomata).   II. Trilobita.
             III. Xiphosura (Poecilopoda).

Of these the first two are extinct, whilst the third is represented by a single recent genus, *Limulus*, the king-crab. This, like its fossil congeners, is marine.

The tracheate division of the Arachnida comprises seven classes :

I. Scorpionida.	V. Araneida.
II. Pseudoscorpionida.	VI. Phalangida.
III. Pedipalpi.	VII. Acarina.
IV. Solifuga.	

The aberrant forms are—

I. Linguatulida.	II. Tardigrada.

           III. Pycnogonida (Pantopoda).

Of these the first class is parasitic and the third is marine.

The Arachnida form a most heterogeneous assemblage ; the group affords shelter to a number of forms whose affinities are by no means clear, and whose structure is in some cases so modified as to possess few or no Arachnid characteristics.

A. THE BRANCHIATE GROUP.

Class III. Xiphosura.

The very remarkable creature which is popularly known as the king-crab, and scientifically as *Limulus*, is an inhabitant of the warmer waters of the east coast of America, the East Indian Archipelago, and the western shores of the Pacific. It is found in from two to six fathoms of water, moving about in the mud or sand, and living chiefly upon certain worms. It may attain a length of several feet.

The body of *Limulus* is enclosed in a thick chitinous cuticle, the dorsal surface of which forms a large shield or carapace. This is produced laterally into two backwardly directed processes, and it covers in the *prosoma*, the anterior region of the body, which bears six pairs of appendages. Anteriorly near the middle line are a pair of simple eyes, and behind and nearer the sides are a pair of compound eyes.

Behind the prosoma, the body is covered in by an abdominal carapace, the anterior part of which shows dorsally faint traces of being composed of seven segments. This region corresponds with the *mesosoma* of other forms; behind it the *metasoma* pre-

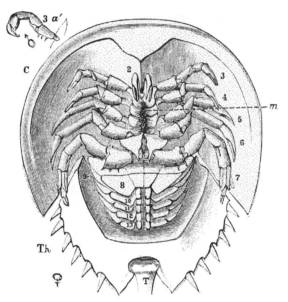

Fig. 222.—Under side of *Limulus polyphemus*, Latr., ♀.

C. Carapace.	3*a'*. Detached pedipalps of ♂.
Th. Abdomen (= meso- and meta-soma).	4-7. Thoracic walking-legs.
T. Telson or spine.	8. Lamellae with genital openings.
2. Chelicerae.	9-13. Branchiferous lamellae.
3. Pedipalps.	*m*. Mouth.

sents a smooth unsegmented surface. The anus is situated at the end of the metasoma, and a long pointed spine extends behind it.

Appendages of *Limulus*.

1. Chelicerae.	7. Plates with genital pores.
2. Leg-like appendages (pedipalpi).	8. ,, ,, branchiae.
3. ,, ,,	9. ,, ,, ,,
4. ,, ,,	10. ,, ,, ,,
5. ,, ,,	11. ,, ,, ,,
6. ,, ,,	12. ,, ,, ,,

The appendages which are found on the ventral surface of the body commence with a small three-jointed pair of chelicerae, placed one on each side of the upper lip, which overhangs the mouth. The next five pairs of appendages are leg-like, and are arranged round the mouth; they consist of six joints—the proximal one or coxa is enlarged, and its anterior border is produced into a process which helps to surround the mouth and form a biting organ. In the female of L. *polyphemus*, the first five of these appendages are chelate, but in the male the first is enlarged but is not chelate; in both sexes the sixth appendage terminates in a number of elongated flattened plates, and this limb is used in the burrowing or digging operations in which the animal delights. All these append-ages are borne on the prosoma; the seventh appendage, or the first mesosomatic, consists of a semicircular plate-like structure hinged on to the body and bearing on its posterior face the two genital pores. This genital plate or operculum folds over and almost covers the five succeeding appendages, which are also plate-like, but, like the former, exhibit traces of a double origin. Each of these, from the eighth to the twelfth, carries a pair of respiratory organs, in the form of branchiae composed of a great number of thin plates like the leaves of a book. Behind the twelfth is the unsegmented metasoma, which bears no appendages.

An internal skeleton or *endosternite*, in the form of a plate of fibro-cartilage, lies between the alimentary canal and the elongated nerve collar. It is not connected in any way with the exoskeleton, but gives origin to a number of muscles. A somewhat similar structure is found in *Apus* and in some other Crustacea.

The alimentary canal, as is usual with the Arthropoda, excepting the Insects, runs in a median line from mouth to anus without convolutions. The mouth is situated some dis-tance behind the anterior end of the body; it leads into a suctorial pharynx, which is lined by chitinous ridges. The pharynx runs forward and widens into a stomach, which turns back, curves posteriorly, and is separated by a valve from the mid-gut; the latter is the absorbent part of the alimentary tract. It extends through the body, and terminates in a short procto-daeum with folded walls. The only gland which opens into the alimentary canal is a large yellow organ, the so-called liver, which communicates with the mid-gut by two pairs of ducts.

The vascular system is very complete, and, unlike many Arthropods, the arteries do not end in irregular lacunar spaces, but are connected with the veins by a definite capillary system. The heart consists of an elongated tube, partially divided into eight chambers, into the anterior end of each of which a pair of ostia open (Fig. 223). The heart ends blindly behind, but in front is continued into a truncus arteriosus; from the base of the latter an artery arises on each side, which passes, together with a vessel from each of the anterior chambers of the heart, into a large collateral artery which runs outside the heart, but parallel to it. The collateral vessels unite behind, and are continued backwards as a single supra-anal vessel. Still

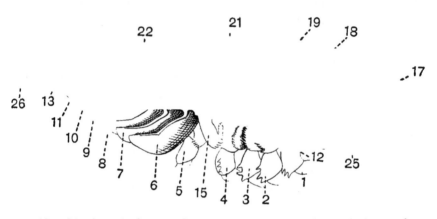

FIG. 223.—*Limulus polyphemus.* Diagrammatic view of left half of body seen from the inside. From Leuckart, partly after Packard.

1-5. The 5 pairs of thoracic limbs.
6. The operculum.
7-11. The gill-bearing mesosomatic limbs.
12. Mouth.
13. Anus.
14. Spine or Telson.
15. Chilaria.
16. Oesophagus.
17. Muscular proventriculus.
18. Intestine.

19. Bile ducts; the openings of two into the intestine are seen.
20. Liver.
21. Heart, with eight venous ostia.
22. Efferent branchial veins leading into pericardium.
23. Frontal artery dividing at
24. Into two marginal arteries.
25. Left aortic arch.
26. Supra-anal artery; the sub-anal artery is shown ventral to this.

The solid black structure beneath the alimentary canal, lying in a blood space, is the nervous system.

farther forward the truncus gives off two cerebral arteries, which pass into a vascular ring which surrounds the oesophagus, and encloses the nerve collar. From this ring a number of vessels arise which supply the appendages, and it is continued backward into a supraspinal vessel in which the ventral nerve cord lies.

The capillaries in which the arteries terminate open into certain reservoirs; passing from these the blood traverses the gills, and is then returned by a system of branchiocardiac canals to the pericardium, and so by the ostia into the contractile heart. The blood has a bluish tinge, due to the presence of haemocyanin, and contains numerous oval corpuscles.

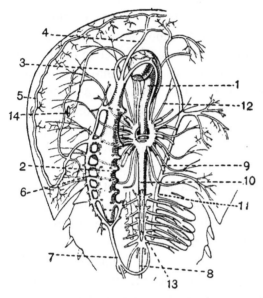

FIG. 224.—Diagram of circulatory system of *Limulus polyphemus*. From Leuckart, after Milne-Edwards.

1. Oesophagus.
2. Heart with 8 pairs of ostia.
3. Aortic arch, the two branches of which unite into a ring surrounding the oesophagus and mouth.
4. Frontal artery.
5. Marginal artery.
6. Collateral artery, running parallel with heart, and supplied by 7 vessels from the heart.
7. Supra-anal artery.
8. Sub-anal artery.
9. One of the 5 pairs of vessels going to the 5 thoracic limbs.
10. Ventral artery surrounding ventral nerve cord and giving off branches to mesosomatic limbs and gills.
11. Longitudinal lateral vein giving branches to gills, the blood is returned from gills to pericardium by the vessels marked (22) in previous figure.
12. Cutaneous nerves which have no arterial sheath.
13. Ventral nerve cord in ventral artery.
14. Lateral eye.

The gills, which are borne upon the posterior face of the five posterior pairs of appendages, consist of a number (150) of very delicate plates, each composed of two thin membranes, in the space between which the blood circulates. The plates lie parallel to one another like the leaves of a book, and the water circulates between them.

A large gland, in its natural state of a brick-red colour, is situated at the base of the leg-like appendages. It consists of a longitudinal portion, from which four lobes are given off, corresponding with the second, third, fourth, and fifth prosomatic limbs. In the adult this structure, which has been termed the *coxal gland*, appears to have no duct, but in the young, both an internal and an external opening have been described. The former opens into a space in the connective tissue which lies between the gland and the ventral blood sinus; the external opening is situated on the dorsal surface of the coxa of the fifth appendage. The chitinous covering of the animal is continued a short distance into the duct. The gland thus appears to have the relationship of a nephridium, and it is worthy of note that it opens upon the same appendage in numerical sequence as does the shell-gland of the Entomostraca, the latter having its orifice on the second maxillae, or fifth appendage.

The nervous system of *Limulus* consists of a supra-oesophageal nerve mass which gives off five nerves, supplying the ocelli, the compound eyes, and the integument in the region of the head (Fig. 223). From the sides of this mass pass a pair of oesophageal commissures, which come together some distance behind the oesophagus; the commissures are, however, united together by a number of transverse connectives situated behind the oesophagus. This elongated oval collar supplies nerves to all the limb-like appendages, and to the operculum which bears the genital orifices. After the fusion of the two circum-oesophageal commissures, the nervous system is prolonged backward as a ventral nerve cord, which during the first half of its course gives off no nerves, but its posterior half supplies five pairs of nerves, which pass to the last five appendages and the surrounding parts.

Both the oesophageal collar and the ventral nerve cord, and many of the more important nerves, are ensheathed in blood-vessels.

Limulus is dioecious, and the male can be distinguished from the female by the thicker and non-chelate character of the second pair of appendages. Both the ovary and the testis are retiform, the network of tubules which compose these glands extending through the pro-, meso-, and meta-soma. The tubules of the testes are in communication with a number of spherical sperm vesicles, which contain immature spermatozoa; the latter, when ripe, are provided with a motile tail. Both the ovaries and the testes communicate with the exterior by means of two ducts, which open independently on the posterior face of the operculum or seventh segment. Fertilisation is external.

The propriety of placing *Limulus* amongst the Arachnida has not been universally accepted, but its resemblance to *Scorpio*, even in points of detail, is striking, and the relationship thus indicated is strengthened by the structure of the fossil Eurypterina, which are to some extent intermediate between these two orders. They have six pairs of thoracic limbs, and the region between the genital plate and the anus is divided into a number of segments gradually diminishing in size.

B. THE TRACHEATE GROUP.

Class I. **Scorpionida.**

Characteristics.—*Arachnida with six pairs of appendages on the prosoma, of which the first and second pairs are chelate, the latter being enlarged. There is a mesosoma of seven segments, and a metasoma of five, terminating in a curved spine armed with a poison gland. There are four pairs of lung-books.*

Scorpions inhabit warm and tropical countries, hiding for the most part under stones, etc., during the day, but running about after dusk. They usually carry their metasoma turned back over the back (Fig. 227). There are several genera: *Scorpio, Buthus, Androctonus*, etc.

Their body is covered by a cuticle, which in some places is thickened and forms stout plates. Six of these have coalesced on the dorsal surface of the prosoma and form a carapace which bears from three to six pairs of simple eyes. Two of these are close together near the median line, the rest are arranged along the anterolateral edge. Behind the carapace are seven short broad terga belonging to the mesosoma. Each of these is connected with its corresponding sterna by soft skin; but the last has to some extent fused with the seventh sternum, forming a complete chitinous ring. The five segments of the metasoma are completely encased in chitin, and so is the postanal poison sting.

Appendages of *Scorpio.*

1. Chelicerae.	7. Plates with genital pores.
2. Pedipalpi.	8. Pectines.
3. Walking-legs.	9. Lung-books.
4. ,,	10. ,,
5.	11.
6. ,,	12. ,,

The first pair of appendages, or chelicerae, are short, three-jointed, and chelate. They are directed forward. The second pair, or pedipalpi, are very much enlarged, and six-jointed; they end in a swollen powerful pair of nippers; the coxae or basal joints of this pair of limbs have a biting process which guards the entrance to the alimentary canal. The remaining prosomatic limbs are walking-legs; they are seven-jointed, and

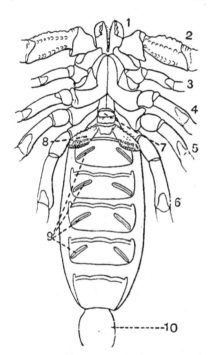

FIG. 225.—Ventral view of pro- and meso-soma of *Buthus afer.* From Leuckart, after Cuvier.

  1. Chelicerae.

  2. Pedipalpi.

3-6. 1st to 4th pair of ambulatory legs.

  7. Genital operculum.

  8. Pectines.

  9. 4 pairs of stigmata on the 3rd-6th mesosomatic region.

10. 1st metasomatic segment.

not chelate. The first and second of these ambulatory appendages have their coxal joint enlarged and turned forward, and the processes thus formed play some part in the ingestion of food. The coxae of the last two pairs of walking-legs are fused together, but those of the left side are separated from those of the right by a small sternum.

The seventh pair of appendages, or first mesosomatic pair,

are probably represented by a small rounded plate with a notch in the middle of its border. This bears on its posterior aspect the genital pores. The eighth pair of appendages are termed the *pectines*; they take the form of an axis, which carries a single row of short processes set like the teeth of a comb; these are probably tactile in function. The remaining segments in the adult scorpion are devoid of appendages, though there are six more pairs of limbs in the embryo; but these do not develope, and are lost.

The third, fourth, fifth, and sixth mesosomatic sterna have each a pair of slit-like openings leading into the lung-books or respiratory organs. These correspond numerically with the last four pairs of branchiae in Limulus.

The endosternite is a free skeletal plate giving attachment to muscles. It lies in the prosoma above the nervous system, and is pierced by a blood-vessel.

The mouth is extremely minute; it leads into a small pharyngeal sac with elastic chitinous walls. Certain muscles

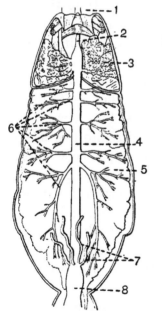

Fig. 226.—Dorsal view of *Buthus occitanus*, the dorsal integument has been removed to show the digestive organs. From Leuckart, after Blanchard.

1. Chelicerae.
2. Oesophagus.
3. Salivary glands.
4. Intestine.
5. Liver.
6. Ducts of liver.
7. Malpighian tubules.
8. Intestine.

act as divaricators of this sac, and thus it functions as a suction-pump, and by its means the scorpion can suck in the juices of its prey, usually spiders or insects. A narrow

oesophagus leads from this sac to a second enlargement, which
receives the ducts of a pair of salivary glands, structures
which are usually associated with a terrestrial mode of life.
The succeeding digestive portion of the alimentary canal remains narrow; it receives four or five ducts which convey the
secretion from a corresponding number of lobes of the liver.
The latter is a considerable gland which takes up a good deal
of space in the wide mesosoma, and even extends into the
narrow metasomatic segments. One or two pairs of delicate
Malpighian tubes are present, and these have been recently
shown in one species to be developementally outgrowths from
the mesenteron; a pair of these tubes are branched. A proctodaeum is present, and ends in an anus situated ventrally at
the end of the metasoma.

The heart in *Scorpio*, as in *Limulus*, consists of a median
tube of eight chambers; it is continued backward in the scorpion
as a posterior aorta which traverses the metasoma. A pair of
valvular ostia open into the anterior end of each chamber,
and a pair of lateral arteries take their origin from the posterior end of each division. The eight chambers lie in the
seventh to the thirteenth segments, the last containing two
chambers. From the anterior end of the heart a truncus
arteriosus leads forward; this vessel gives off two lateral
branches, which embrace the oesophagus and then unite into a
median artery which runs backward above the nerve cord.
In front of this ring the anterior aorta divides into two lateral
vessels and a median one, these supply the appendages of the
prosoma, the brain, and other organs. After passing through
the body, the blood collects in a ventral reservoir and passes
thence to the lung-books. Here it is oxygenated, and is then
conveyed by special veins back to the pericardium, and so
into the heart. The blood in *Androctonus* is, when oxygenated,
of a deep indigo blue, coloured by haemocyanin; its corpuscles
are oval, and are remarkably large.

The four pairs of oval stigmata, which are obliquely placed
on the sterna of the ninth, tenth, eleventh, and twelfth segments, lead into sacs, into the lumen of which project numerous
lamellae. These lamellae, of which there may be as many as
130, arise from a definite axis, which is for the most part

adherent to the wall of the sac, only a short portion being free. Between the thin walls which constitute each lamella, the blood flows, whilst the air circulates between the lamellae.

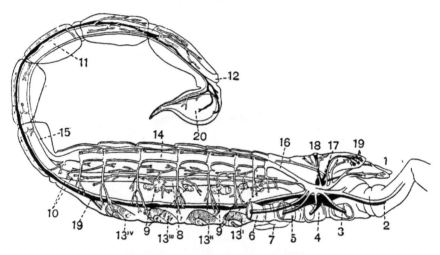

FIG. 227.—View of internal anatomy of *Buthus,* showing digestive, circulatory, respiratory, and nervous systems.   From Leuckart, after Newport and Blanchard.

1. Chelicerae.	12. Anus.
2. Pedipalpi.	13^I-13^{IV}. Lung sacs.
3-6. Ambulatory limbs.	14. Heart.
7. Pectines.	15. Posterior aorta.
8. Mesenteron.	16. Anterior aorta.
9. Lobules of liver, with ducts entering mesenteron.	17. Brain.
10. Malpighian tubules, portions of.	18. Median eyes.
11. Proctodaeum.	19. Lateral eyes.
	20. Poison gland.

A pair of coxal glands occupy a position in the prosoma near the base of the fifth and sixth appendages; they appear to be actively secretory, but apparently in the adult have no outlet.   In the embryo, however, according to Laurie, this gland originates as a tube which opens to the exterior at the base of the fifth appendages, and internally into a coelomic space.   The coxal gland of *Scorpio,* like that of *Limulus,* thus seems to be of the nature of a nephridium, but the part which it plays in the excretion of nitrogenous waste matter, and whether it shares this function with the Malpighian tubules, is a subject still requiring investigation.

The nervous system of *Scorpio* comprises a supra-oesopha-geal ganglion which sends nerves to the central and marginal eyes, and a circum-oesophageal collar, which gives off the ventral nerve cord (Fig. 227). The nerves to the first six pairs of appendages, as well as to the genital plate, the pectines, and the two following segments, arise from the collar or the sub-oesophageal ganglion. The ventral cord for some distance bears no ganglion, the first being situated in the eleventh segment and supplying the third pair of lung-books. Behind this there are six pairs of ganglia, one in each segment, the

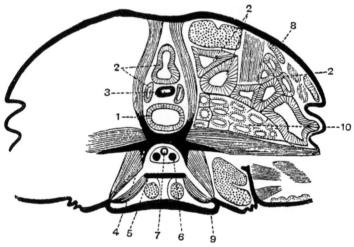

FIG. 228.—Transverse section through the body of *Euscorpius italicus* in the region of the endosternite. After Lankester.

1. Alimentary canal.
2. Caeca of gastric gland (liver).
3. Anterior aorta.
4. Endosternite.
5. Supraneural plate of endosternite.
6. Right ventral nerve cord.
7. Supraneural blood-vessel.
8. Chitinous tergum.
9.   ,,   sternum.
10. Right coxal gland.

The dotted areas represent sections of various muscles.

last being in the fourth metasomatic segment. The sense organs are the simple eyes ·borne on the carapace, one pair near the middle line, and a small group on each side near the edge, and the pectines, which may be tactile in function.

The poison glands are two in number, both situated in the postanal spine. They are provided with a muscular tunic, the contractions of which serve to express the poisonous secretion. Each gland has a duct, and the ducts open by separate orifices close to the apex of the spine. The poison is fatal to most small animals, such as spiders and insects;

and owing to the way the metasoma is usually carried, over the head, it can readily be injected into any prey which has been caught and held by the pedipalpi.

The generative glands, like those of *Limulus*, consist of a network of tubules; these are hollow, and the cells lining their lumen give rise to the reproductive cells. The testes consist of two pairs of longitudinal tubules united by cross branches, on each side of the body; the efferent duct leads to the external opening on the first mesosomatic somite; its distal end is modified to form an intromittent organ, and it bears accessory glands.

The ovary is single, and consists of three longitudinal tubules united by transverse ducts; it lies ventral to the liver, and is to some extent embedded in that organ. The ovary is situated in the mesosoma, and the oviducts run forward, one on each side, to open upon the genital operculum. The ova are formed by the growth of cells lining the tubules, and they project from the surface in the form of follicles. They are fertilised *in situ*, at any rate, in one species. After going through the early stages of segmentation, they pass into the oviduct, and there complete their developement. The young are born in a condition resembling their parents, the scorpion being viviparous.

CLASS II. **Pseudoscorpionida.**

CHARACTERISTICS.— *The abdomen is not divided into meso- and meta-soma, but is broad and flat, and consists of ten or eleven somites; the animals are all minute; they breathe by tracheae; the pedipalpi are chelate.*

In many respects the Pseudoscorpions resemble minute scorpions, but their abdomen is not produced into a tail, nor does it terminate in a poison gland. They breathe by tracheae, which open externally by two pairs of stigmata situated in the third and fourth abdominal somites. Like spiders, they are provided with spinning glands, which lie partly above the liver in the prosoma, and which open at the tip of the moveable joint of the chelicerae.

26

Coxal glands are present in the form of closed tubes, which may be bent once; these lie in the region of the last

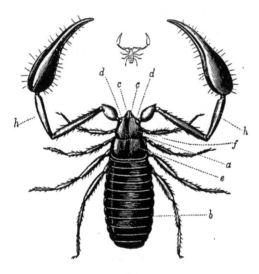

FIG. 229.—*Chelifer sesamoides*, Savigny.

a. Thorax (prosoma).

b. Abdomen.

c. Chelicerae.

d. Eyes.

e. Hinder segment of thorax.

f. Front segment of thorax.

h. Pedipalpi.

three pairs of legs, and at present no opening to the exterior has been described.

The ventral ganglia have fused into a circular mass, from which nerves radiate to the various organs.

The intestine is folded in a single loop, a very rare thing in Arachnids. The rectum has a diverticulum in which, as is the case with spiders, the excrement accumulates.

The genital opening is between the second and third abdominal sterna. The females carry the eggs about attached to the abdomen, and the young live for some time in a small web formed from the secretion of the spinning glands.

There are several genera of Pseudoscorpions: *Chelifer*, *Obisium*, *Chthonius*, and *Chernes*, all have a wide distribution. *Chelifer cancroides* is known as the book-scorpion, as it is found among old books and papers. Other species live under bark, or in moss in damp places; they feed mostly on mites and small insects. They can run rapidly, sideways or backward, and are often transported long distances by clinging with their pedipalpi to flying beetles, etc.

CLASS III. **Pedipalpi** (Scorpion-spiders).

CHARACTERISTICS.—*The pedipalpi are clawed or chelate, and the anterior pair of legs are prolonged into long antenniform processes. Abdomen of eleven or twelve segments, distinct from thorax; there are two pairs of lung sacs.*

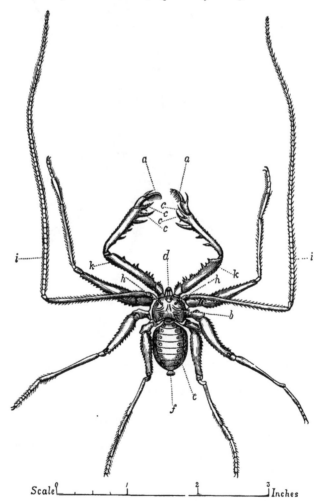

FIG. 230.—*Phrynus medius*, Koch.

a. Moveable fangs or claws on fourth joint of pedipalpi (*k*), forming with claws (*c*) on third joint a modified didactyle claw.
b. Thorax.
d. Chelicerae.
e. Abdomen.
f. Button at end of abdomen.
h. Eyes.
i. Long antenniform legs.

This group includes the two genera *Phrynus* and *Thely-phonus*; they are confined to the tropical and subtropical regions of both the Old and the New World. They are of considerable size.

The abdomen is distinctly marked off from the thorax; it consists, in *Phrynus,* of eleven segments, and does not

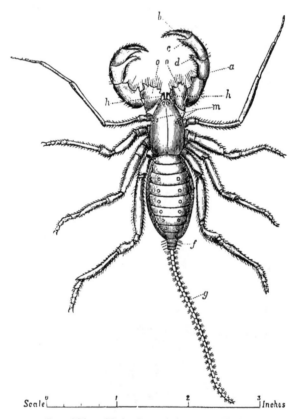

Scale ⌊ᵒ┊_____┊_____┊₁_____┊₂_____┊₃ Inches

FIG. 231.—*Thelyphonus giganteus,* Koch.

a. Pedipalpi.
b. Moveable fang or claw forming with
c. Claw on 4th joint, a didactyle claw.
d. Claw on 3rd joint.
f. Segmental elongation of abdomen supporting tail.

g. Tail.
h. Chelicerae.
m. Eyes.
o. First palpal joint, with characteristic denticulations.

exhibit any distinction of meso- and meta-soma; in *Thely-phonus,* however, the tenth, eleventh, and twelfth segments are small, and the twelfth bears a long-jointed filament representing the telson.

Appendages of *Pedipalpi.*

1. Chelicerae.
2. Pedipalpi.
3. 1st pair of legs, antenniform.
4. 2nd pair of legs.
5. 3rd ,,   ,,
6. 4th ,,   ,,

The chelicerae are clawed, and probably contain poison glands; the pedipalpi are clawed in *Phrynus*, and chelate in *Thelyphonus*. The elongated anterior legs function as antennae, which they resemble in structure. The other legs are seven-jointed. A pair of closed coxal glands lying between the second and fourth pairs of legs have recently been described.

The genital orifice is ventral, at the anterior end of the abdomen, and the respiratory sacs open, one pair on the

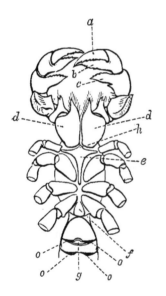

Fig. 232.—*Thelyphonus*, portion of under side.

a. Moveable claw on pedipalp.

b. Fixed claw on pedipalp.

c. Claw on third palpal joint.

d. Coxal joint of pedipalp.

e, f. Sternal plates.

g. Orifice of sexual organs.

h. Place of labium.

o. Orifices of respiratory organs.

posterior edge of both the second and the third abdominal segments. *Phrynus*, like the scorpions, produces its young alive. There are eight eyes: two large ones near the centre of the cephalothorax, and six small ones round its margin.

Thelyphonus seems to be nocturnal in its habits, and passes the day under pieces of bark, etc. When irritated, it is said to carry its tail erect like the scorpions, and the chief interest of the group lies in their forming an intermediate link between the Scorpionida and the Araneida.

Class IV. Solifuga.

CHARACTERISTICS.—*The head and abdomen are distinct from the thorax, which consists of three segments; the chelicerae are chelate, and the pedipalpi leg-like. The abdomen consists of nine segments. Respiration is carried on by tracheae.*

The best-known species of this group is *Galeodes*, which is chiefly found in warm places in the Old World. It is an animal of considerable size; the body of an Indian species is described as being the size of a thrush's egg. *Galeodes*

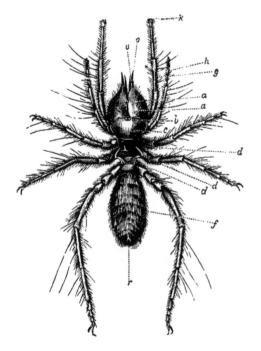

FIG. 233.—*Galeodes araneoides,*
Pallas.

a. Chelicerae.

b. Eyes.

c. Head.

d. Thorax.

f. Abdomen.

g. Pedipalpi.

h. Palpiform legs.

k. Digital joint (capsule).

o. Shear-like points of falx (end joint of chelicera).

r. Anus.

araneoides has a body two inches long, and with the legs may extend over ten inches.

In the division of its body into head, thorax, and abdomen, and in its method of breathing, *Galeodes* approaches the Insecta. The body and limbs are both well covered with thick-set hairs.

Appendages of *Galeodes.*

1. Chelicerae, very strong.
2. Pedipalpi, limb-like.
3. 1st pair of walking-legs.
4. 2nd pair of walking-legs.
5. 3rd   ,,   ,,
6. 4th   ,,   ,,

The chelicerae are very strong, and project in front of the head; the lower limb is moveable, and works vertically against the fixed upper half. The pedipalpi are limb-like. The anterior pair of legs belong to the head; their basal joint serves as a cutting (maxillary) process, at the side of the mouth, an arrangement similar to that found in *Limulus* and *Scorpio*. Like the pedipalpi, they are directed forward, and like them may serve as tactile organs.

The three posterior legs, which are borne by the three free thoracic segments, are clawed. Another point in which these Arachnids resemble Insects is the presence of a well-formed tarsus. The abdomen is composed of nine or ten segments.

The basal joint of the fourth pair of legs bears five curious

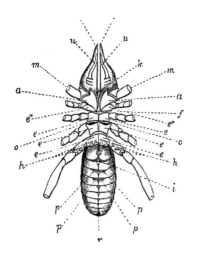

Fig. 234.—Under side of *Galeodes arane-oides*, with legs and palpi truncated.

a. Coxa.
e. Basal leg-joints.
e^x.
e^x. 1st joint of 1st pair of legs.
f. Labium.
h. Racket-shaped appendages.
i. External orifice of genital organs.
k. Tongue.
m. Pedipalpi.
o. Openings of respiratory organs at base of 2nd pair of legs.
p. Orifices leading to tracheae at fore margin of 2nd and 3rd segments of abdomen.
r. Anus.
t. Fixed upper fangs or claws of chelicerae.
u. Moveable lower fangs or claws of chelicerae.

racket-shaped processes, pointing backwards, of unknown function.

The tracheae open to the surface by three pairs of stigmata; the anterior pair are on the thorax, the other two open behind the second and third abdominal segments.

The stomach is provided with lateral caeca; the anus is posterior, and the genital orifice opens on the first segment of the abdomen. Paired coxal glands are found in the thorax; they are much coiled, but apparently have no opening to the exterior.

A pair of poison glands are found in the thorax, closely applied to the walls of the stomach. The ducts are coiled, and open near the bases of the chelicerae ; muscle fibres are described near the orifice, and the contraction of these is believed to eject the poison into the wounds caused by the chelicerae.

Two simple eyes occur in the head. The animals are nocturnal, and spend the day under stones or rubbish, or in holes in the ground. They are extremely voracious, and live on large insects, such as beetles and grasshoppers; they will even attack scorpions, and small vertebrates such as lizards and mice. The female lays about fifty eggs at a time, and the young are hatched in an immature condition.

Class V. **Araneida** (Spiders).

CHARACTERISTICS.—*Abdomen unsegmented, soft, and stalked ; chelicerae with poison glands, pedipalpi leg-like, two or four lung sacs, and four or six spinnerets.*

Epeira diademata is one of the largest species of British spider, and at the same time is a very common one, being found in its wheel-shaped web in most gardens and woods; it therefore serves as a convenient type of the order.

The head is united with the thorax, but the abdomen is constricted at its base, and being swollen and unsegmented, gives a very characteristic appearance to the members of this group.

Appendages of *Epeira.*

1. Chelicerae.
2. Pedipalpi.
3. 1st pair of walking-legs.
4. 2nd pair of walking-legs.
5. 3rd   ,,   ,,
6. 4th   ,,   ,,

The chelicerae are two-jointed, the terminal joint being curved and pointed, and capable of folding down in a sub-chelate fashion; at the base of these appendages lie a pair of poison glands, and these pour their secretion through a duct which opens to the exterior near the tip of the terminal joint. This poison is capable of killing insects and seriously affecting larger animals.

The pedipalpi have a cutting blade-like basal joint which takes part in mastication, as is the case in *Limulus*, *Scorpio*, and *Galeodes*.   They are usually six-jointed, and are frequently clawed, but never chelate.   In the male as it grows older the terminal joint of the pedipalps grows larger, and after the final moult it appears as the *palpal organ*, whose presence may modify the shape of the adjacent joints.   The palpal organ varies in different species, but it consists essentially of a hollow sac which communicates with the exterior by a duct, opening at the tip of the segment.   The use of these organs is to deposit the spermatozoa in the receptaculum seminis of the female.

The four legs are seven-jointed, the terminal joint bearing two or three claws, and in some species a number of short hairs, which aid them in walking up walls and on ceilings.

The abdomen is separated from the thorax by a constriction, it is unsegmented and soft, and is larger in the female than in the male.   Near its base on the ventral surface is the unpaired genital opening, and on each side of this lies the opening of a pulmonary sac; in some species there are two pairs of these openings, the posterior pair leading in some species (*Mygale*) into a second pair of pulmonary sacs, in others (*Argyroneta*) into a tracheal system.

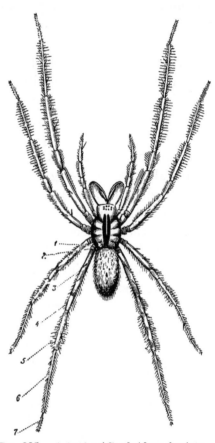

Fig. 235.—A Spider (*Cambridgea fasciata*, Koch).   Adult male.

1-7. Seven joints of leg.

Near the posterior end of the abdomen are found the spinnerets, these are four or six in number, according to the species.   In *Epeira diademata*, one of our largest spiders, there

are four conical two-jointed spinnerets, whose apices meet
together when at rest; if these be separated, two smaller
single-jointed spinnerets are disclosed. Each of these projections is very mobile, and they can be separated and approximated
with ease. They are perforated at their apices by a number
of minute pores, amounting to about 600 in all, through which
the "silk" leaves the body. This silk is secreted by an enormous number of glands which lie in the abdomen; these in the
species in question are of five distinct kinds. Each kind of
gland has not only a definite relation to the spinnerets, some
supplying two pairs and others only one, but the threads
secreted from the various kinds of glands differ in quality, and
are used for different purposes. Thus the lines in the concentric threads of the well-known wheel-like webs differ from
the radial lines in being sticky, and so serving to hold the
captured flies, etc., and these again differ from the threads
which form the cocoon, or from those used to bind up captured
prey; and each of these various kinds of thread is the product
of one definite set of glands.

Between and a little behind the posterior spinnerets the
anus is situated.

In *Epeira* the two stigmata leading to the pulmonary sacs
are on a level with the genital orifice. The cavities of the two
pulmonary chambers are in communication with one another
across the median line. Each cavity is to a great extent
occluded by a number (about sixty) of horizontally placed
lamellae, which are attached as a rule to the wall of their
sac by their anterior and lateral edges, but present a free
edge posteriorly. These lamellae have very thin chitinous
walls, which are prevented from collapsing by the presence of
a number of little cellular pillars. Between the dorsal and
ventral wall of each lamella the blood circulates, whilst the
air passes in the slit-like spaces between the neighbouring
lamellae.

In addition to the pulmonary sacs, *Epeira* also has a
tracheal system. A single stigma opens just in front of the
anterior spinnerets, and leads into a median sac; from this
four tracheae emerge. These are chiefly distributed to those
organs which lie behind the pulmonary sacs. In other species,

as in *Argyroneta*, the tracheal stigmata are situated just behind the pulmonary, in the same position as the second pair of lung sacs in the Tetrapneumones.

The heart in *Epeira* is separated from the dorsal integument as well as from the intestine by some of the lobes of the liver; it lies in a pericardium, and is confined to the abdomen. The heart is described as giving off an anterior and a posterior aorta, and four pairs of lateral vessels. Three pairs of lateral ostia admit the blood which has collected in the pericardium into the heart; this contracts, and forces it through the blood-vessels. The anterior aorta splits in the thorax, and each half bends backwards and downwards to supply the legs and neighbouring parts. From the summit of each bend a cephalic artery runs forward to supply the organs in the head. The course of the lateral vessels in the abdomen cannot be easily followed, as the vessels have very thin walls, and soon lose themselves in the tubules of the liver, etc. The blood is ultimately collected in various sinuses, one of which is continuous with the spaces in the lamellae of the lungs. From these organs it is returned to the pericardium, and thence to the heart.

The blood is colourless, and contains relatively few large round corpuscles, but many amoeboid ones.

The mouth is guarded on each side by the maxillary process of the pedipalpus, and in front and behind by an upper and lower lip; it is a transverse slit and leads into the pharynx, a narrow tube lined with chitin which passes perpendicularly upwards and forms a right angle with the next section of the alimentary canal, the oesophagus. This tube is also lined with chitin, it is encircled by the nerve ring and opens into an expansion known as the sucking-stomach. The lumen of the latter is triangular in cross section, and its walls are strengthened by chitinous plates; strong muscles unite the dorsal wall of this organ with the chitinous integument of the tergum, and similar muscles pass between its ventro-lateral walls and the endosternite; the contraction of these muscles enlarges the lumen of the organ, and the juices of the prey are thus sucked into the digestive canal of the spider.

The sucking-stomach communicates with the very small

true stomach, this gives off a remarkable series of caeca, two of which pass forward and end blindly near the poison glands. In some species, as *Tegenaria domestica* and *Agelena labyrinthica*, these two caeca fuse together and form a ring above the

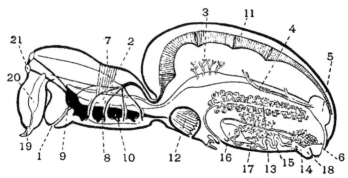

Fig. 236.—Semi-diagrammatic view of a Spider (*Epeira diademata*), to show internal organisation. After Warburton.

1. Mouth.	11. Dorsal vessel or heart, with lateral
2. Suctorial stomach.	ostia.
3. Liver ducts.	12. Lung-sac.
4. Malpighian tube.	13. Ovary.
5. Stercoral pocket.	14. Acinate and pyriform silk-glands.
6. Anus.	15. Tubuliform silk-gland.
7. Dorsal muscle of stomach.	16. Ampulliform silk-gland.
8. Lower portion of stomach.	17. Aggregate or dendriform silk-gland.
9. Cerebral ganglion, giving off nerves	18. Spinnerets or mammillae.
to the eyes.	19. Falx (distal joint of chelicera).
10. Thoracic ganglion, giving off nerves	20. Poison gland.
to the legs.	21. Eye.

cerebral ganglion.   The stomach gives off four pairs of lateral caeca, each of these passes into the base of a leg and then turns back again and ends blindly between the ventral nerve mass and the ventral integument.   These caeca in *Epeira diademata* end blindly, but in some species they fuse together and form a common cavity.   Thus the stomach and its caeca may in the Araneida form a very complex system of anastomosing loops, some of which project a short way into the legs.

From the stomach the intestine passes backward; it traverses the constricted stalk between the cephalothorax and abdomen, and then takes a curved course lying beneath the heart, and finally opens into a large rectum.   The intestine gives off numerous tubules which bifurcate into a large number of small ducts which ultimately end blindly; these constitute the " liver," a very extensive glandular organ which occupies a large part of the cavity of the abdomen. The secretion which it pours into the intestine is chiefly of use in digesting proteids.

The cloaca or " stercoral pocket " is an extensive chamber which occupies about one-sixth of the abdominal space ; the faeces seem to accumulate in this in considerable masses. It opens by the anus, which lies behind the posterior spinnerets.

The Malpighian tubules, the excretory organs of the spiders, are four long fine white ducts, closed at their free end and opening at the other into the intestine just where the latter passes into the cloaca. Their secretion contains guanine or some allied body.

A pair of coxal glands are present in *Epeira*, but in a very degenerated state ; in *Tegenaria* they are larger, and show some trace of a duct, and in the young, just-hatched spiders, a duct can be traced which opens to the exterior on the coxal joint of the first pair of legs ; the position of this opening is thus farther forward in the Dipneumones than is usually the case in Arachnids, but in a specimen of another genus, *Dysdera rubicunda*, a second opening has been described on the third pair of legs, the usual position.

Two poison glands of a pyriform shape are found just under the integument in the anterior end of the cephalothorax, their ducts traverse the chelicerae and open at the tip of these appendages. In *Amaurobius* the poison is said to be secreted and stored up through the winter, consequently this genus is especially venomous during the spring.

The nervous system in *Epeira* lies chiefly in the posterior half of the cephalothorax ; compared with some other Arachnids it exhibits great concentration (Fig. 236). The supra-oesophageal ganglion is large, and supplies nerves to the eyes and to the chelicerae, it is connected by lateral commissures with the sub-oesophageal nervous system. The ventral chain of ganglia, which are distinct in many Arachnids, in the Araneida are fused into one large ganglion situated in the cephalothorax. In *Epeira* this lies under the sucking-stomach and upon the ends of the gastric caeca, it is star-shaped, with numerous rays, of which the four middle ones on each side are stouter than the anterior and posterior. The rays are each continued into nerves; of these the first pair is continuous with the circum-oesophageal commissures, the second supplies the pedipalpi; the next four stout rays supply the four legs, and the posterior pair is continued back into the abdomen, the two nerves lying so close together as to be often mistaken for one.

The sense organs of spiders are limited to eyes which are always simple, and to taste organs. The number and arrangement of the eyes are of great use in classification, the various families having a different and definite arrangement of these organs. In *Epeira* the visual organs are situated on the anterior part of the cephalothorax; there are four large eyes arranged quadrilaterally, and on each side of the square are two smaller eyes, the members of each pair being connected by a ridge of chitin.

The male, as is usually the case in spiders, is smaller than the female, and whilst the latter usually sits in the centre of the circular web, the male is to be found hidden under the leaves of a neighbouring bush.

The ovary is situated in the abdomen; it lies ventrally, and is surrounded by the caeca of the liver and spinning-glands. It consists of two hollow tubes which unite posteriorly, and are continued backward as a thread; anteriorly each ovary is continuous with an oviduct; during the breeding season the ovaries are much swollen, and the ova form projections on the walls of the organ (Fig. 236). The oviducts open into a uterus, and this opens into a recess on the ventral wall of the abdomen. The external genital armature of the female is called the *epigynium*; it is often of a very complex character, and its nature is of considerable use in determining the species of a spider. In addition to the opening of the oviducts, it also receives the external ducts of the two spermathecae; these ducts may be short and wide, or long and coiled. There are also a pair of internal ducts which lead from the spermathecae to the oviduct, and along which the spermatozoa pass to fertilise the ova before they are laid. The epigynium is rendered more complex by the presence of numerous structures which serve to guide the palpal organs of the male into the external openings of the spermathecae; it only exists in its complete form in the mature female, and appears after the ninth or last moult.

The testis in the male consists of two caeca, placed ventrally in the anterior part of the abdomen ; each passes into a vas deferens, and the two vasa deferentia unite together in a small vesicle which opens to the exterior just behind the level of the pulmonary sacs. In *Tegenaria guyonii* the semen is deposited on a web constructed for the purpose, and taken up by the palpal organs and by them introduced into the spermatheca of the female, an operation not unattended by danger, as the female, when hungry, has been known to seize and devour the small male. The number of eggs laid varies from above one thousand (*Argiope cophinaria*) to two (*Oonops pulcher*) ; they are generally deposited in cocoons. The Lycosidae carry their cocoons attached to the abdomen of the mother, and when the young are hatched they are borne on her back till the first moult. *Theridion* carries the cocoon between the base of the legs, and if it be lost searches about diligently till it is found.

The Araneida are divided into two sub-orders:

1. TETRAPNEUMONES, with four lung sacs, and as a rule four spinnerets.
2. DIPNEUMONES, with two lung sacs and six spinnerets.

Sub-order 1. TETRAPNEUMONES.

These are large hairy spiders which do not weave .webs, but burrow in the earth, lining their tubular tunnels with a thick web. They sit at the entrance of these holes waiting for their prey, which, in the case of the gigantic South American *Mygale avicularia*, often takes the form of small birds. Other members of this sub-order are *Cteniza* and *Nemesia*, the trapdoor spiders, found in Southern Europe ; these genera close the entrance of their tubular home with a lid or trap-door. Both these genera have two pairs of spinnerets, of which the posterior pair are very long and bend up over the abdomen. *Atypus*, the only genus of Tetrapneumones found in Britain, has six spinnerets.

The endosternite, which was well developed in *Limulus* and

Scorpio, is also found in *Mygale*. Coxal glands are also present, in *Mygale* they seem to be closed, but in *Atypus* they open on both the first and third pairs of legs.

Sub-order 2. DIPNEUMONES.

This sub-order contains the majority of spiders, which are classified in very numerous families. *Epeira diademata* belongs to the EPEIRIDAE, a family the members of which make the well-known circular webs with radiating lines. Another family, the LYCOSIDAE, or wolf-spiders, includes the *Lycosa tarantula* of Italy and Spain. The THOMISIDAE, or crab-spiders, have a flat round abdomen and rather short legs; they frequently run sideways, and build no webs. The AGELENIDAE make horizontal webs, prolonged at one point into a tunnel, in which the spider sits until some prey becomes entangled in the web, when it rushes out, kills it, and binds it up. This family includes the *Agelena labyrinthica* and *Tegenaria derhamii*, the commonest household spider. *Argyroneta aquatica*, which also belongs to this family, makes a bell-shaped, water-tight nest, attached to some submerged water-plant. This it fills with air, carrying it down from the surface in bubbles which it has entangled between its spinnerets and the posterior pair of legs.

CLASS VI. **Phalangida.**

CHARACTERISTICS. — *Arachnida with their abdomen not constricted off from the cephalothorax; four pairs of very long and slender legs; they breathe by tracheae only; chelicerae chelate.*

The **Phalangida**, sometimes called the **Opilionina,** and popularly known as harvestmen, resemble spiders in appearance, but have no constriction between the abdomen and thorax, and are almost always of a sombre gray, brown, or blackish hue.

Appendages of *Phalangium.*

1. Chelicerae.	4. 2nd pair of walking-legs.
2. Pedipalpi.	5. 3rd ,, ,,
3. 1st pair of walking-legs.	6. 4th ,, ,,

The chelicerae are chelate, and the pedipalpi are not modi-

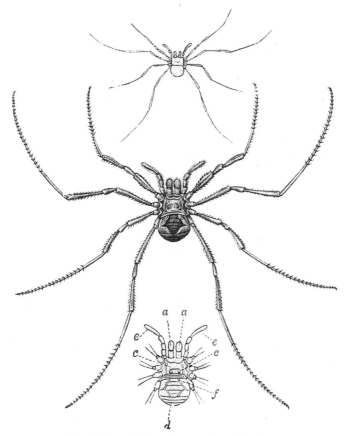

FIG. 237.—*Phalangium copticum*, Savigny.

a. Chelicerae.	*e.* Pedipalpi.
c. Eyes.	*f.* Junctional line of thorax and ab-
c. (to right) Thorax.	domen.
d. Abdomen.	

fied for purposes of fertilisation, but are short and limb-like. The four pairs of legs are very long and thin, and very easily

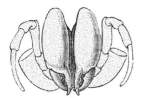

FIG. 238.—Chelicerae and pedipalpi of *Phalangium copticum*, seen from in front.

break in two; the proximal joint of the first two pairs of legs is produced into a process, the coxa or maxilla, which assists in eating, and the third leg also sends in a process towards the mouth, recalling the condition of things in *Scorpio.*

The abdomen consists of six to ten segments; its sterna project forward, and, as the anterior one bears the genital pore, this orifice is situated at the level of the third pair of legs. Just behind this are the two stigmata which open into the tracheae. These are at first wide, but they soon narrow and subdivide into small branches; the whole system is strengthened by a spiral thickening of chitin.

The oesophagus is wider than in spiders, and the Phalangids eat solid food; the stomach is provided with a number of caeca, and there are a pair of Malpighian tubules; the anus is terminal. The coxal glands are well developed; they consist of a coiled tube, which it is suggested has been mistaken for the Malpighian tubules; this opens at one end into a ventrally placed sac, and at the other to the exterior at the base of the third pair of legs.

The heart is a long dorsal vessel with three chambers and three pairs of ostia. The nervous system consists of a bilobed supra-oesophageal ganglion and a concentrated ventral mass. There are no spinning-glands.

The Phalangids have no external indications of sex. Both

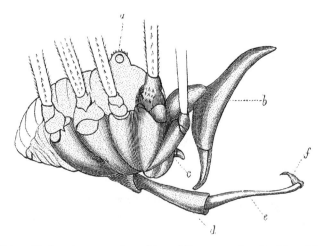

Fig. 239.—*Phalangium cornutum*, Linn. (Harvestman). Profile, with legs and palpi truncated.

a. Eye eminence.	d. Sheath of penis protruded.
b. Chelicerae.	e. Penis.
c. Portion of mouth apparatus.	f. Glans.

ovary and testis are in the form of ring-shaped glands; the ova are said to be fertilised in the ovary. The testes sometimes produce ova as well as spermatozoa. The female is provided with a large ovipositor, and the male with a large penis; both of these structures are usually retracted within a sac.

The females lay their eggs under stones or in crevices in the ground during the autumn; the adults of both sexes then die, so that until the eggs hatch out in the following spring no Phalangid is to be seen. They are largely nocturnal in their habits, and feed on small insects, spiders, etc., but at times they are cannibal. Their long thin legs enable them to run over grass and hay, and to steal with a "gliding spring" upon their prey. There are about twenty-four species found in Britain.

CLASS VII. Acarina (Mites).

CHARACTERISTICS.—*Abdomen unsegmented and fused with thorax. The oral appendages are adapted for piercing, sucking, or biting. Tracheae usually present.*

The mites comprise an immense number of species, which vary greatly both in appearance and in habits; as a rule they are minute in size with stout roundish bodies. Many of them are parasitic on either plants or animals, others live on cheese, etc., and some are predaceous.

Appendages of *Acarina*.

1. Chelicerae.   4. 2nd pair of walking-legs.
2. Pedipalpi.   5. 3rd   ,,   ,,
3. 1st pair of walking-legs.   6. 4th   ,,   ,,

The chelicerae are clawed or chelate, or they may form piercing stylets, in which case they are protected by a sheath formed by the base of the pedipalpi. The rest of the pedipalpus projects as a tactile palp, or may be clawed. The four pairs of legs vary in different species, they usually end in a pair of claws, but these may be replaced by a sucking-disk. The claws are often wonderfully adapted to fit round the hairs of animals upon which the mites may be living as ectoparasites.

The alimentary canal is often provided with salivary glands; the stomach may give off numerous caeca, which in some cases are forked. The stigmata are two in number, situated near the hind pair of legs, the tracheae may be long and fine, interlacing with the viscera, but in some species (*Nothrus*) they are short and thick, and in *Hoplophora* they are still shorter and end in swollen vesicles.

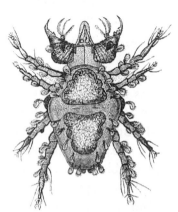

FIG. 240.—*Cheyletus flabellifer* (Book Mite, quite unconnected with books).

As a rule a circulatory system is absent, but the GAMASIDAE, or beetle mites, are described as having a two-chambered heart with two pairs of ostia.

A coxal gland has been described, but no opening to the exterior has yet been found. The nervous system is very concentrated, one or two pairs of eyes may be present or these organs may be entirely absent.

The male and female generative glands are very much alike; as in Phalangids, they form a ring. Their external opening is on the anterior end of the abdomen or between the last pair of legs. The Acarina are oviparous, the young are hatched with three pairs of legs and undergo a number of ecdyses. *Sphaerogyna ventricosa* is said to give birth to mature mites which are fertilised as soon as born.

The Acarina are divided into many families, some of which may be mentioned. The DERMATOPHILI comprise one genus, *Demodex*, which is found living in hair follicles both in domesticated animals and in man. It has a suctorial rostrum, four pairs of rudimentary legs, and a long annulated abdomen.

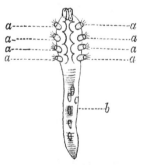

FIG. 241.—*Demodex folliculorum*, Simon. Under side.

a. Rudimentary legs.
b. Abdomen.

The SARCOPTIDAE are microscopic mites which live parasitically in the skins of mammals, and give rise to diseases known as the mange or the itch. The female *Sarcoptes scabei* makes a tunnel in the skin and sits at the inner end, pushing the eggs towards the exterior as they are laid.

The PHYTOPTIDAE are gall mites, and make amongst others the curious conical excrescences which are so common on the leaves of lime trees, maples, etc. In many of them the posterior pair of legs are replaced by bristles.

The HYDRACHNIDAE or water-mites are usually brightly coloured. They often hang on to water-insects, and may be frequently found attached by their rostrum to *Dytiscus, Nepa,* etc. One species, *Atax bonzi,* is common in the mantle-cavity of *Unio.*

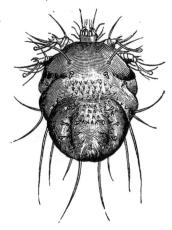

FIG. 242.—*Sarcoptes scabei* (the Itch Mite), female. After Meguin.

The TROMBIDIIDAE are brightly-coloured mites, whose larvae

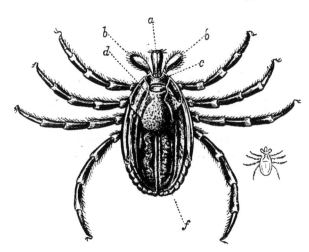

FIG. 243.—*Ixodes aegyptius*, Savigny.

a. Chelicerae.	*d.* Thorax.
b. Pedipalpi.	*f.* Abdomen.
c. Head.	

often live parasitically on spiders or insects. Some of them, as *Tetrarhyncus telearius,* commonly known as the red spider spin webs under leaves, in which whole colonies of the mites dwell.

The IXODIDAE or ticks are blood-sucking mites which live on undergrowth in forests ; the females crawl upon man,

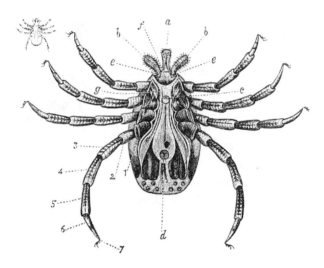

| FIG. 243.—The same (*Ixodes aegyptius*) ; under side.

a. Chelicerae.
b. Pedipalpi.
c. Genital aperture.
d. Anal aperture.

e, f. Maxillae and labium soldered together.
g. Sternum.
1-7. Joints of leg.

cattle, or even snakes, and attach themselves by their strong rostrum.   They suck blood to such an extent that their bodies sometimes swell to the size of a small walnut.

The TYROGLYPHIDAE include *Tyroglyphus siro*, the cheese or flour mite, and several other species which feed on vegetable substances.

CHAPTER XXI

CHORDATA

CHORDATA
- HEMICHORDATA
  - Enteropneusta—*Balanoglossus.*
  - Cephalodiscida—*Cephalodiscus.*
  - Rhabdopleurida—*Rhabdopleura.*
- UROCHORDATA (Tunicata)
  - Larvacea—*Appendicularia.*
  - Thaliacea—*Doliolum, Salpa.*
  - Ascidiacea — *Ciona, Botryllus, Pyrosoma.*
- CEPHALOCHORDATA—*Amphioxus.*
- VERTEBRATA.

   I. HEMICHORDATA (Enteropneusta, etc.)
  II. UROCHORDATA (Ascidians or Tunicata).
 III. CEPHALOCHORDATA.
 IV. VERTEBRATA.

CHARACTERISTICS.—*The phylum Chordata consists of coelomate animals characterised by (i.) the possession of gill-slits or apertures leading from the pharynx to the exterior, these may persist through life, or they may be but temporary openings existing only in the embryo ; (ii.) by the presence of a skeletal rod, the notochord, which is formed in the dorsal middle line by cells budded off from the hypoblast of the embryo ; (iii.) by a dorsal nervous system, which, with a few exceptions, is tubular and does not form a nervous ring round the alimentary canal; (iv.) by the segmentation of the mesoblast, which, although often obscured in later life, is always found in the earliest stages.*

From some points of view the Chordata is the most important of the various phyla which make up the animal kingdom. The first three groups are either degenerate or contain but few species; the fourth group, however—the Vertebrata—has, from the size of its numerous members and the great variety of form they present, and above all from the fact that it culminates in the genus *Homo*, acquired a great importance, and it was formerly customary to oppose this group of Vertebrata to the remainder of the animal kingdom, which were classed together under the general heading Invertebrata. This division of the animal world was to a great extent supported by the fact that much more was known about the anatomy and developement of the comparatively conspicuous Vertebrata than about the much less accessible and often minute Invertebrata, but in late years our knowledge has been much increased in the latter direction, and it is now recognised that the group Vertebrata is, from a strictly morphological standpoint, of no greater interest than any of the other large groups which compose the animal kingdom.

CLASS I. **Hemichordata.**

CHARACTERISTICS.—*The anterior end of the body forms an elongated proboscis, at the base of which the mouth opens. The region surrounding and immediately behind the mouth is termed the " collar," and this is succeeded by a trunk containing the generative organs. A series of pores or gill-slits are generally present. A diverticulum of the pharynx projects forward into the proboscis-stalk and forms the notochord. The sexes are separate.*

The Hemichordata are a group recently constituted by Bateson to contain the genus *Balanoglossus*. Harmer's investigations on the structure of *Cephalodiscus* have shown that that genus must be assigned a place near *Balanoglossus*, and still more recently Fowler's researches have shown that *Rhabdopleura* must also be included in the group. The two last-named forms were previously placed in the neighbourhood of the Polyzoa; at present we know nothing of their embryology. The group now comprises the three sub-classes (i.) Enteropneusta, (ii.) Cephalodiscida, (iii.) Rhabdopleurida, each represented by a single genus.

Sub-class I. ENTEROPNEUSTA.

CHARACTERISTICS.—*Soft worm-like animals with ciliated ectoderm. The trunk, succeeding the collar, is long and somewhat flattened. The alimentary canal is straight, and the anus opens at the end of the trunk. There are numerous gill-slits, which increase in number during life.*

There are several species of *Balanoglossus* described.

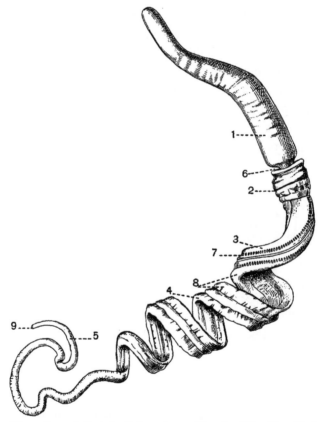

FIG. 244.—View of an adult male *Balanoglossus Kowalevskii.* Magnified one and a half times. After Bateson.

1. Proboscis.	6. Mouth.
2. Collar.	7. Gill-slits.
3. Trunk.	8. Lateral edge of the body where it
4. Region of Testes.	becomes flattened.
5. Tail.	9. Anus.

B. clavigerus and *minutus* are found in the Bay of Naples, *B. salmoneus* and *robinii* off the coast of Brittany. The species which is in the main described below is *B. kowalevskii,* which, together with *B. brooksii,* is found on the east coast of the United States.

Balanoglossus differs from other members of the phylum with which it is grouped in the fact that its body is covered externally with cilia, and in this respect it resembles a Nemertine. The ciliated epidermis contains a considerable number of unicellular mucous glands, which secrete a very copious mucus. This secretion serves to line the tubes in the sand and mud in which the animal lives. In one species the mucus hardens into a definite tube, in the walls of which grains of sand are embedded. The body of *Balanoglossus* has usually a very characteristic odour.

The body-cavity or coelom consists of a single unpaired portion occupying the proboscis, a paired portion in the collar, and a paired portion in the trunk. These are derived from five coelomic pouches of the archenteron of the embryo. The original cavities of these are, however, much obliterated by the growth of stellate connective tissue cells (Fig. 245). The spaces

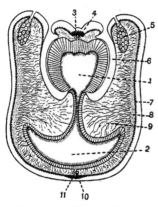

FIG. 245.—Section, partly diagrammatic, through the respiratory part of the trunk of *Balanoglossus.* The whole course of a gill-slit could not be seen in a transverse section, as the slits run obliquely.

1. Dorsal portion of alimentary canal.
2. Ventral portion of alimentary canal.
3. Dorsal nerve cord, in the skin.
4. Dorsal blood-vessel.
5. Ovary.
6. Gill-slit.
7. Nerve plexus at the base of the skin.
8. Longitudinal muscle fibres.
9. Connective tissue cells filling up the coelom.
10. Ventral blood-vessel.
11. Ventral nerve in the skin.

between the cells communicate with the exterior in the proboscis by a ciliated proboscis pore,—in one species this is paired,—and the right and left collar cavities open to the exterior by a right and left ciliated collar pore. These latter have been compared to the "brown tubes" of Amphioxus. Underlying the epidermis is a well-marked layer of longitudinal muscle fibres. The muscles of the trunk show no trace of metameric repetition.

The mouth opens ventrally between the base of the proboscis and the anterior rim of the collar. It leads into a straight alimentary canal. The anterior half of the digestive tube is partially divided into a dorsal and ventral half by two lateral horizontal folds. The gill-slits open into the dorsal half. Behind the region of the gill-slits the intestine is characterised by a ciliated dorsal and ventral groove. In some species paired hepatic caeca open at intervals into this region, and in at least one species these diverticula are described as opening on to the exterior. A median dorsal and ventral mesentery support the intestine.

In the region of the collar the alimentary canal gives off a diverticulum, which grows forward and supports the proboscis. The cells of this diverticulum become vacuolated, and assume an appearance which is common in the notochordal tissue of higher animals. The lumen of the diverticulum disappears, except at its base, and the whole structure, which extends through but a short region of the body, has been regarded as a notochord. This interpretation is supported both by its mode of origin and its structure. On the other hand, the chief blood-vessel is dorsal to it, instead of being ventral to it, as in the higher Chordata.

The paired gill-slits are visible externally as small pores situated in the dorso-lateral grooves of the anterior portion of the trunk. Each gill-slit opens internally into the dorsal thick-walled section of the alimentary canal. They increase in number throughout life, new slits arising behind those already existing. When they first appear they are circular in outline, but, as is the case in Amphioxus, a tongue-like process grows down from the dorsal surface and reduces the opening to a U-shaped aperture. Between neighbouring gill-slits and in the tongue-shaped process there is a skeletal rod, the whole forming a branchial skeleton comparable with that of Amphioxus. The gill bars receive a supply of blood from the dorsal blood-vessel, and water passes through the gill-slits from the alimentary canal to the exterior. The slits are arranged obliquely, so that in a single transverse section it is impossible to see the whole course of any one. In the section (Fig. 245) the gill-slit has been drawn diagrammatically.

The posterior edge of the collar projects backward over a few of the anterior gill-slits, the number varying in different species. This fold is termed the opercular fold, and the recess covered in by it has been regarded as a rudimentary atrial cavity.

The vascular system comprises a dorsal vessel which extends from the tail to the heart. The latter is a thin-

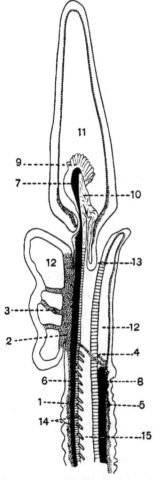

Fig. 246.—Diagrammatic longitudinal vertical section through the anterior end of *Balanoglossus minutus.* Slightly altered, after Bateson.

1. Dorsal nerve cord.
2. Central nervous system (*i.e.* in the collar region).
3. Cord running from the central nervous system to the skin.
4. Ring of nervous tissue round the collar.
5. Ventral nerve cord.
6. Dorsal blood-vessel.
7. Heart.
8. Ventral blood-vessel.
9. Proboscis gland.
10. Notochord.
11. Cavity of proboscis.
12. Cavity of collar.
13. Mouth.
14. Gill-slits.
15. Oesophagus.

walled pulsating vesicle lying in the base of the proboscis, just dorsal to the anterior end of the notochord. The heart is connected with a plexus of capillaries ramifying in a glandular structure, which forms a cap to the anterior end of the notochord. This is termed the proboscis gland. From the posterior end of the collar a well-developed ventral vessel passes backward, supported like the dorsàl vessel in a median mesentery. These chief vessels are placed in communication with one another by plexuses of capillaries in the skin and in the walls of the alimentary canal, and the skin plexus is specialised into a more or less definite circular vessel connecting the dorsal and ventral vessel in the opercular fold or the posterior edge of the collar. The blood is said to be free from corpuscles, but the fluid which occupies the remnants of the coelom contains amoeboid corpuscles. The course of the blood is forward in the dorsal, and backward in the ventral vessel.

The nervous system consists of a dorsal and ventral cord, which lie in the skin, and extend from the anus to the posterior edge of the collar; at this level the ventral cord divides into two strands, which pass round the alimentary canal and join the dorsal cord (Fig. 246). The dorsal cord in the region of the collar has lost its connection with the epidermis, and by a process of delamination, aided by invagination at its ends, has come to form a partially tubular cord. This is the portion of the nervous system in which the cellular elements are to a great extent aggregated. Its posterior end receives the dorsal nerve and the two branches of the ventral nerve. In some species three nerves arise from this central nervous system and pass towards the dorsal skin, these three nerves have been compared to the dorsal roots of spinal nerves in Vertebrates. Anteriorly this central nervous system is continuous with a well-marked sub-epidermic plexus of nerve fibrils which exists in the proboscis (Figs. 246 and 247); a similar network lies in the skin of the trunk, and is continuous with the dorsal and ventral nerves.

The various species of *Balanoglossus* are all dioecious. Both ovaries and testes consist of sacs which open directly on to the epidermis, from which they are probably derived. These sacs occur in the region of the gill-slits, and open externally to the latter; they are also found serially repeated along that part of the trunk which succeeds the branchial region. The ova do not escape singly through the external orifice, but the whole follicle breaks away and then disintegrates.

The systematic position of *Balanoglossus* has given rise to much divergence of opinion amongst zoologists. Bateson's

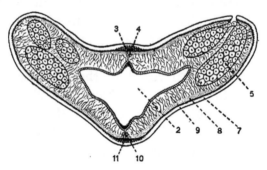

FIG. 247.—Section through *Balanoglossus* behind the region of the gill-slits.

2. Alimentary canal supported by a dorsal and ventral mesentery and showing the dorsal and ventral grooves lined by long cilia.
3. Dorsal nerve cord in the skin.
4. Dorsal blood-vessel.
5. Ovarian follicles, opening to the exterior.
7. Nerve plexus at the base of the epidermis.
8. Longitudinal muscle fibres.
9. Connective tissue cells filling up the coelom.
10. Ventral blood-vessel.
11. Ventral nerve cord in the skin.

recent researches on the embryology of this form, however, justify us in placing it amongst the Chordata, but it has affinities in at least two other directions. Certain species, in the course of their developement, pass through a larval stage termed the *Tornaria*, which shows the closest resemblance to the characteristic *Bipinnaria* larva of the Echinodermata. The presence of this larva in the ontogeny of these two groups seems to point to some connection between their remote ancestry, and this to some extent is emphasised by the fact that the Echinodermata, like *Balanoglossus*, are ciliated externally. Again, the structure of the body-wall, its external ciliation, the form of the body, the absence of segmentation in the muscles, the structure of the generative organs, and perhaps the proboscis, are all features which recall the similar parts in the Nemertines.

Sub-class II. CEPHALODISCIDA.

CHARACTERISTICS.—*The pharynx is pierced by one pair of gill-slits. The alimentary canal has a neural flexure, so that the mouth and anus are approximated. The collar bears twelve tentacular plumes, which are hollow; their cavities are continuations of the collar cavity.*

This sub-class contains one species, *Cephalodiscus dode-calophus*, which, owing to Harmer's researches, has been

FIG. 248.—Ventral view of *Cephalodiscus dodecalophus*. After M'Intosh.

    1. Tentacular plumes.
    2. Bud at the end of pedicle, two others are also shown.
    3. Proboscis.

associated with *Balanoglossus,* and placed in the group Hemi-chordata. *Cephalodiscus* is an organism which reproduces by budding, and it has a certain superficial resemblance to some Polyzoa, but it shows closer affinities to *Balanoglossus* in its internal anatomy. It resembles the last-named genus in the division of its body into three regions,—the proboscis, collar, and trunk,—and in the arrangement of its coelomic cavities and the possession of proboscis- and collar-pores. Moreover, it is provided with gill-slits, with a diverticulum of the alimentary

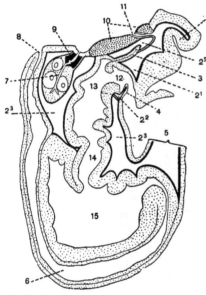

FIG. 249.—Longitudinal section of an adult Cephalodiscus, supposed to be taken sufficiently on one side of the middle line to allow of the representation of one of the ovaries, and of one of the proboscis pores. After Harmer.

1. Proboscis.
2^1. Body-cavity of proboscis.
2^2. Body-cavity of collar.
2^3. Body-cavity of trunk.
3. Notochord (really visible only in a median section).
4. Operculum.
5. Pedicle or stalk cut through.
6. Intestine.
7. Ovary.
8. Anus.
9. Pigmented oviduct.
10. Central nervous system.
11. One of the proboscis pores.
12. Mouth.
13. Pharynx or branchial region of gut.
14. Oesophagus.
15. Stomach.

canal (notochord) growing into the proboscis stalk, and with a nervous system which closely resembles that of *Balanoglossus*.

Cephalodiscus has only been found once. It was dredged in the Straits of Magellan by the " Challenger " from a depth of 245 fathoms. The various " polypides," with their budding pedicles, live inside a coenoecium or tubular case secreted around them, in much the same way as Appendicularia secretes its " house."

Sub-class III. RHABDOPLEURIDA.

CHARACTERISTICS.—*Gill-slits and proboscis pore absent. The intestine has a neural flexure. There are two tentacular plumes having the same relation as in the Cephalodiscida.*

· *Rhabdopleura normani* has been dredged in the Hardanger Fjord, off the Shetlands, and off one of the islands of the Tristan d'Acunha group. It consists of an irregularly branching colony attached to foreign bodies. The zooids are very minute; they are connected by a stem, and the whole is ensheathed in a tubular investment, probably secreted by the proboscis (epistome). A pair of processes bearing tentacles, into which the cavity of the collar space is continued, exist,

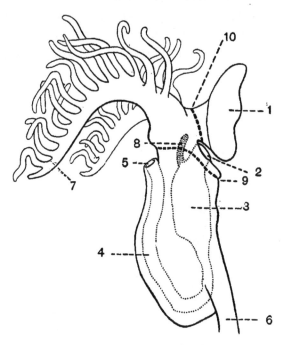

Fig. 250.—Side view of *Rhabdopleura normani*. Partly diagrammatic. After Lankester and Fowler.

1. Proboscis (epistome).
2. Mouth.
3. Stomach.
4. Intestine.
5. Anus.
6. Stalk.
7. Right tentacular arm, only one row of tentacles is shown on each arm.
8. Notochord.
9. Dotted line indicating the division between the regions of the body and of the collar.
10. Dotted line indicating the division between the regions of the collar and of the proboscis.

The animal is represented as transparent to show the alimentary canal and notochord.

and closely resemble the similar structures in *Cephalodiscus*. The zooids creep up the tube in which they are enclosed by means of the proboscis, and are retracted by means of muscle cells. The relations of the notochord, a solid diverticulum of the pharynx, correspond with those of *Balanoglossus* and *Cephalodiscus*.

Class II. Urochordata (Tunicates or Ascidians).

CHARACTERISTICS.—*Hermaphrodite Chordata, with free-swimming larvae, which usually become sessile and degenerate adults. These are either solitary or form colonies. In the adult the nervous system is, with few exceptions, reduced to a single ganglion, the elongated nervous system and the notochord being confined to the larval stages. A metamorphosis usually occurs.*

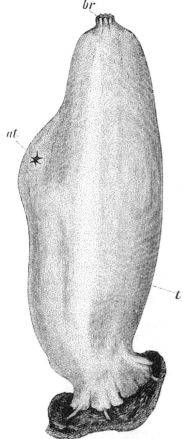

FIG. 251.—*Ascidia mentula*, from the right side. After Herdman.

br. Branchial aperture.
at. Atrial aperture.
t. Test.

The simple Ascidians—*Ascidiae simplices*—have a more direct life-history, and to some extent a less modified structure than many of the other members of Urochordata. *Ciona intestinalis*, which may serve as a type of this group, is abundant in the Mediterranean, and is found widely distributed in temperate seas. It is a hyaline transparent creature, when young presenting the appearance of opalescent glass, but as it grows old it is apt to become overgrown with foreign organisms, etc., and presents an opaque appearance. It attains a length of 5 or 6 inches.

At one end the *Ciona* is attached to a rock or some other foreign substance, at the other end the mouth is situated, surrounded by eight lobed processes, this may be regarded as the anterior end of the animal. About an inch behind the mouth, and on that side of the animal which may be regarded as the dorsal, another aperture surrounded by six

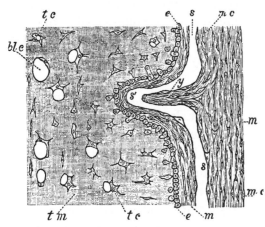

FIG. 252. — Diagrammatic section of part of mantle and test of an Ascidian, showing the formation of a vessel and the structure of the test. After Herdman.

m. Mantle.
e. Ectoderm.
tc. Test cell.
tm. Matrix.
blc. Bladder cell.

s, s'. Blood sinus in mantle being drawn out into test.
mc. Mantle cells.
y. Septum of vessel.

lobes is found; this is the atrial pore, and through this the water which has passed through the perforated pharynx and purified the blood, the waste matter from the intestine, and the generative cells are all discharged.

The whole animal is enveloped in a hyaline *tunic* or *test*, which is a cuticular excretion of the ectoderm, and into which ectodermal cells and blood-vessels wander. The test is turned in for a short distance at both the oral and atrial openings; at the posterior end it is produced into various processes and lobes which fit into crevices of the rock, etc., and serve to fix the animal; the true skin or dermis is continued into these processes. The gelatinous substance which composes the test of *Ciona* contains numerous cells split off from the ectoderm; in the genus in question many of these cells soon die, but in others they live for some time, forming fusiform or stellate cells, often pigmented, or they develope a large vacuole (Fig. 252), or in some genera they secrete calcareous spicules. The test is undoubtedly a cuticle secreted by the ectoderm, but it is kept alive to a certain extent by the cells which wander into it, and by the blood-vessels which make short incursions into it. The test contains in many instances a chemical substance identical or closely allied to cellulose, a substance rarely met with in animals, but almost universal in plants. Kowalevsky has recently shown that in some species the cells which wander into the test arise from the mesoderm, and pass through the ectoderm on the way to their final resting-place in the test.

The ectoderm, which secretes the test, forms a single layer of cubical cells; beneath this is a layer of muscles arranged in large longitudinal bands, with few anastomoses, and numerous small transverse bands, which anastomose freely. Nerves and blood sinuses also ramify in the skin.

The general disposition of the organs of the body in a simple Ascidian is as follows : the mouth leads into a large branchial sac, which extends throughout four-fifths of the body; this is pierced by very numerous slits, which, instead of opening directly to the exterior, communicate with a large chamber, the atrial cavity, which completely surrounds the branchial sac, except along the middle ventral line, and which opens to the outside world through the atrial pore. The atrial cavity is produced originally by an invagination from the exterior, and is consequently lined throughout by ectoderm. Behind the branchial sac, occupying the posterior fifth of the animal, is a closed space, in which the stomach, heart, and generative organs are enclosed; the intestine and ducts of the reproductive glands run up the dorsal surface of the branchial sac, and open into the atrial chamber, the latter near its pore.

A constant current of water, maintained by the action of the cilia lining the pharynx, enters the mouth and passes out through the slits in the branchial sac into the atrial cavity, and out at the atrial pore, carrying with it the waste matter

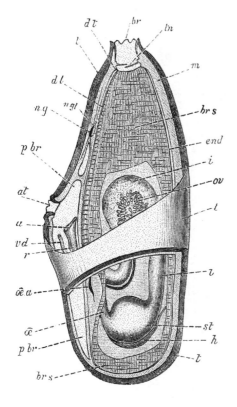

FIG. 253.—Diagrammatic dissection of *Ascidia mentula* to show the anatomy of a simple Ascidian. After Herdman.

at. Atrial aperture.
br. Branchial aperture.
a. Anus.
brs. Branchial sac.
dl. Dorsal lamina.
dt. Dorsal tubercle.
end. Endostyle.
h. Heart.
i. Intestine.
m. Mantle.
ng. Nerve ganglion.
œ. Oesophagus.
œ.a. Oesophageal aperture.
ov. Ovary.
pbr. Peribranchial cavity.
r. Rectum.
st. Stomach.
t. Test.
tn. Tentacles.
vd. Vas deferens.
ngl. Subneural gland.

from the alimentary canal and also the generative products when ripe.

A little distance within the oral opening is situated a ring of tentacles, which project across the mouth and possibly function as a filter; the area between this ring of tentacles and the branchial sac is termed the prebranchial zone. The branchial sac is a large space with perforated walls; it is attached to the mantle or body-wall along the median ventral line, and the large ventral blood-vessel, which communicates behind directly with the heart, is situated along this line of attachment. The rest of the branchial sac is separated from the mantle by the peribranchial or atrial chamber, which opens to the exterior by the atrial pore. The rectum and

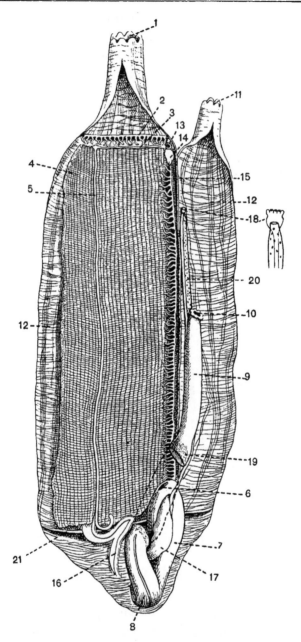

Fig. 254.—View of *Ciona intestinalis* lying on its right side. Both the branchial and the atrial cavities have been opened by a longitudinal incision to display the internal organs.

1. Branchial aperture.
2. Tentacles.
3. Peribranchial grooves.
4. Perforated walls of branchial sac.
5. Endostyle.
6. Oesophageal opening leading from the branchial sac to the stomach, rather diagrammatic.

the genital ducts run for some distance along the dorsal middle line, and with these runs the dorsal blood-vessel. The walls of the branchial sac are also connected with the mantle at irregular intervals by blood-vessels, which pass from the bars to supply the substance of the mantle. The branchial bars run at right angles to one another, and leave between them small ciliated slits or stigmata, through which the water taken in at the mouth escapes into the peribranchial chamber, the current being maintained by the cilia. The transverse bars contain blood-vessels communicating with the dorsal and ventral blood-vessels. Connecting two adjacent transverse bars are a number of small longitudinal bars, which divide the space up into the stigmata. There are two kinds of transverse bars, the primary and the secondary, the latter being the smaller. In addition to the small longitudinal bars, there are also some large ones situated in a different plane from the others, farther inside the branchial sac; these are connected by a short vessel with the transverse bars at their points of intersection (Fig. 256), and at the same spot they give off little finger-shaped processes, which hang into the lumen of the sac.

A ciliated groove, known as the *hypopharyngeal groove* or *endostyle*, runs along the ventral line of the branchial sac (Fig. 254). The lips of the groove project into the cavity of the sac, and it is lined by cells which are partly glandular and partly ciliated. At the posterior end it is continued into a *cul-de-sac*, which has a slight spiral twist; anteriorly its lips are continuous with the *peripharyngeal* bands, two ciliated ridges which encircle the anterior end of the branchial sac

7. Stomach.
8. Intestine showing typhlosole, part of it removed to show subjacent structures.
9. Rectum.
10. Anus.
11. Atrial aperture.
12. Inner surface of mantle, showing longitudinal and transverse muscle fibres.
13. Dorsal tubercle.
14. Subneural gland and brain.
15. Cut edge of branchial sac.
16. Heart.
17. Ovary.
18. Pore of vas deferens. The openings of the oviduct and the vas deferens are shown enlarged to the right.
19. Testicular follicles on intestine; the greater part of them have been removed with the intestine.
20. Oviduct.
21. Septum shutting off that part of the body-cavity which contains the heart, stomach, and generative organs.

and form the posterior limit of the prebranchial zone (Fig.

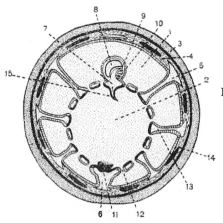

FIG. 255.—Transverse section through the branchial region of a *Ciona intestinalis*. After Roule.

1. Atrial chamber.
2. Branchial chamber.
3. Test.
4. Mantle.
5. Vessel passing from a branchial bar into the mantle.
6. Ventral blood-vessel at base of endostyle.
7. Dorsal blood-vessel at base of dorsal lamina.
8. Rectum.
9. Oviduct.
10. Vas deferens.
11. Ciliated endostyle.
12. Branchial slit, stigma.
13. Branchial bar connected with mantle.
14. Muscle fibres.
15. Dorsal lamina.

254). The anterior of these two ridges is the most prominent; it forms a simple circle; the posterior is, however,

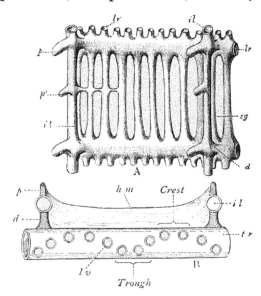

FIG. 256.

A. Part of branchial sac of *Ascidia* from inside.

B. Transverse section of same, drawn to different scale.

tr. Transverse vessel.
d. Connecting duct.
hm. Horizontal membrane.
il. Internal longitudinal bar.
lv. Fine longitudinal vessels.
p, p′. Papillae.
sg. Stigmata.

continuous with the dorsal lamina, a structure which runs

along the dorsal middle line of the branchial sac. In *Ciona* the dorsal lamina is divided up into a number of *languets*, or tongue-like processes, which hang into the lumen of the sac (Fig. 254). One of these is given off at the level of each transverse vessel, and in this and some other respects they resemble the finger-like processes borne on the longitudinal bars. The dorsal lamina extends to the end of the branchial sac, and terminates close to the mouth of the oesophagus.

Ciona lives upon minute organisms, etc., which float into the branchial sac with the water which is continually streaming through the body. These small particles become entangled in strands of mucus, and are directed to the dorsal lamina by the action of the cilia borne by the peripharyngeal groove and the endostyle or hypopharyngeal groove; the languets of the dorsal lamina, or epipharyngeal band, guide the food into the entrance of the oesophagus. The mucus which serves to entangle the food particles is probably partly secreted by the glandular cells of the hypopharyngeal groove, but probably also to a great extent by a gland which is situated under the central nerve ganglion.

This glandular structure, usually known as the *subneural gland*, is a compact body with few ramifications; its duct opens by a flattened mouth, which sometimes is folded in the most complicated fashion, but in *Ciona* is curved into a simple horse-shoe, situated dorsally near the posterior margin of the prebranchial zone. This gland has been regarded by some writers as homologous with the hypophysis cerebri of the Vertebrate brain. Some authorities have regarded this gland as a renal organ, but its function is more probably to secrete the mucus in which the food particles become entangled.

The mouth of the oesophagus is an oval slit situated at the dorsal side of the branchial sac close to the end of the dorsal lamina, it leads into a short transparent tube, the oesophagus, which soon expands into the spherical stomach. There are no muscle fibres in the wall of the oesophagus, nor indeed in the whole length of the alimentary canal, except in the rectum; the food is propelled onward by means of the cilia which line the digestive tract. The stomach is large, and its posterior half is covered by the follicles of the testis; it leads into the intestine, the first part of which is also covered by the same structures (Fig. 254). The intestine makes a slight twist round the ovary, and is then continued into the rectum, which runs straight along the dorsal wall of the branchial sac

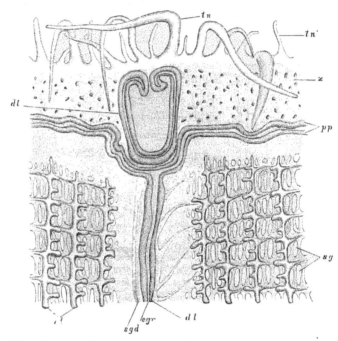

FIG. 257.—Mouth of the subneural gland and neighbouring parts in *Ascidia mentula*. After Herdman.

egr. Epibranchial groove.
pp. Peribranchial groove.
tn. Tentacles.
sg. Stigmata.
dt. Opening of duct of subneural gland.

dl. Dorsal lamina.
sgd. Duct of subneural gland.
il. Internal longitudinal bars.
z. Peribranchial zone.

and opens into the atrial cavity. The internal surface of the first part of the intestine is much increased by the existence of a typhlosole which projects into its lumen.

The heart is contained in a pericardium situated ventral to the oesophagus in the angle between the posterior limit of the branchial sac and the stomach (Fig. 254). The pericardium contains a corpusculated fluid, its cavity is derived from the original coelom of the larva. The heart is a cylindrical tube bent into the shape of a V, the angle pointing dorsalwards. One arm of the V is continuous with a ventral or branchio-cardiac vessel or sinus which runs along the line of attachment of the hypopharyngeal groove; the other arm is continued into a cardio-visceral vessel, which breaks up into many splits or sinuses amongst the viscera. From these splits in the tissues of the various organs the blood passes into a

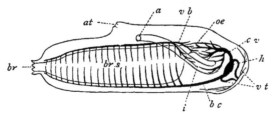

FIG. 258.—Diagram of circulation in a simple Ascidian. Herdman.

br. Mouth.
at. Atrial aperture.
a. Anus.
brs. Branchial sac.
oe. Oesophagus.
h. Heart.

i. Intestine.
bc. Branchiocardiac or ventral vessel.
cv. Cardiovisceral vessel.
vb. Viscerobranchial or dorsal vessel.
vt. Vessels to test.

viscero-branchial vessel situated at the base of the dorsal lamina. From this the blood passes into the vessels in the lateral bars of the branchial sac which carry the blood down to the ventral vessel, and it thus returns to the heart. A peculiarity of the circulation in Ascidians is that its course is from time to time reversed, after contracting for a certain number of times in one direction the heart stops and then recommences in the other, thus reversing the course of the circulation.

The vessels arising from each end of the heart give off branches into the test; these subdivide, and their smaller branches usually end in spherical cavities (Fig. 252). They probably serve to nourish the test and keep it alive, and in some species they may possibly play some part in respiration. The blood contains amoeboid and spherical corpuscles.

The cavity of the heart and of the blood sinuses connected with it is directly derived from the blastocoel or segmentation cavity of the embryo; this is an interesting fact, taken in conjunction with the similar origin of the space of the heart in the lamprey, and possibly in other Vertebrates. The nature of the space in which the stomach, ovary, and pericardium lie, is at present a matter of some uncertainty; it contains a corpusculated fluid.

The free-swimming larvae of the Tunicata are provided with a tubular nervous system derived from the epiblast. This has an enlarged anterior end into which two unpaired sense organs—a dorsal eye and a ventral auditory sac—project. The rest of the nervous system is traversed by a canal, and in its course along the tail of the larva it gives off paired nerves to the segmentally arranged muscles. In the larva a noto-chord also is present underlying the nervous system. When

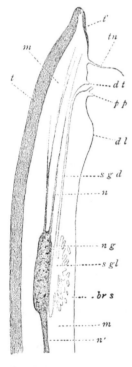

Fig. 259.—Diagrammatic section through the anterior dorsal part of *Ascidia mentula*, showing the relations of the nerve ganglion, subneural gland, etc. After Herdman.

tn. Tentacle.
t. Test.
m. Mantle.
dt. Opening of subneural gland.
pp. Peribranchial band.
t'. Test turned in at mouth.
dl. Dorsal lamina.
sgd. Duct of subneural gland.
sgl. Subneural gland.
ng. Brain.
n. Anterior nerve.
n'. Posterior nerve.
brs. Branchial sac.

the larva fixes itself, the notochord and the greater part of the nervous system atrophy. In *Ciona* the latter is reduced to an oblong ganglion, situated just below the ectoderm and above the subneural gland, in the angle between the oral and atrial cones. The ganglion is solid, and gives off numerous nerves, chiefly from its angles; these exhibit a rather marked asymmetry, the nerves of the left side dividing close to their origin, whilst those of the right remain for some distance un-branched. The specialised sense organs of the larva have atrophied.

Ascidians have no nephridia, and their nitrogenous waste

matter is not as a rule expelled from the body, but is stored up in certain cells in which the renal products continue to accumulate throughout life. In *Ciona* the renal cells are found close under the epidermis, in the neighbourhood of the opening of the vas deferens; they are of an orange colour, and contain granules which give micro-chemical reactions characteristic of uric acids and urates. It is the accumulation of these cells which forms the orange-red ring around the external opening of the vas deferens. The cells have no ducts, but since they are separated from the surrounding water by but a thin layer of epithelium, it has been thought that their excreta may diffuse out. In other genera of simple Ascidians, the renal cells are arranged around a vesicle into which they pour their secretions; the cavities of these vesicles are closed, and the nitrogenous waste matter does not leave the body. Like the lumen of the genital glands and of the pericardium, the cavities of these vesicles are possibly derived from the coelom. The vesicles are usually of a brown colour, and are found in the loop of the intestine, extending along its wall.

The Ascidians are hermaphrodite. In *Ciona* the testes consist of numerous follicles branching over the stomach and intestine; these are whitish in colour, and their presence gives this part of the alimentary canal a peculiar appearance; the larger of the follicles gradually unite to form a narrow duct, the vas deferens, whose wall is continuous with that of the testis. The vas deferens runs along by the side of the rectum, and ultimately opens into the atrial cavity about an inch in front of the anus. The cells lining the cavity of the follicles of the testis give rise to the spermatozoa. The latter are minute, with small heads and long tails; they apparently ripen and are discharged from the body at the same time as the ova.

The ovary is a comparatively compact gland situated in the general perivisceral space, in a loop of the intestine; the ova are modified cells lining the walls of its cavity. The walls of the ovary are continuous with the oviduct, which runs alongside the vas deferens and opens into the atrial cavity close behind the pore of the male duct. The oviduct may be distinguished from the vas deferens by its large size,

and often by the presence of ova in it; the vas deferens is
a very narrow duct of an opaque white colour.

The structure of an adult Ascidian, such as *Ciona*, shows
but slight resemblance to that of the more typical Chordata,
but if the life-history of one of these creatures be followed out,
it is seen that the larval stages are much more highly differentiated than the adult, and that in all essentials they conform
to the type of the Chordata.   The free-swimming tadpole

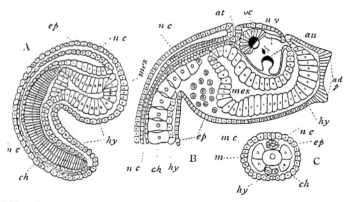

FIG. 260.—Stages in the embryology of a simple Ascidian.   After Kowalevsky.

A. Embryo showing body and tail and completely formed neural canal.
B. Larva just hatched: the tail is cut off.
C. Transverse section of tail of larva.
adp. Adhering papillae of larva.
at. Epiblastic atrial involution.
au. Auditory organ of larva.
ch. Notochord.
ep. Epiblast.
hy. Hypoblast.
nc. Neural canal.
oc. Ocular organ of larva.
m. Muscle cells of tail.
mes. Mesenteron.
mc. Mesoderm cell.
nv. Cerebral vesicle at anterior end of neural canal.

larva of the simple Ascidians is provided with a tail, in which
is a skeletal rod of supporting tissue, the notochord.   Unlike
the same structure in other Chordata, this rod is confined to
the tail.   Above this lies the tubular nervous system, which
gives off several pairs of nerves to the muscles of the tail; these
latter show some traces of metameric repetition.   Anteriorly
the nervous system is enlarged, and a median eye and auditory
sac are connected with it.   The origin of the various organs
in the larva closely resembles that of the other Chordata.
The gill-slits in the larva do not exceed a few pairs in
number, and it has been maintained that the numerous
stigmata opening through the walls of the branchial sacs

do not correspond with so many gill-slits, but are rather sub-divisions of a single original pair.

The Urochordata are divided into three orders : (i.) LARV-ACEA, (ii.) THALIACEA, and (iii.) ASCIDIACEA.

Order 1. LARVACEA.

The Larvacea include but four genera, the best-known of which is the genus *Appendicularia*. The members of the group are of extreme interest, since they have undergone no degeneration, but retain those features of the active larval condition which are lost in the other members of the group.

Appendicularia, like the other Larvacea, is a minute free-

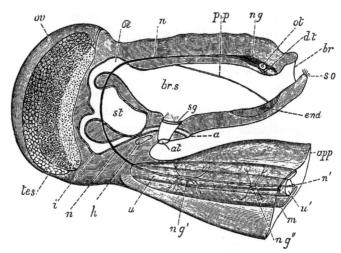

FIG. 261.—Semi-diagrammatic view of *Appendicularia* from the right.
After Herdman.

a. Anus.	*ot.* Otocyst.
at. One of the atrial apertures.	*ov.* Ovary.
app. Tail.	*pp.* Peripharyngeal band.
brs. Branchial sac.	*ng.* Cerebral ganglion.
br. Branchial aperture.	*ng'.* Caudal ganglion.
dt. Dorsal tubercle.	*ng".* Enlargement of nerve cord in tail.
end. Endostyle.	*so.* Sense organ (tactile) on lower lip.
h. Heart.	*sg.* Ciliated aperture in pharynx.
i. Intestine.	*st.* Stomach.
m. Muscle band in tail.	*tes.* Testes.
n. Nerve cord in body.	*u.* Notochord.
n'. Nerve cord in tail.	*u'.* Its cut end.
oe. Oesophagus.	

swimming organism, found near the surface of the water in all parts of the ocean. Its body is divided into a short trunk and a long tail which is attached ventrally to the body and is bent forward. The mouth leads straight into the branchial sac, which opens directly to the exterior by two ciliated apertures; there is no atrial chamber. The hypopharyngeal groove is short, and there is no dorsal lamina. There is an oesophagus stomach and intestine which ends in the anus situated just in front of the ciliated pores. The main nervous ganglion is in connection with an otocyst and a pigment spot or eye, it gives off posteriorly a nerve cord, which passes by the side of the alimentary canal into the tail and then runs along the left side of the notochord. In the tail the nervous axis enlarges into several ganglia, which give off nerves to the surrounding parts. The notochord is confined to the tail, and the muscles in the same region are broken up into segmentally repeated bands. The heart is said to be composed of two cells. The testes and ovaries are at the posterior end of the body, and open directly to the exterior.

Appendicularia, like the other members of the order, possesses the power of secreting with extraordinary rapidity a temporary gelatinous covering or test, which corresponds with the test of *Ciona*. This test is, however, soon cast aside, but another one may be formed shortly afterwards.

The Larvacea do not reproduce by budding, and their developement is direct.

Order 2. THALIACEA.

This order contains certain free-swimming forms, which in the adult state are not provided with a tail with notochord, etc. Some members of this order, as *Doliolum*, are single animals with a complicated alternation of generations in their life-history. The sexual form is hermaphrodite, and the ovum gives rise to tailed larvae which grow into asexual forms differing slightly from the sexual generation. These asexual forms reproduce by budding. The buds develope into the sexual generation, which is complicated by being polymorphic; there being three forms, of which only one has functional generative organs.

Other members of this group present still greater compleṣ\[x]ities, since the sexual generations form colonies. · The Salpidae are the most striking. example of this. The solitary

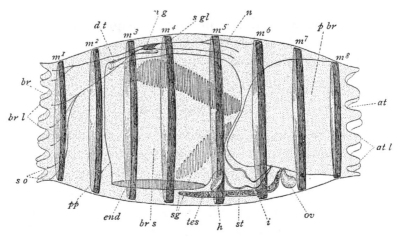

FIG. 262.—*Doliolum denticulatum*, sexual generation from the left side. After Herdman.

m^1-m^8. Muscle bands.	*at.* Atrial aperture.
ng. Nerve ganglion.	*ov.* Ovary.
sg. Stigmata.	*i.* Intestine.
sgl. Subneural gland.	*st.* Stolon of buds.
pbr. Peribranchial cavity.	*h.* Heart.
atl. Atrial lobes.	*tes.* Testis.
so. Sense organs.	*brs.* Branchial sac.
brl. Branchial lobes.	*end.* Endostyle.
dt. Dorsal tubercle.	*pp.* Peripharyngeal band.
n. Nerves.	*br.* Branchial aperture, mouth.

form of this family gives rise by a process of budding to a chain or stolon of embryos. These become detached in larger

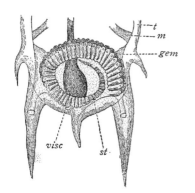

FIG. 263.—Posterior part of the solitary form of *Salpa democratica-mucronata*, showing a chain of embryos nearly ready to be set free. After Herdman.

gem. Young aggregated Salpae forming the chain.

st. Stolon.

m. Muscle band.

t. Test.

visc. Visceral mass.

or smaller groups, and each of them when mature produces sexual organs. Their ova when fertilised give rise to the

29

solitary forms. The sexual forms are always protogynous, the ovum reaching maturity before the spermatozoon ; thus self-impregnation is prevented.

All the members of this order are pelagic, and, as is usual with those animals which live at the surface of the sea, they are very transparent, or of a bluish hue.

Order 3. ASCIDIACEA.

This order includes some forms, such as *Ciona*, which are fixed, and rarely reproduce by budding; but it also includes certain free-swimming pelagic colonies of Ascidians, and certain fixed colonies the individuals of which have no separate tests, but are embedded in a common mass, which increases by gemmation. In the free-swimming forms the individuals are arranged in a cylinder with their mouths on the outer side and their atrial openings abutting on the hollow of the cylinder. There is but one genus of these pelagic forms, *Pyrosoma*, which includes, however, many species.

INDEX

The names of Genera are printed in Italics.

29*

THE END

Printed by R. & R. CLARK, *Edinburgh*

Lightning Source UK Ltd.
Milton Keynes UK
UKHW021455021218
333216UK00010B/816/P

9 780428 719470